The
Complete Book of
Locks and Locksmithing

3rd Edition

The
Complete Book of
Locks and Locksmithing

3rd Edition

C. A. Roper
Bill Phillips

TAB BOOKS
Blue Ridge Summit, PA

NOTICES
Biaxial® Medeco Security
Kaba® Lori Corporation
Primus® Schlage Lock Company
Simplex® Simplex Access Controls Corporation

THIRD EDITION
SECOND PRINTING

© 1991 by **TAB Books**.
TAB Books is a division of McGraw-Hill, Inc.

Library of Congress Cataloging-in-Publication Data

Roper, C. A. (Carl A.)
    The complete book of locks and locksmithing / by C. A. Roper and
Bill Phillips.
      p.   cm.
    Includes index.
    ISBN 0-8306-7522-1   ISBN 0-8306-3522-X (pbk.)
    1. Locksmithing.   I. Phillips, Bill, 1960-   . II. Title.
TS520.R58 1990
683′.3—dc20   90-43543
                                                CIP

TAB Books offers software for sale. For information and a catalog, please contact TAB Software Department, Blue Ridge Summit, PA 17294-0850.

Questions regarding the content of this book should be addressed to:

Reader Inquiry Branch
TAB Books
Blue Ridge Summit, PA 17294-0850

Acquisitions Editor: Kimberly Tabor
Book Editor: Stephen Moro
Production: Katherine G. Brown
Book Design: Jaclyn J. Boone

Contents

Acknowledgments

*T*his book could not have been written without help from hundreds of individuals and companies. While there are far too many to mention each by name, we would like to single out a few who offered us a great deal of assistance. They include the following:

A-1 Security Manufacturing Corp.; Abus Lock Co.; Adams Rite Manufacturing Co.; Alarm Lock Systems, Inc.; Arrow Lock; Belwith International Ltd.; Black & Decker U. S. Power Tools Group; Roger A. Bush; Charles W. Chandler; Dominion Lock Co.; Framon Mfg. Co., Inc.; Ilco Unican Corporation; Kwikset Corporation; Lock Corporation of America; M.A.G. Eng. & Mfg., Inc.; M.K. Morse Co.; William D. McInerney; Medeco Security Locks; Bert Michaels; Milwaukee Electric Tool Corp.; The National School of Locksmithing and Alarms; Porter Cable Corp.; Preso-matic Lock Co. Inc.; R&D Tool Company; Rofu International Corporation; Schlage Lock Company; Securitech Group, Inc.; Securitron; Security Engineering, Inc.; Sentry Door Lock Guards; Simplex Access Controls Corp.; Skil Corporation; Slide Lock Tool Co.; Star Key & Lock Mfg.; Donald Edward Stewart; Kimberly Taber; Taylor Lock Co.; Trine Consumer Products Division, Square "D" Co.; and Stan Willis.

C.A. Roper offers a very special thank-you to his wife, Lynda, and to his children.

Bill Phillips offers special thanks to the following special people: Gloria J. Glenn; Daniel M. Phillips; Michael M. Phillips; and Janice M. Phillips.

Introduction

*T*his completely revised and updated handbook is written for everyone who is interested in locks and keys. Students, security professionals, do-it-yourselfers, and collectors will all find a lot of fascinating and useful information within these pages.

This third edition contains many new photographs and drawings, an expanded glossary, and new chapters about current locksmithing tools and locking devices. It also includes a new chapter on collecting locks and keys.

Other chapters contain practical information on installing and servicing mechanical and electrical locking devices. With the information in this book, you can do your own locksmithing work or even start your own business.

We have tried to include everything you need to know. We've included detailed information about all of the basic lock mechanisms—combination, warded, lever, disc, pin-tumbler, high-security, and electric. Chapters are devoted to interior and exterior locksets, with special emphasis on the most popular brands. While locksmithing requires surprisingly few tools, each tool you will need is described, and many are illustrated. Instructions are given on how to make many of your own tools, as well as how to choose and buy factory-made tools.

A new chapter in this edition, Chapter 22, focuses on buying and using key duplicating machines. The photographs, drawings, and instructions in Chapter 22 will help you get the most for your money when you buy a key duplicating machine.

Do-it-yourselfers will welcome the chapters on rekeying locks and gaining entry into locked cars and buildings. Those chapters, as well as many others in this book, contain practical step-by-step instructions.

Professional locksmiths will find useful information in the chapters on high-security mechanical locks, master keying, electrical access and exit control systems, and home and business services.

We could go on and on describing the tremendous amount of information contained in this book. After you've looked at the table of contents, flipped through the pages, and read a couple of chapters no further explanation will be necessary.

We enjoyed writing this book, and hope you enjoy reading it. Let us hear from you if you have any questions or comments. Send your letter to:

Bill Phillips, c/o TAB BOOKS, Blue Ridge Summit, PA 17294-0850.

I

A short history of locks and keys

*L*ocks and keys have been around for centuries. They have evolved with man as a natural part of his existence, signifying his continual need for protection of life and property. Every civilization in history understood the need for security, and their understanding spawned a persistent search for more reliable locks, and more sophisticated keys.

EGYPT

Long before the glories of Greece, Rome, and the Greco-Roman empire, the Egyptians developed a reliable lock that operated in essentially the same way our locks do today. They invented the pin-tumbler lock and key. Old paintings inside pyramids give us a general idea of what the lock was like (FIG. 1-1). The internal form of the lock was similar to that reinvented in the mid-1800s by Linius Yale, Sr., in the United States.

The lock body was made of wood and mounted securely to the door. The bolt passed through the lock body and into a bracket mounted on the wall. In the locked position, hardwood or iron pins dropped into holes in the bolt, fixing it securely in the lock body. The key was no more than a length of wood fitted with pins that matched those in the lock body. It was inserted through the hollow end of the bolt (the bolt hole, forerunner of the keyhole) and raised to disengage the pins from the bolt. The bolt was then free to slide. The size, pattern, number, and length of the key pins varied, even as they do today.

GREECE

Most Greek doors pivoted at the center and were secured by bolts from the inside. In the few cases where locks were used, they were very primitive. The key was a large crescent-shaped affair, resembling a modern sickle. To work the bolt, the owner inserted the key into the keyhole and gave it a twist. The tip of the key aligned with a hole in the bolt and moved it left or right, locking or unlocking the door.

ROME

Unfortunately, most Roman locks have been destroyed by time, but the few keys and locks that remain give testimony to the Roman genius. The basic lock was of the warded type. (A

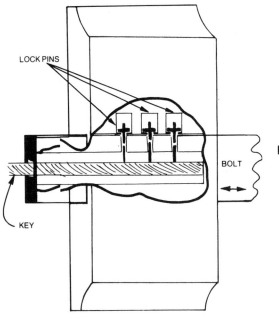

LOCK PINS

BOLT

KEY

1-1 An ancient Egyptian lock. Raising the key brought the pins to the shear line and released the bolt. The same principle was used some 400 years later on the Yale lock.

ward is a projecting ridge in a lock or on a key designed to permit only the correct key to be inserted in a lock.) Many locks are still made on this principle. In addition, the Romans made ingenious attempts to disguise the keyway as ornamentation or to hide it altogether. One famous Roman lock is in the shape of a fish. The keyway is revealed when one of the fins is turned.

Roman innovations included the use of wards in their own iron case, the metal key, the spring-loaded bolt, the spring-operated padlock, and the first true (i.e., removable) padlock. One curious aspect of Roman lock lore is that the keys were designed to be worn as rings because togas did not have pockets (FIG. 1-2).

After the fall of Rome, Europe was swept by Barbarian hordes moving south and east from the Rhineland and Scandinavia. Brute force was the order of the day, and locks meant little. Men put their faith in arms and fortresses.

This period passed as the Barbarians learned from their victims and settled down into the routines of farming and fishing. The Vikings who settled in Brittany evolved a new kind of civilization, with a king at the apex and power radiating downward from him into various levels of nobility. This social pattern, known as *feudalism*, gave the nobility access to wealth and property, and the art of locksmithing again flourished.

EUROPE

In Europe, keys were made that could move about a post and shift the position of a movable bar (the locking bolt). In its various forms it worked quite well. The first obstacles to unauthorized use of the lock were various internal wards. Medieval and Renaissance craftsmen made warded locks and improved upon them. Some were very complex, using many interlocking wards and complicated keys. Since many of these wards could be easily bypassed by almost anyone, newer methods had to be devised.

Later, chests were made entirely of metal; the locksmith reasoned that if one bolt was secure, then a chest with eight, ten, twelve, or even fifteen locking bolts would be more secure. Thus, craftsmen created elaborate internal mechanisms to allow the many bolts to be shifted by one or two separate keys. Levers, springs, ratchets, and pinions were employed to do this job. The locksmiths also installed separate locks for separate bolts; the locks had false keyholes and safety devices, such as springloaded knives that would injure or kill the thief. Hidden keyholes were extremely popular among the merchant and upper classes.

In France, the treatise *The Art of the Locksmith* (1767) was published, describing examples of the *tumbler* or *lever lock*. Exactly who invented it is unknown, but credit is long overdue. As locksmithing advanced, multiple-lever locks came into being. Two, three, six, and more levers were used. Each lever had to be lifted; when all were in proper alignment, the bolt could be moved, opening or closing the lock.

ENGLAND

The English also worked on newer and better locking devices. Incentive was given in the form of cash awards and honors to those who could successfully open these newer and more complex locks. In the forefront of lock designing were three Englishmen: Robert Barron, Joseph Bramah, and Jeremiah Chubb.

Robert Barron patented the double-action tumbler lock in 1788. Like others before it, this lock employed a series of lever tumblers pegged at one end (FIG. 1-3). One side of these tumblers engaged a notch in the bolt. The key bore against the opposing surface. Notches on the key corresponded to the individual tumblers. The width of the key-bearing surface varied with each tumbler: a wide tumbler required a correspondingly deep notch in the key; a narrow tumbler required a shallow notch. When the key and tumbler stack matched, all tumblers would move in unison and release the bolt.

Barron's patent involved the key-bearing surface on the tumbler. Earlier locksmiths were content with a simple notch on the lower edge of the tumbler stack. When the tumblers were raised high enough, the bolt would release. Barron pierced the tumbler stack so that the key controlled both up and down movement of the individual tumblers. This refinement worked in conjunction with a more complex gating; the bolt would remain latched until all tumblers were at the same height. Because the tumblers move in two directions, this lock is

1-2   Roman keys were worn as rings.

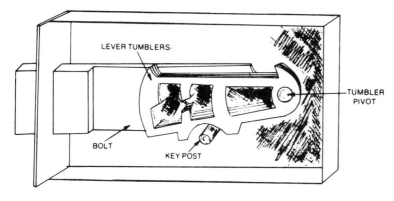

LEVER TUMBLERS·

TUMBLER PIVOT

BOLT

KEY POST

1-3 An 18th-century lever-tumbler mechanism. The width of the tumblers corresponded to top slots in the key.

described as a *double-action* type. Barron added up to six of these double-lever actions to his lock and thought it virtually impossible to open except with the proper key. He soon found out differently.

Another Englishman, Joseph Bramah, wrote *A Dissertation on the Construction of Locks*, which exposed the many weaknesses of existing so-called thiefproof locks and pointed out that any of them could be picked by a good lock specialist or criminal with some training in locks and keys. Bramah admitted that Barron's lock had many good points, but he also revealed its major fault: the levers, when in the locked position, gave away the lock's secret. The levers had uneven edges at the bottom; thus, a key coated with wax could be inserted into the lock and a new key could be made by filing where the wax had been pressed down or scraped away. Several tries could create a key that matched the lock. Bramah pointed out that the bottom edges of the levers showed exactly the depths the new key should be cut in order to clear the bolt. Bramah suggested that the lever bottoms should have a smooth surface, and that the lever slots should be cut unevenly. Then only a master locksmith could open the lock.

Bramah's lock, patented in 1784, employed a series of notched lever tumblers that were aligned by corresponding notches in the key. The novelty was in the way the levers were arranged. Earlier locks were built on the pattern shown in FIG. 1-3; Bramah's lever tumblers were mounted vertically. Bramah used radial tumblers and a barrel key (FIG. 1-4). The pin on the side of the key rotated the lock.

Though this lock could be picked, the job was beyond the average thief. Each tumbler had to be aligned and the control piece had to be turned in the right direction.

Jeremiah Chubb added refinements to the Barron lock in an attempt to make it more secure. One of Chubb's improvements was a metal "curtain" which fell across the keyhole when the mechanism began to turn, making the lock difficult to pick. He also added a detector lever that indicated whether the lock had been tampered with. A pick or an improperly cut key would raise one of the levers too high for the bolt gate. This movement engaged a pin that locked the detector lever. The detector could be cleared by turning the correct key backwards and then forwards.

During this period, many robberies were committed against property and person. In 1817, the Portsmouth, England dockyard was robbed; the British Crown offered a reward to anyone who could devise a lock that could be opened *only* by its own key—one that was impossible to open by lockpicking or the use of a false key. A year later Chubb patented his lock and won the prize money.

This lock got much attention. One convict, a former locksmith, claimed he could open the best of locks. He was guaranteed $250 and his freedom if he could open Chubb's lock.

I-4 A Bramah radial lever lock (circa 1790).

The British Government and Chubb supplied the convict with a lock, key blanks, appropriate lockpicking implements, and some dismantled locks to work on. After working for some three months, he finally gave up and served out his sentence. England announced that it finally had an unpickable lock.

Hobbs, an American locksmith, made a kind of minor career picking English locks. The manufacturers responded by developing new locks and manufacturing techniques. Many locks—over 3000—were patented in England alone during the 18th and 19th centuries. Key designs were as varied and ingenious as the locks they fitted. Perhaps the most interesting was the detachable bit key. The bits—the part of the key that worked the lock—could be disassembled and rearranged in different combinations. Only the owner knew the proper combination; if the key were lost or stolen it would be useless unless bitted correctly. Other keys had projecting pins like the Bramah key, and they were intended for rotating locks. Others were flat with a wide tip, or socketed for detachable bits. Another type had a detachable tip. Elaborately notched keys were not uncommon, and were used with extremely complex locking wards. Each ward had a distinctive shape location in the lock. A false key might turn the lock a few degrees, only to be blocked by additional wards.

The swivel key was another example of 19th century ingenuity. The lever, or the biting portion of the key, and the barrel, or shank, were moved independently of each other. Each half was turned in opposite directions to open the lock.

AMERICAN LOCKS

Early American lock bolts were mounted on the inside of the door, and could be opened from the outside by means of a latchstring. Hence the phrase, "the latchstring's always out."

At night the string would be pulled inside, "locking" the door. Of course, someone had to be inside to release the bolt. An empty house was left unlocked. As the country was settled, theft increased; local merchants and blacksmiths soon got into the lock trade. Small cabinet and cupboard locks were very popular. The lady of the house (or, in large establishments, the butler) was in charge of the household keys. The doors were unlocked in the morning and the cabinets and cupboards were opened to dispense the day's supplies and immediately relocked. At night the doors were secured.

In the 1850s two inventors, Andrews and Newell, were granted patents on an important new feature—removable tumblers. The tumblers could be disassembled and scrambled to make, in effect, a new lock. The keys had interchangeable bits that matched the various tumbler arrangements. After locking up for the night, a prudent owner would scramble the key bits. Even if a thief got possession of the key, it would take him hours to stumble on the right combination. In addition to removable tumblers, this lock featured a double set of internal levers.

Newell was so proud of this lock that he offered a reward of $500 to anyone who could open it. A master mechanic took him up on the offer and collected the money. This experience convinced Newell that the only secure lock would have its internals sealed off from view. Ultimately, the sealed locks appeared on bank safes in the form of combination locks.

As we have seen, Hobbs picked the famed English locks with ease. Until Hobbs' time, locks were opened by making a series of false keys. If the series was complete, one of the false keys would match the original. Of course, this procedure took time. Thousands of hours might pass before the right combination was found. Hobbs depended upon manual dexterity. He applied pressure on the bolt while manipulating one lever at a time with a

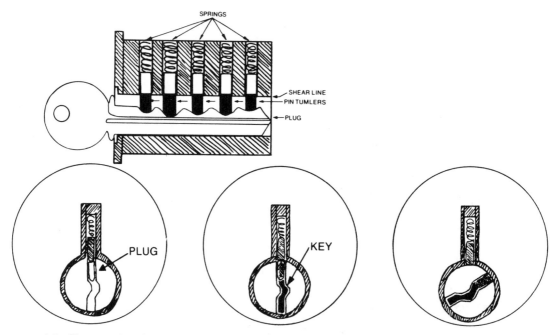

1-5   The operation of the Yale cylinder lock. The key raises the pin tumblers to the shear line and allows the plug to rotate.

small pick inserted through the keyhole. As each lever tumbler unlatched, the bolt moved a hundredth of an inch or so.

Until the early 19th century, locks were made by hand. Each locksmith had his own ideas about the type of mechanism—the number of lever tumblers, wards, and internal cams to put into a given lock. Keys contained the same individuality. A lock could have 20 levers and weigh as much as five pounds.

Linius Yale changed all this. An inventor of milling machinery, Yale devised a simple, safe, and compact key-operated lock that could be made on automatic machinery. This lock was based upon the principle first developed by the Egyptians. Yale improved upon this principle and developed a keying system that is still used today. His son continued in the trade, making further refinements on his father's work. He developed a simplified cylinder with a rotating internal plug core that advanced lock design to new heights. Other refinements included the solid case and the small keyway that made life difficult for pick artists. The son also developed a series of dies and cutters to mass produce all parts of the lock.

The major structural difference between the Yale lock and the Egyptian prototype was the revolving plug that replaced the sliding bolt. A tailpiece on the back of the plug worked the bolt mechanism. Using machine tools, the Yale family was able to produce these locks with the same dimensions and tolerances. The keyway was ridged to accept a grooved key blank. Only the correctly cut key would turn the plug and open the lock. This lock meant that thieves had to start over and devise new methods and tools. Figure 1-5 illustrates the basic Yale mechanism.

Competitors were quick to copy the Yale cylinder. The reliability of these locks revolutionized the industry. Some of the early lock companies are still in business.

2

Tools of the trade

*H*aving proper tools and supplies are crucial to performing locksmithing tasks. How much money you'll need to spend on those items depends on two things: how many tasks you want to be able to perform, and how many of the necessary items you're willing to make.

You will need some tools and supplies to put the information you learn from this book into practice. Without practice, it isn't possible to learn locksmithing. Proficiency in the trade requires both knowledge and skills.

Some of the items you'll need can be easily made in a home workshop. Others can be purchased locally. But some items must be purchased from locksmith supply houses. (A list of suppliers is provided in Appendix B.)

WORKBENCH

Whether you practice locksmithing in a shop or at home, you'll need a workbench. You can temporarily use a table or desk, but a workbench is more practical and comfortable to use. If you have basic woodworking tools and skills, you should have no trouble making your own workbench.

The workbench should be:

Long enough to ensure adequate work space.
Strong enough to support a key machine at one end, out of the way.
Solid enough to keep the key machine in alignment.
High enough to allow you to work without stooping.
Wide enough to store parts and supplies; 30 inches is the comfortable maximum.
Lit from overhead and behind, or from the sides. (Never have the light too close to the bench.)

Workbench location

The location of the workbench should also be considered. If possible, place it in a well ventilated area, away from general traffic, near frequently used equipment and supplies, with an easy access from several directions. Many locksmiths place storage bins near their workbenches to hold tools and supplies (FIG. 2-1).

2-1 Arco bins are made of heavy polypropylene for long maintenance-free service.

Examples of workbench designs are shown in FIGURES 2-2 and 2-3. No one type of workbench is ideal for everybody; individual tastes and needs differ.

A vise should be installed at one end, out of the way. It should be mounted on a swivel base, be sturdily built, and have jaws at least 3 inches wide. The jaws should open to at least 5 inches.

Common tools and supplies

Many common tools are used in locksmithing. You probably already have most of them. You can find the rest at virtually any hardware store.

Allen wrench set

Bezel nut wrench

Bolt cutters, 16'

C-clamps (FIG. 2-4)

Center punches

Coping saw and blades

Cut-all knives and blades

Dent puller

Dial caliper (FIG. 2-5)

Disc grinder

Drill bits (assorted sizes and types)

Drill, electric (preferably ½" chuck) (FIG. 2-6)

Drill, cordless (FIG. 2-7)

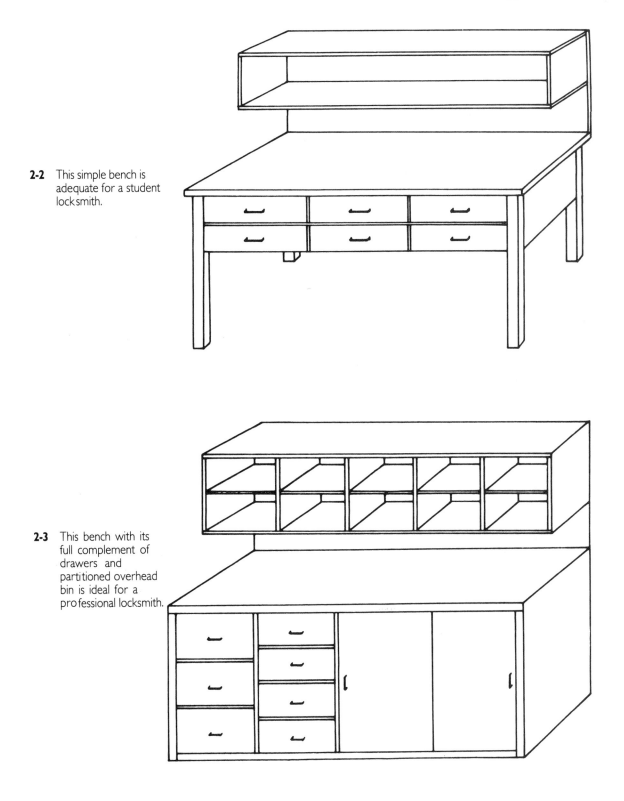

2-2 This simple bench is adequate for a student locksmith.

2-3 This bench with its full complement of drawers and partitioned overhead bin is ideal for a professional locksmith.

2-4   Stanley C-clamps are light and durable.

Extension cord (at least 50′)
Files (assorted types and sizes) (FIG. 2-8)
Flashlight
Bench grinder with wire wheel
Hacksaw and blades (FIG. 2-9)
Hammers (ball-peen and claw) (FIG. 2-10)
Handcleaner
Hollow mill rivet set
Lubricant (such as WD-40)
Masking tape
Measuring tape (at least 12′)
Nails and screws (assorted)
Plastic glue
Pliers (adjustable, cutting, and locking lever)
Rivet assortment

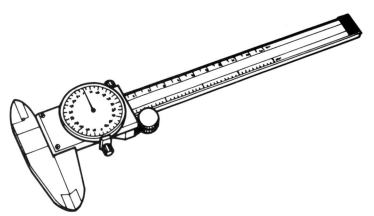

2-5   A dial caliper is great for measuring pins, key blanks, cuts in keys, and plugs. (courtesy Ilco Unican Corp.)

2-6 An electric drill is an important tool for installing locks. (courtesy Black & Decker)

2-7 A cordless drill is useful when no electricity is available. (courtesy Milwaukee Tool Co.)

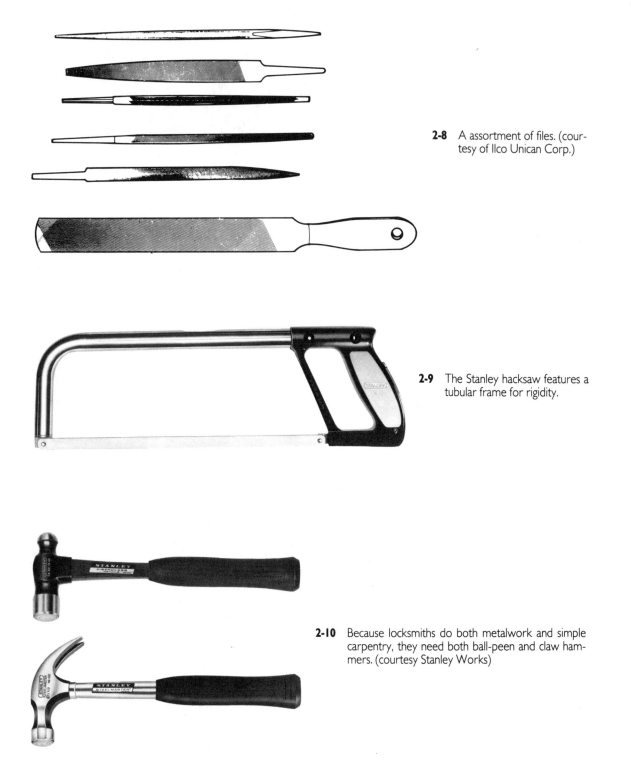

2-8 A assortment of files. (courtesy of Ilco Unican Corp.)

2-9 The Stanley hacksaw features a tubular frame for rigidity.

2-10 Because locksmiths do both metalwork and simple carpentry, they need both ball-peen and claw hammers. (courtesy Stanley Works)

Rubber mallet
Scratch awl
Sandpaper and emery cloth
Screwdriver assortment (FIG. 2-11)
Scribers
Snap ring pliers (assorted sizes)
Socket sets, ½' & ¼'
Storage trays (FIG. 2-12)
Tap set
Tool boxes
Vise
Vise grips
Wood chisel set (FIG. 2-13)
Wrenches (adjustable and pipe)

2-11 Electrician's screwdrivers (top) are designed for precision work; standard screwdrivers are used on the heavier jobs (bottom). (courtesy Stanley Works)

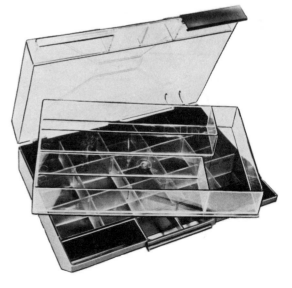

2-12 Parts cabinets and storage boxes are a necessity. (courtesy Raaco)

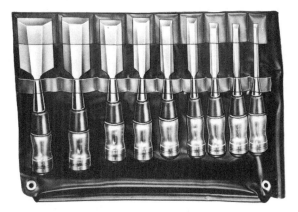

2-13   A Stanley wood chisel set.

Special tools and supplies

Many of the following tools and supplies are used primarily for locksmithing tasks and aren't normally sold in hardware stores. Later chapters explain more about the various items in this list.

Automobile entry tools

Broken-key extractors (FIG. 2-14)

Cam assortment

Cam screw assortment

Chrysler shaft puller

Circular hole cutter kit (FIG. 2-15)

2-14   Broken-key extractors are useful for removing broken key parts from locks. (courtesy A-1 Security Manufacturing Corp.)

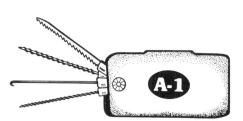

2-15 Circular hole cutter kits are useful for installing locksets. (courtesy Skil Corporation)

2-16 A manual code key cutter can be easily transported to allow for code-key cutting anywhere. (courtesy Ilco Unican Corp.)

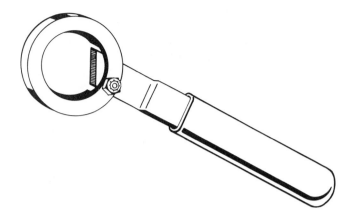

2-17 A cylinder removal tool makes it easier to remove stubborn mortise cylinders. (courtesy A-1 Security Manufacturing Corp.)

Code books
Code key cutting machine, electric.
Code key cutting machine, manual (FIG. 2-16)
Cylinder removal tool (FIG. 2-17)
Depth gauge
Disc-tumbler assortment
Door handle clip tool (FIG. 2-18)
Face caps
Face cap pliers
Flat spring steel stock
Flexible light, portable
Key blank assortment
Key duplicator machines, electric (FIG. 2-19)
Key layout board

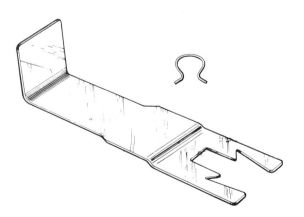

2-18 A door handle clip tool helps remove the retainer clip that secures the door handle to a door. (courtesy A-1 Security Manufacturing Corp.)

2-19 Different key duplicating machines are used to duplicate different types of keys. (courtesy Framon Manufacturing Corp.)

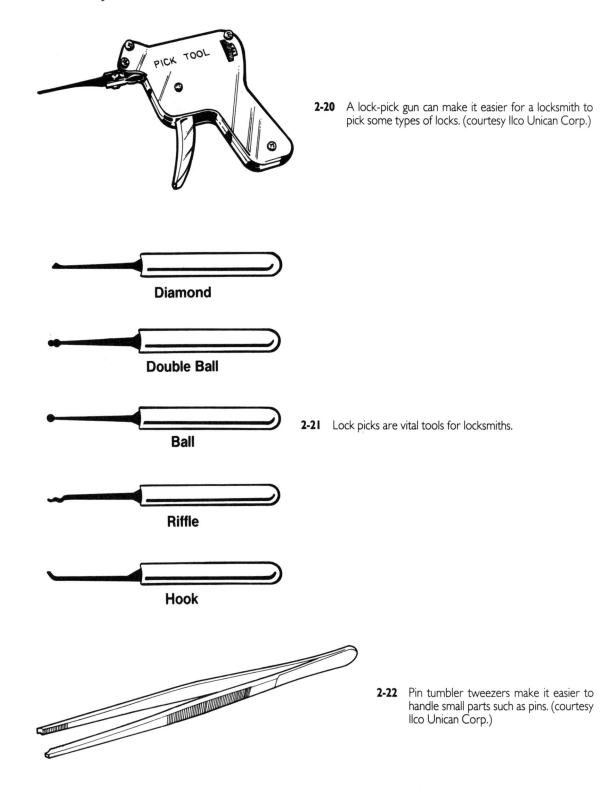

2-20   A lock-pick gun can make it easier for a locksmith to pick some types of locks. (courtesy Ilco Unican Corp.)

Diamond

Double Ball

Ball

2-21   Lock picks are vital tools for locksmiths.

Riffle

Hook

2-22   Pin tumbler tweezers make it easier to handle small parts such as pins. (courtesy Ilco Unican Corp.)

Key marking tools

Lever tumbler assortment

Lock-pick gun (FIG. 2-20)

Lock-pick set (FIG. 2-21)

Lock plate compressor

Lock reading tool

Pin kits

Pin tray

Pin-tumbler tweezers (FIG. 2-22)

Plug followers (FIG. 2-23)

Plug holders (FIG. 2-24)

Plug spinner (FIG. 2-25)

Punch and die set in single and double D configurations

Retainer clip tool (FIG. 2-26)

Retainer ring assortments

Shims (FIG. 2-27)

Spindle assortment

Spring assortment

Steering wheel pullers

Trim removal tool

Tubular key decoder (FIG. 2-28)

Tubular key lockpicks (FIG. 2-29)

Tubular key lock saw (FIG. 2-30)

Torque wrenches (FIG. 2-31)

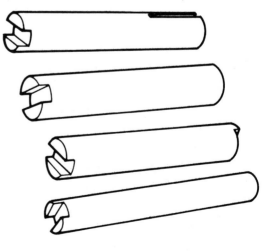

2-23 Plug followers are useful when disassembling and assembling pin tumbler cylinders. (courtesy Ilco Unican Corp.)

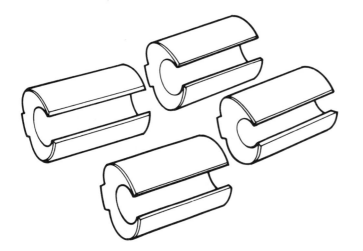

2-24 Plug holders aren't vital, but are useful when working on cylinders.

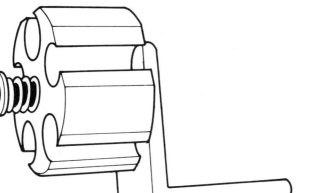

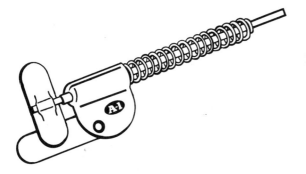

2-25 A plug spinner quickly rotates a plug clockwise or counterclockwise; it's used when a lock has been picked in the wrong direction. (courtesy A-1 Security Manufacturing Corp.)

2-26 A retainer clip tool is used for automotive lock work. (courtesy A-1 Security Manufacturing Corp.)

2-27 Shims are used when disassembling pin tumbler cylinders. (courtesy Ilco Unican Corp.)

2-28 A tubular key decoder helps a locksmith determine the bitting of tubular keys. (courtesy A-1 Security Manufacturing Corp.)

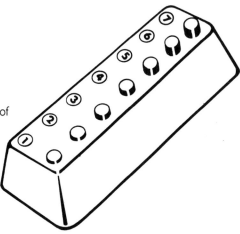

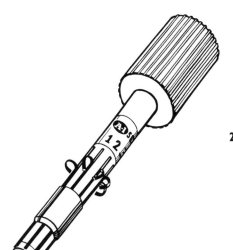

2-29   Tubular key lockpicks are vital to locksmiths who work on tubular key locks. (courtesy A-1 Security Manufacturing Corp.)

2-30   A tubular key lock saw is used to drill out tubular key locks. (courtesy A-1 Security Manufacturing Corp.)

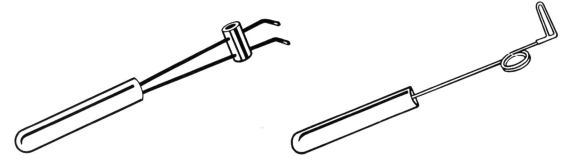

2-31   Torque wrenches come in various shapes. (courtesy A-1 Security Manufacturing Corp.)

2-32 Wedges are used to help gain entry into locked automobiles.

Wafer tumbler assortment

Wedges (FIG. 2-32)

It isn't necessary to have all the tools in both of the above lists to practice the material in this book. As you study each chapter, you'll learn which tools you'll need. But with all of these tools and supplies, you will have enough equipment to start a locksmithing business.

3

Types of locks
and keys

The locking devices commonly used today can be grouped into a number of general types. These types include the following

- Warded locks
- Mortise locks
- Automobile locks
- Key-in-knob locks
- Surface-mounted auxiliary locks
- Lever locks
- Luggage locks
- Padlocks
- Combination locks

An example of each type is shown in FIG. 3-1. Within these few major types all design variations fall. Some locks, such as a combination door lock, represent a melding of two different types.

In this chapter we will consider the various types of locks and keys that are in common use today (FIG. 3-2).

Warded locks are either of the *rim* or *mortise* type, and are among the oldest locks still being made (FIGS. 3-3 and FIG. 3-4).

Lever and *wafer* locks are related, although they were introduced to the public at different times.

A variety of *padlocks* exist—warded, wafer, disc, and pin tumbler. Smaller padlocks have a single ward and take very simple keys. The common railroad lock is a padlock. It is opened with an old-style barrel key and uses a simple lever mechanism. Some railroad locks work dependably even after being exposed to the elements for fifty years. Padlocks can also be combination locks.

The simplicity of the cylinder lock made possible its mass production. The cylinder is a single rotating unit firmly encased within a solid metal housing. The creation of the Yale cylinder lock encouraged several variations: the key-in-knob lock, the double lockset (FIG.

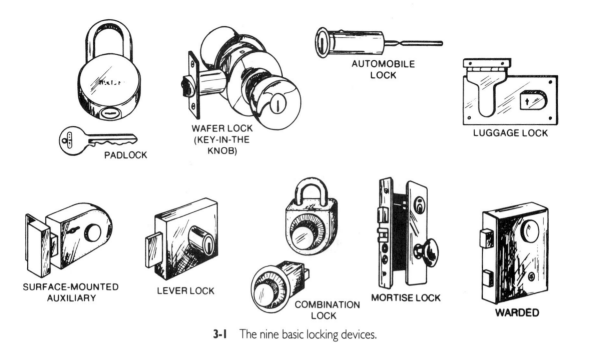

3-1   The nine basic locking devices.

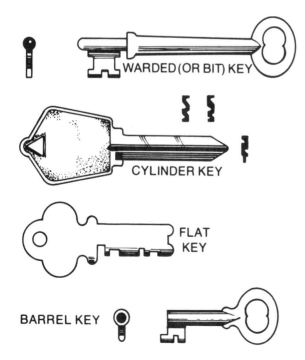

3-2   Four basic key types in use today.

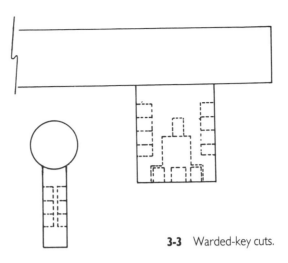

3-3 Warded-key cuts.

3-5), and the desk lock. A variation of the cylinder lock used in coin-operated machines is the Ace circular lock (FIG. 3-6).

A combination lock consists of a series of interconnecting wheels rotating about a central core that is controlled in its revolutions by an outside combination knob. The number of revolutions of the knob necessary to open the lock is pretty much standard: a 3-2-1 sequence is most common; some high-security locks may require a 4-3-2-1 rotation. Combination locks are of two basic types: the hand and the key-change variety. A kind of hand combination lock not dependent upon internal wheels is the pushbutton lock, shown in FIG. 3-7.

Surface-mounted auxiliary locks include deadlatches, surface-mounted cylinders (when used separately from another lock unit), chain door guards, rim latches, surface bolts, and chain bolts (FIG. 3-8).

3-4 Typical warded locks. (courtesy Taylor Lock Company)

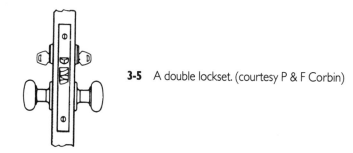

3-5   A double lockset. (courtesy P & F Corbin)

The automotive lock employs the sidebar principle. When the key engages the tumblers, it aligns them so that a sidebar built into the lock drops into place, allowing the key to turn to open the lock. The sidebar principle is also used on certain pin-tumbler cylinders.

KEY BLANK IDENTIFICATION

Three parts of the key blank are used to match a given key with a key blank:

- □ The *head shape* of the key.
- □ The *length* of the key.
- □ The *millings* (grooves) of the key.

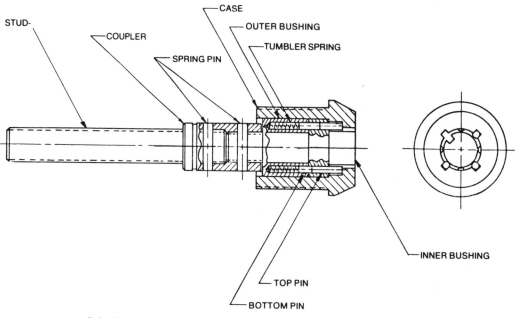

3-6   A cross section of an Ace circular lock. (courtesy Desert Publications)

3-7 A Simplex pushbutton combination lock.

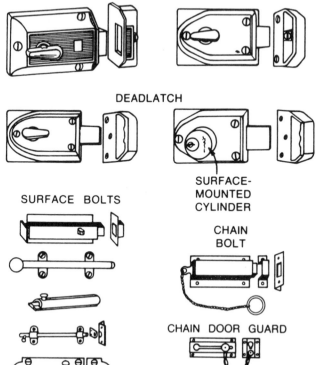

3-8 Surface-mounted auxiliary locks.

The head shape will usually pinpoint the manufacturer, since most manufacturers like to give their keys an individual look. The length and millings are very critical. To determine the length, lay the key over the illustration in the key catalog and align the shoulders. The length is then observed at a glance because the key catalog drawings are the size of the original keys. To match the grooves, stand the key directly over the cross-section that appears under each blank. The grooves must match *exactly* to be considered the same.

Key manufacturers put out catalogs that identify their key blanks, their sizes, and a cross section of each different blank. In addition, they may have a cross-reference section which refers to other manufacturers' keys that are applicable to each of the keys in that particular catalog. These cross-reference sections are valuable, as they obviate reference to a wide variety of catalogs in order to determine the manufacturer of a given key blank. Merely refer to the key type and cross-reference it to the one that you have.

A problem that arises here is that some manufacturers have literally *hundreds* of key blanks, and the cross-reference section becomes unwieldly at times. A catalog that carries the most popular and commonly used blanks is best to use. About 95 percent of the keys you will have to duplicate will be found in such catalogs. Other catalogs can be used for general reference for the other 5 percent, but are not absolutely necessary to have for your shop.

KEY BLANK EXAMPLES

On the following pages are illustrations of various types of key blanks. Figure 3-9 shows key blanks from the Taylor Lock Company. These cylinder blanks will fit a variety of American locks. Notice that in Fig. 3-10 the blank is further laid out to incorporate a master key system and from the cross section, you can identify the various keyways that will fall into this particular master key system.

Figure 3-11 shows another keyway system with the applicable blanks. Notice that within this key series there are three different key lengths, which allows for a wider use of the master key system. The length also allows for more pins to be used in the cylinders. This means a greater variety in the number of possible key combinations available to the system.

Foreign automobiles (and, naturally, their keys) are increasing in popularity, so it is necessary to have a well-stocked variety of foreign lock keys in the locksmith shop. Figure 3-12 provides illustrative details on a variety of the more common key blanks in use. You should have these types, and also a slightly wider supply of other foreign auto key blanks available to service the needs of your customers.

Automotive Key Blanks

The duplication of automotive keys is more prevalent in recreational areas and very large cities. Lost or stolen keys mean duplication and/or rekeying on a weekly basis for many locksmiths. It becomes imperative that as a locksmith you have in your shop an ample and wide selection of key blanks on hand in this particular area. Keep a selection of the most common and popular automotive blanks in your van plus one or two others that might be required. (Note: A standard selection of automotive key blanks should be in *every* locksmith shop, but in large cities and in recreational areas such as beaches, parks, etc., the need is greater than usual. For this reason, in these areas, the number and variety of blanks will be greater than in others. Always consult with your distributor or manufacturer's representative to obtain a more accurate view of the types and quantities of blanks that are necessary; don't buy a larger selection than required just because you think you may *some day* need it.

Figure 3-13 is a list of domestic automotive blanks, identified by the number and the vehicle they accompany. These are, by far, the most common ones for which you will be cutting keys for. (This is a representational selection of popular keys that require frequent

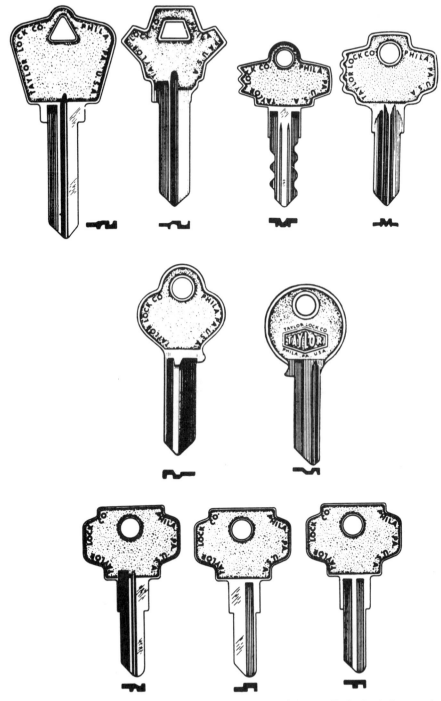

3-9 Representative sampling of some standard key blanks. (courtesy Taylor Lock Company)

3-10 Six pin cylinder key and various associated keyways. Keyway M is used only for the master key. (courtesy Taylor Lock Company)

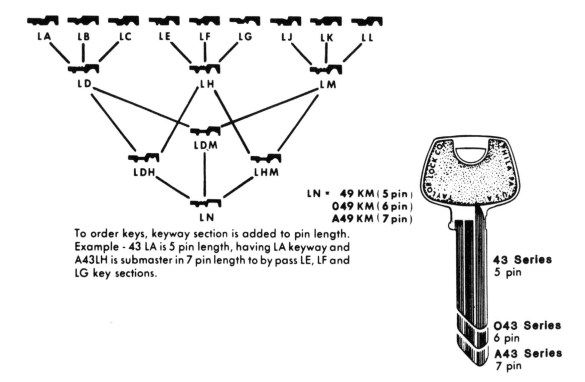

LN = 49 KM (5 pin)
   049 KM (6 pin)
   A49 KM (7 pin)

To order keys, keyway section is added to pin length. Example - 43 LA is 5 pin length, having LA keyway and A43LH is submaster in 7 pin length to by pass LE, LF and LG key sections.

43 Series
5 pin

O43 Series
6 pin

A43 Series
7 pin

3-11 Larger master key system which provides a greater variety of key uses and control features through sub-masters and individual use keys. (courtesy of Taylor Lock Company)

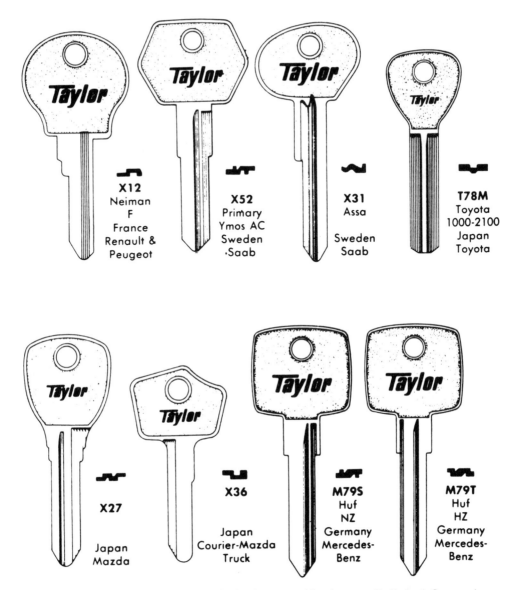

3-12 Variety of popular key blanks for foreign automobiles. (courtesy Taylor Lock Company)

duplication; it is not all-inclusive.) Refer to your key catalog for illustrations of all the automotive keys and start to develop a working memory of them. You should be able after a period of time to look at a key that is already cut (or even a blank) and determine with relative accuracy what make of automobile the key goes to, and where the key blanks are located on your key blank board. Figure 3-14 shows some of these key blanks. Figure 3-15 lists the most common foreign keys by number of automotive type, while FIG. 3-16 shows a variety of these keys.

AD1	Audi 100 and 100LS—1971 on	LU1	GM LUV Ign./Door & Trunk/Glove—1973 on
CP1	Capri Ign. 1971 on—Ign./Trunk 1975 on	MZ1	Mazda Door/Trunk—1971 on
CP2	Capri Door/Trunk 1971-74	MZ2	Mazda—Supplements MZ1
CP3	Capri—Supplements CP2	MZ3	Mazda, Courier Trucks 1971-73
CP4	Capri—Fiesta	MZ4	Mazda/Courier R100/1200-Door/Trunk 1970-72
		MZ5	Mazda—Supplements MZ4
5DA1	Datsun		
5DA2	Datsun 1970-72, Mazda 1971 on, Eng. 1969 on	RP1	Renault—Peugeot
DA3	Datsun—1970 on	5RP2	Renault-Peugeot Ignition 1971 on
DA4	Datsun & Subaru—1970 on	TO1	Toyota
LDC1	Long Head Dodge Colt, Arrow, Chall., Sapporo, Jap. Opel—1970 on	TO2	Toyota Ignition—1969 on
		UN2	Union (Hillman—Vauxhall—English Ford)
5FT1	Fiat Ignition—1967 on	UN3	English 1960 on, Volvo Door/Trunk 1970
FT2	Fiat Trunk 1967 on—Pantera-Ferrari	UN4	Union—MG—Triumph
FT3	Fiat—Supplements FT2	6VL1	Volvo 1960 to 1965-6 Pin
HN1	Honda Ignition 1970 on, MG Ignition 1972	VW1	Volkswagen
HN2-3	Honda MASTER (Codes #1001-1700, 2001-2700) —1976	VW2	Volkswagen 1965-70
		VW3	Volkswagen Beetle—1971 on
HN4	Honda Trunk 1973-76	VW4	Volkswagen VW411-412-DA-RA-SC, Fox, Audi 4000/5000, Porsche, Jetta
HN5	Honda Cycles—1977 on	VW5	Volkswagen Bus—1971 on

3-13   Automotive key blanks by key number and associated automotive use. (courtesy Star Key and Lock Co.)

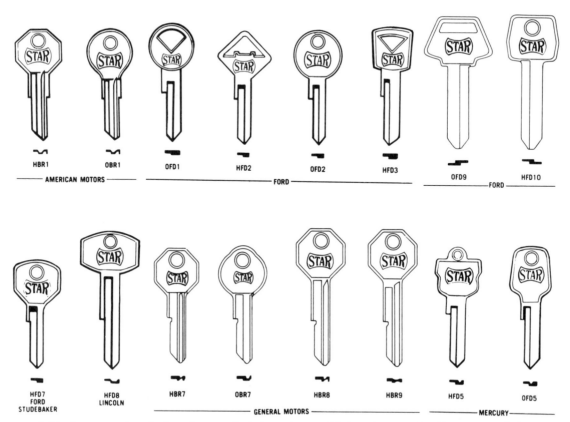

3-14   Representative selection of commonly duplicated automobile keys. (courtesy Star Key and Lock Co.)

HBR1	American Motors Ignition/Door to 1959		HFD1	Ford Ignition/Door—1952 to 1954
OBR1	American Motors Trunk/Glove to 1959, GAS locks		OFD1	Ford Trunk/Glove—1952 to 1954
OBR1DB	Briggs & Stratton—GAS & utility locks		HFD2	Ford Ignition/Door—to 1951
			OFD2	Ford Trunk/Glove—to 1951
HBR2	GM Ignition/Door—all years to 1966		HFD3	Ford Ignition/Door—1955 to 1958
OBR2	GM Trunk/Glove—all years to 1966		OFD3	Ford Trunk/Glove—1955 to 1958
HBR3	American Motors Ignition/Door—1960 on		HFD4	Ford Ignition/Door—1952 to 1964
OBR3	American Motors Trunk/Glove—1960 on		OFD4	Ford Trunk/Glove—1952 to 1964
OBR4	GM Long Head Ignition/Door—all years to 1966		HFD5	Mercury Ignition/Door—1952 to 1964
			OFD5	Mercury Trunk/Glove—1952 to 1964
HBR5	GM Ignition/Door "A"—1967		HFD6	Ford, Linc., Merc. Ign./Trunk Square Head Master—1952 to 1964
HBR5M	GM Ignition/Door Master "A/C"—1967/68			
OBR5	GM Trunk/Glove "B"—1967		OFD6	Ford, Linc., Merc. Ign./Trunk Round Head Master—1952 to 1964
HBR6	GM Ign./Door & Trunk/Glove Master—hexagon head—1967		HFD7	Ford Ign./Door to 1951—Studebaker 1947 on
			HFD8	Lincoln Ignition/Door—1952 to 1964
OBR6	GM Ign./Door & Trunk/Glove Master—oval head—1967		HFD9	Ford Ignition/Door—1965/66 only (double side)
			OFD9	Ford Trunk/Glove—1965/66 only (double side)
HBR7	GM Ignition/Door "C"—1968		HFD10	Ford Ignition/Door—1965 on (double side)
OBR7	GM Trunk/Glove "D"—1968		OFD10	Ford Trunk/Glove—1965 on (double side)
HBR8	GM Long Head Ign./Door Master "A/C"—1967/68/71/72 (Ign. only 1975/76/79/80)		HPL1	Chrysler, Plymouth, Dodge Ign./Door 1949-55
			OPL1	Chrysler, Plymouth, Dodge Trunk/Door 1949-58
HBR9	GM Long Head Ign./Door & Trunk/Glove Master "E/H"—1969/73/77/81		HPL2	Chrysler, Plymouth, Dodge Ign./Door to 1948
			OPL2	Chrysler, Plymouth, Dodge Trunk/Glove to 1948
OBR9	GM Oval Head Ign./Door & Trunk/Glove Master "E/H"—1969/73/77/81		HPL3	Chrysler, Plymouth, Dodge Ign./Door 1956-59
			OPL4	Chrysler, Dodge Trunk/Glove 1959-65
HBR9E	GM Ign./Door "E"—1969/73 (Ign. only—1977/81)		OPL5	Plymouth Trunk/Glove 1959-65
OBR9H	GM Trunk/Glove "H"—1969/73 (Door also—1977/81)		HPL6	Chrysler, Plymouth, Dodge Ign./Door 1960-67
HBR10J	GM Ign./Door "J"—1970 (Ign. only—1974/78/82)		OPL7	(Master) Long GM to 1966 & Chrysler Trunk 1959-65
OBR10K	GM Trunk/Glove "K"—1970 (Door also—1974/78/82)		OPL8	Chrysler Trunk/Glove—1966/67
HBR11	American Motors Ign./Door—1970 square head—1960 on		HPL68	Chrysler Ign./Door Master 1956 on—hexagon head
			OPL68	Chrysler Trunk/Glove Master—1966 on
OBR11	American Motors Trunk/Glove—1970 ellipse head—1960 on		OPL70	Chrysler Ign./Door & Trunk/Glove Master—1966 on
			HPL73	Chrysler Ign./Door Master 1956 on—diamond head
HBR12A	GM Ign./Door "A"—1967/71 (Ign. only—1975/79)		HYA4	Kaiser Frazer Ign./Door—1949/50
OBR12B	GM Trunk/Glove "B"—1967/71 (Door also—1975/79)		HYA5	Studebaker Ign./Door—1940 on
OBR13	GM Ign./Door & Trunk/Glove Master "A/B"—1967/71/75/79			
OBR13S	GM Ign./Trunk—fits most years—1967 on—FOR LOCKSMITHS ONLY—NOT FOR DUPLICATION			
HBR14C	GM Ign./Door "C"—1968/72 (Ign. only—1976/80)			
OBR14D	GM Trunk/Glove "D"—1968/72 (Door also—1976/80)			

3-15 Most popular foreign automotive keys listed by key number and automotive use. (courtesy Star Key and Lock Co.)

Look-alike keys

As the name implies, you will find key blanks that are "look-alike" keys—blanks made by a manufacturer other than the original that look just like the original blank. These provide you with a great advantage; as look-alikes, they are quickly identified with a specific lock manufacturer; thus, you can quickly set the proper key for duplication.

Not all keys made are look-alikes. This brings up an important point—comparison charts or cross-reference key charts. They go by either name but the function is the same, and they are an important reference tool in your shop. You may have a customer who brings in a key for duplication, and for one reason or another, you cannot put your finger on the specific key blank that is required in order to duplicate it. Figure 3-17 is a manufacturers' key comparison chart for automotive and house keys.

From this chart you have, say, two keys that require duplication—one for the house and the second for the family car. The house key has a National/Curtis number of IN1, whereas, the car key is 1127DP. Here's how to quickly determine whether you have the particular blanks for these two keys or not. (As an up-and-coming locksmith, you carry a wide assortment of the most commonly used keys; plus, you have a cross-reference key chart listing.)

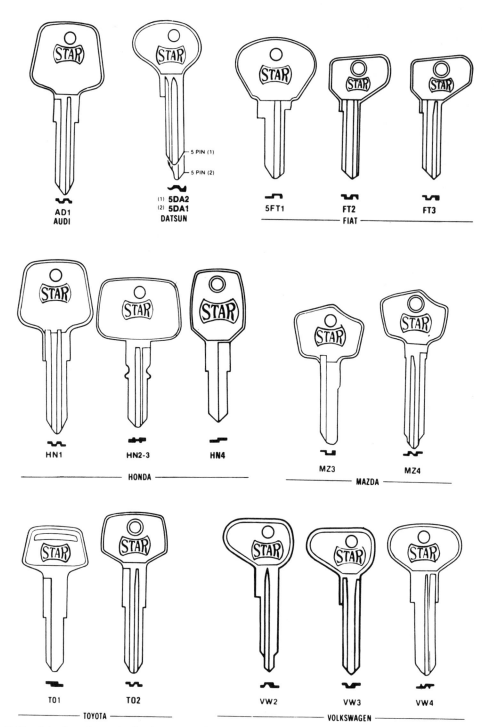

3-16 Selection of popular foreign automotive key blanks that require frequent duplication. (courtesy Star Key and Lock Co.)

AUTOMOTIVE

STAR	Ilco	Natl/Curtis	Make
HBR2	H1098LA	B10	GM Ignition to 1966
OBR2	01098LA	B11	GM Trunk to 1966
HBR3	H1098NR	B24,RA2	Amer. Mtrs. Ign. 1960 on
OBR3	1098NR	B4,RA1	Amer. Mtrs. Trunk 1960 on
HBR5	H1098A	B40	GM Ignition "A"-1967
HBR9E	P1098E	B44	GM Ignition "E" - 1969 & 1973 (Ign. only-1977/81)
OBR9H	S1098H	B45	GM Trunk "H" - 1969 & 1973 (Door also-1977/81)
HBR10J	P1098J	B46	GM Ignition "J" - 1970 (Ignition only-1974/78/82)
OBR10K	S1098K	B47	GM Trunk "K" - 1970 (Door also-1974/78/82)
HBR11	1970AM	RA4	Amer. Mtrs. Ignition - 1970 square head - 1960 on
OBR11	S1970AM	RA3	Amer. Mtrs. Ignition - 1970 ellipse head - 1960 on
HBR12A	P1098A	B48	GM Ignition "A" - 1971 & 1967 (Ign. only-1975/79)
OBR12B	S1098B	B49	GM Trunk "B"-1971 & 1967 (Door also-1975/79)
HBR14C	P1098C	B50	GM Ignition "C" - 1968/72 (Ign. only-1976/80)
OBR14D	S1098D	B51	GM Trunk "D" - 1968/72 (Door also-1976/80)
HFD4	1127DP	H27	Ford Ignition 1952-64
OFD4	1127ES	H26	Ford Trunk 1952-64
HFD10	1167FD	H33,H51	Ford Ign.Double Side 1965 on
OFD10	S1167FD	H32,H50	Ford Trk.Double Side 1965 on
OPL4	1759L/1764S	Y141,B29-31	Chrysler Trunk 1959-65
HPL68	1768/69/70CH	Y152/150/148/146	Chrysler Ign.Master 1956 on
OPL68	S1768/69/70CH	Y151/149/138	Chrysler Trk.Master 1966 on

HOUSE

STAR	Ilco	Natl/Curtis	Make
5CO1	1001EN	CO7	Corbin X1-67-5
5DE3	D1054K	DE6	Dexter 67
5EA1	1014F,X1014F	EA27,EA50	Eagle 11929 - Harloc
5HO1	1170B	HO3/HO1	Hollymade - Challenger K1
5IL1	1054F	IN1	Ilco 1054F - Keil 159AA
5IL2	1054K	IN3	Ilco - Dexter - Weslock
5IL7	X1054F	IN21,IN18	Ilco X1054F
5KE1	1079B	K2	Keil 2KK
5KW1	1176	KW1,PT1	Kwikset 1-1063 - Donner - Petco
5LO1	1004	L1	Lockwood B308 - 5 Pin
MA1	1092	M1	Master 1K,77K
5RO4	R1064D	NA5, NA6	Rockford 411-31A
5RU2	1011P	RU4	Russwin 981B
5SA1	01010	S4	Sargent 265U
5SE1	1022	SE1	Segal K9 - Norwalk - Star
5SH1	1145	SC1	Schlage 35-100C(923C)
SH2	1307A	SC6	Schlage 35-180(920A)
SH6	1307W	SC22	Schlage 35-200(927W)
5TA4	1141GE	T4,T7	Taylor 111GE
5WK1	1175, 1175N	WK1/WK2	Weslock 4425
5WR2	1054WB	WR2,WR3	Weiser 1556 - Falcon
4YA1	999B	Y145,Y220	Yale 9½
5YA1	999	Y1	Yale 8 - Acrolock

3-17 Key comparison chart for automotive and residential key blanks. (courtesy Star Key and Lock Co.)

Go to the cross-reference list and find that the IN1 goes to the Star blank 5IL1 (Ilco house lock). The 1127DP is an Ilco key, for which you do have the Star blank, which is HFD4, going to a Ford auto ignition. Rapidly looking at the key blanks on the way, you now quickly select the two proper keys. Within minutes you have another satisfied customer, thanks to the cross-reference listing.

The cross-reference listing is available from your key blank representative, or you can spend between $20 and $50 and get a complete cross-reference book. This is both an advantage and a disadvantage. With the full book, you may have to look in a number of sections, whereas with the smaller individual cross-reference breakout from your key distributor/manufacturer, you are more likely to find the required key quickly.

A copy of a cross-reference listing is a prime requisite for every shop. Remember, go first with the listing obtained from your key manufacturer; the book can come later, if it is ever really needed.

4

The warded lock

*T*he warded lock is the oldest lock still in use and is found in all corners of the world. It employs a single or multiple warding system. Because of its simple design, its straightforward internal structure, and its easily duplicated key, this lock is an excellent training aid for locksmiths. This same simplicity means that warded locks give very little security. Use these locks only in low-risk applications such as storage sheds and rooms where absolute security is not essential.

At one time, warded locks were used on most doors. These locks are still found in abundance in older buildings still standing in decayed metropolitan neighborhoods, such as the center city of Philadelphia, Market Street in San Francisco, the Old Town section of Chicago, and the East Side of New York City.

The oldest of these buildings have the cast iron locks on the doors; some date back to the last century. Later types were made of medium gauge sheet metal. The casing consists of two stampings, the cover plate and the backplate. The latter mounts the internal mechanism and forms the sides.

The warded lock derives its name from the word *ward*, meaning *to guard*. The interior of the lock case has protruding ridges or wards that help to protect against the use of an unauthorized or improperly cut key. Normally, there are two interior wards positioned on the inside of the cover and backing plate, directly across from each other.

This lock is sometimes mistakenly referred to as a *skeleton-key* lock. The proper and full name is the *warded-bit key* lock.

TYPES

Two types of warded locks are currently in use, the *surface-mounted* or *rim* lock and the *mortised* lock (FIG. 4-1). While both types are similar in structure and size, the degree of security they give varies. The internal mechanisms of both operate on the same principle, but the mortise lock may have several additional parts. Differences between these locks are as follows:

Surface-mounted (rim) lock	Mortised ward lock
Mounted on door surface.	Mounted inside of door.
Secured by screws in the doorface.	Secured by screws in the side of the door at the lock faceplate.
Door can be any thickness.	Door must be thick enough to accommodate.
Thin case.	Fairly thick case.
Short latchbolt throw.	Up to 1″ latchbolt throw.
Lock from either side.	Locked from either side.
Strike can be removed with door closed.	Strike cannot be removed with door closed.
Very restricted range of key.	Restricted range of key changes.
Very weak security.	Weak security.

CONSTRUCTION

The basic interior mechanism is drawn in FIG. 4-2. Since the relative security of any lock lies in the type of key used, the number of key variations possible, and the amount of access to the locking mechanism afforded by the keyhole, the warded lock is the least secure.

If a lock were designed to have no more than 10 different key patterns (changes), and 1000 locks were made, 10 different keys would open all 1000 locks. By the same token, one key would open the lock it was sold with and 99 others. Furthermore, it is often possible to cut away parts of a key to pass (negotiate) the wards of all 1000 locks. From this you can see that lock security is related to the kind and number of key changes built into the system when it is initially designed.

The keyhole is an access route to the interior mechanism of the lock. The larger the keyhole, the easier it is to insert a pick or other tool and release the bolt.

In theory, each warded lock may be designed to accept 50, or even 100 slightly different keys. In practice, these locks tend to become more selective as they age and wear. The lock

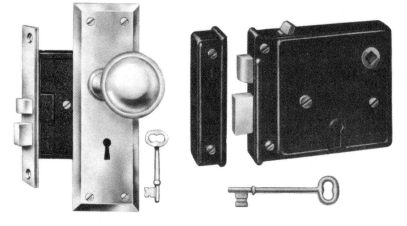

4-1 Warded locks are available in mortised (left) and surface-mounted varieties. (courtesy of Taylor Lock Co.)

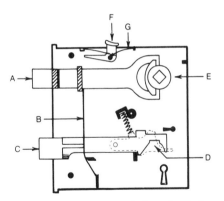

4-2 The internals of a typical warded lock.

may respond to the original key or to one very much like it, but other keys that would have worked when the mechanism was new no longer fit.

While this may seem fine and well for the lock owner, excessive wear increases the potential of key breakage within the lock and of jambs in the open, partly open, or closed position. It can also mean that a new lock will have to be installed.

Most surface-mounted and mortised locks are intended to be operated from both sides of the door. Keyholes and doorknob spindle holes extend through both sides of the lock body. Occasionally you will encounter a surface-mounted lock with a doorknob spindle and keyhole only on one side. The other side is blanked off.

A lock of this type can be modified to accept a key from the other side. This modification entails cutting a keyhole through the door and lock body and may require some filing on the key. Figure 4-3 illustrates the differences in keys. Note the additional cut on the left-hand key.

4-3 The key on the right is for a lock activated from one side of a door; the key on the left can pass the wards from either side.

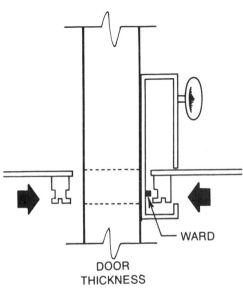

WARD

DOOR
THICKNESS

OPERATION

The key must be cut to correspond to the single or multiple side and end wards, that have been designed into the lock. After the key passes these wards, it comes into contact with the locking mechanism. The cuts on the key lift the lever to the correct height and throw the dead bolt into the locked or unlocked position.

Turning the doorknob activates the spindle and, so long as the dead bolt is retracted, releases the door.

Figure 4-4 depicts various keyhole control features that allow only certain types of cut keys to enter the keyhole. Figure 4-5 shows a key entering a keyhole. Notice that the key has the appropriate side groove to allow it to pass through the keyhole and into the lock. If you were to file off this obstruction (called a *case ward*), any key thin enough to pass could enter. (Some ward bit keys are quite thick.) By the same token, a very thin key is able to pass whether or not the side ward is present (FIG. 4-6). The common skeleton key is a prime example of this; it is thin enough to pass most case wards, but it will not necessarily open the lock!

Figure 4-7 shows a key engaging the bolt. While there is only one set of wards in this particular lock, because this ward is set where it is, it gives more security than a lock with no ward, or one with a ward that has been worn down to almost nothing.

REPAIR

Warded locks are not repaired to any great extent since replacement is usually cheaper. You should, however, have a supply of spare parts for these locks. The most frequent failure is a broken spring. Over a period of time, the spring may crystallize where it mates with the bolt. In addition, the wards can break or wear down into uselessness.

Replace a broken spring with a piece of spring stock, cut to length and bent to the correct angle. Some spring stock must be tempered before use; other springs come already tempered. If tempering is necessary, heat the spring cherry red and quench it in oil. You can

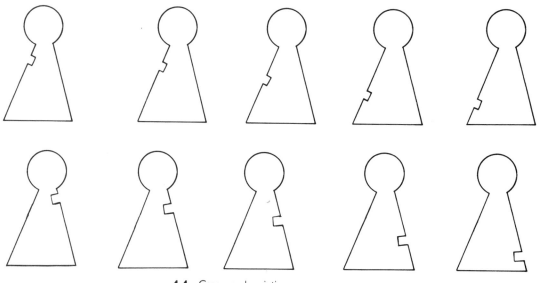

4-4 Case ward variations.

4-5   A slot milled on the edge of the key allows the
key to pass the case ward.

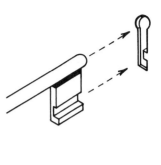

4-6   A skeleton key is one that has been ground down
to defeat the case ward.

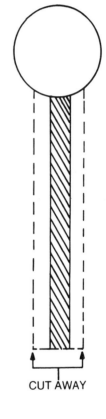

CUT AWAY

4-7   The single ward is better than none.

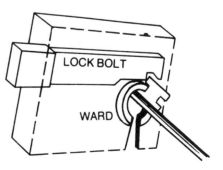

LOCK BOLT

WARD

save time by purchasing standard springs already bent into a variety of shapes that are designed to fit almost all locks.

Worn or broken wards on locks with cast cases can be repaired by drilling a small hole in the case and forcing a short brass pin into the lock case. The best technique for brittle cases is to braze a piece of metal on the case at the appropriate spot and file it down to the appropriate size. Wards on locks with sheet metal cases can be renewed by indenting the case with a punch ground to a fairly sharp point. If the factory has already punched out the wards, it would be best to braze a piece of metal at the proper spot.

Since most of these locks are inexpensive and offer minimal security, you should remind the customer that the cost of repair may far exceed the cost of the lock. Purchasing a new and more secure lock has definite advantages, and since you are already there, the homeowner's cost would then be less than ordering one and having you make a second trip to install it.

Should the homeowner decide to take your advice and purchase a new lock, ask to keep the broken one; it is of no use to him and parts are always nice to have. Sooner or later you will have to repair another lock of the same type; having the correct parts at hand will save you time. Furthermore, you have made a sale, and by being allowed to keep the old lock, have obtained parts at no cost.

Removing paint from these locks is also a form of repair. A major cause of lock failure is the home painter. He does not take the time to remove the lock or to cover it with masking tape. Some paint usually gets into the mechanism. To clean a paint-bound lock, follow this procedure:

Remove the lock from the door. (Run a sharp knife around the edges, so the new paint will not be cracked and broken.) Disassemble the lock. Using a wire brush, scrape the paint from the parts. In extreme cases, you may have to resort to a small knife to do the job or else soak the individual parts in paint remover. Dry each part thoroughly. Check for rust and worn parts and replace as needed. Assemble and mount the lock.

Use paint remover only as a last resort, since it leaves a residue that attracts dust and lint. When you use paint remover, you must clean each part before assembly.

Locks that are difficult to operate usually have not been lubricated in a long time, if ever. Never use oil to lubricate a lock. The professional approach is to use a flake or powdered graphite. Apply the lubricant sparingly. Remember, a little bit goes a long way, and this is especially true of graphite. Should you overuse it, you may have to explain why there is a dark patch on the carpet that cannot be cleaned . . . graphite stains are almost impossible to remove.

WARDED KEYS

Warded keys are made of iron, steel, brass, and aluminum. Iron and aluminum keys have a tendency to break or bend within a relatively short time; steel and brass keys can outlast the lock. The warded key has seven parts, as shown in FIG. 4-8. The configuration of the box, length of the shank, and the relative thickness of the shoulder are not critical to the selection or the cutting of the key.

Types of warded keys

Warded keys come in various types: the *simple* warded key, the *standard* warded key, the *multicut* key, and the *antique* key. Simple warded keys are often factory-made precut keys which fit several different keyholes. Multicut keys, on the other hand, are designed for specific locks. The standard warded key is usually mass-produced, but it has more precut ward and end cuts than the simple warded key. Standard warded keys can be easily

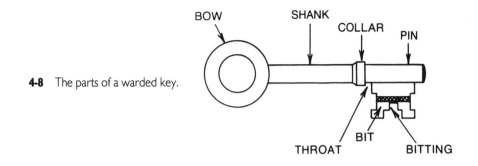

4-8 The parts of a warded key.

converted into master keys by cutting. The antique key may have several kinds of cuts: ward cuts, end cuts, and even side (or bullet) groove cuts extending the length of the bit. Antique keys usually go to older locks, but these keys are still manufactured.

Selection of key blanks

A key blank is a key that has not been cut or shaped to fit a specific locking mechanism. When selecting a blank for a duplicate warded key, the following should be considered:

Pin size The pins of both keys (original and duplicate) must be the same diameter. If your eyesight cannot correctly determine whether or not they are the same size, you can use either calipers or a paper clip. Use the calipers to compare the diameter of the original key with the diameter of the duplicate. You can use a paper clip that has been wrapped around the original key to check the diameter of the duplicate.

Length From the collar to the end of the pin, both keys should be approximately the same length. This is important because this portion of the key enters the keyhole and operates the lock.

Height The height of the bitting (cuts in the bit) must be the same on both keys. If the bitting in the duplicate is higher, it should be filed down; if it is lower, another blank should be selected.

Bow The bows need not be identical, but generally should be closely matched.

Width The width of the bittings should be approximately the same. If the bitting on the duplicate is too wide, the extra thickness may prevent the duplicate from entering the lock.

Thickness The thickness of the bits should be the same. If the original key bit is tapered, the bit of the duplicate should also be tapered. You may have to select a blank with a thick bit that you can file down to the correct taper.

If you don't have a micrometer, Table 4-1, showing standard wire diameters, can help you determine the approximate diameter of warded key blanks. You can take a standard piece of wire with a known diameter and compare it with any key blank. Also, use a drill to determine the approximate diameter of warded key blanks. Insert the pin of a blank into the hole that matches the blank's diameter.

Duplicating a warded key by hand

Select the proper key blank.

Wrap a strip of aluminum, approximately 1½ inches × 2¼ inches or 2½ inches about the pin and bit of the original key with one edge against the collar (FIG. 4-9).

Table 4-1   Standard Wire Gauges
(inches and millimeters). 1 mm = 0.03937"; 1" = 25.4 mm.

Standard Wire Gauge Number	Inches	Millimeters	Standard Wire Gauge Number	Inches	Millimeters
4/0	0.400	10.16	7	0.176	4.47
3/0	0.372	9.45	7½	0.168	4.27
2/0	0.348	8.84	8	0.160	4.06
0	0.324	8.23	9	0.144	3.66
1	0.300	7.60	10	0.128	3.25
2	0.276	7.01	11	0.116	2.95
2½	0.264	6.71	12	0.104	2.64
3	0.252	6.40	13	0.092	2.34
3½	0.242	6.15	14	0.080	2.03
4	0.232	5.89	15	0.072	1.83
4½	0.222	5.64	16	0.064	1.63
5	0.212	5.38	17	0.056	1.42
5½	0.202	5.13	18	0.048	1.22
6	0.192	4.88	19	0.040	1.02
6½	0.184	4.67	20	0.036	0.91

Clamp the original key (wrapped in the aluminum) into a vise, bitting edge up. Ensure that the aluminum fits snugly about the key bit and pin.

Cut off excess aluminum around the bit of the key. Remove the aluminum strip and smoke the key with a candle.

After properly smoking the key, place the strip back on the bit and reclamp.

Using a warding file, cut down the aluminum strip—*not the original key*—until it is in the shape of the original key. Since the aluminum is easily bent, use the file only in one direction—away from you. The stroke should be firm and steady at first. As you file closer to the cuts of the original key, the strokes should be shorter and lighter. When the file just barely touches the original key and starts to remove the candle black, *stop and go no further*. If the candle black is removed and the shiny surface of the original key is revealed, you know that you have filed too deeply.

Fit the aluminum strip onto the key blank.

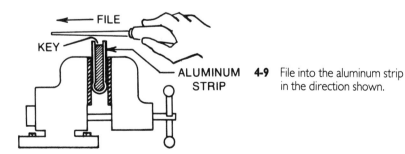

4-9   File into the aluminum strip in the direction shown.

File down the exposed areas on the bit until it matches the outline of the aluminum strip. Be careful not to cut into the strip. Again, use shorter and lighter strokes as you get closer to finishing each cut.

If the original key has a side groove that matches a keyhole ward, this too must be cut. Use another strip of aluminum. If the original key has two grooves you must wrap the strip around the bit so that both grooves are covered.

Using a scriber, scratch the metal strip to indicate the top and bottom of the groove(s). Fit the strip onto the key blank and mark the positions of the groove(s) on both ends of the bit. By connecting the marks with lines, you know exactly where to file.

To determine the depth of a groove cut, put one edge of a metal strip into the groove of the original key and scribe the depth of the groove on it. This mark will be your depth guide when filing the groove on the duplicate key.

The impression method

The term *impression* implies a duplicate key that is made from the lock itself as opposed to copying the original key. Impressioning is a skill that sets the locksmith apart from the individual who merely duplicates keys in the local five-and-dime store. It is a skill that is an extremely important and valuable asset.

Impressioning is the ability to decipher small marks made on a smoked key blank that has been inserted into a lock and turned. Interpretation of the marks tells the locksmith what cuts to make, where to make them, and how deep they must be. The advantage of impressioning is that you need not disassemble a lock or remove it from a door to make a key.

The first cut allows that blank to enter the keyhole. Refer to the previous section for information on what is required to properly make such a cut. Since you do not have the original key, smoke the end edge of the blank and insert it into the keyhole so the edge comes into contact with the case ward. Scribe mark the top and bottom of the ward on the blank. Remove the key; the candle black that was removed indicates the depth of the cut and the scribed marks show the position of the cut on the blank. Transfer the scribe marks to the near end of the bit and draw lines connecting the two pairs of marks. As in the previous section, use a small piece of metal to make a depth gauge so you will not file too deeply. Mount the blank in the vise and cut the ward slot. When you're finished, the key should pass the case ward.

The next step is the preparation of the blank for impressioning the internal wards. Recall that the key must pass certain side wards. When the lock is assembled, how can you be sure exactly where to cut the key so it will pass these wards? You can't. Thus, you will have to prepare the key blank. Two methods may be used; the first is to coat the key with a thin layer of wax. This is unprofessional and can harm the lock; the wax may clog the mechanism and require that the lock be disassembled and cleaned *at your expense*—you can't charge a customer for your mistake. The professional method is to smoke the key. The smoking must be thick enough to form a stable marking surface. If, for example, the blank is thicker than it should be, enough blackening must be present to give true readings. A thin coat will speckle as you turn the key and send you on wild goose chases.

The technique for impressioning the key is as follows:

Insert the key into the lock and turn it with authority.

Remove the key. You will notice one or more bright marks where the blackening has been removed. These marks indicate obstructions and you will have to file the appropriate cut for the key to pass.

Blacken, insert, turn, and remove the key; if a mark is present below the point you filed, the cut is too shallow. Carefully file it deeper.

When the cut is deep enough, use emery paper and clean off the fine burrs left by the file. This is the professional way.

Reinsert the key and turn it again. It should pass the wards. If it tends to stick slightly, a quick pass over the cut lightly with the warding file will alleviate this problem. Note: The cut should be square on all sides. Think of each ward cut as a miniature square, perfect and even on all sides.

Smoke the key, insert it, and turn it. The edges of the cuts should be shiny. The brighter the spot, the greater the pressure at that point. As the individual cuts are filed deeper, the bright spots grow dimmer. As this happens, you know that you are close to the point where you should stop filing.

Each time you insert the key to test the depth of the cut, you must be more cautious. You are working "in the blind;" a small overcut means the depth is permanently wrong on the blank and you will have to start over. This is the reason for making a few light, but firm, strokes as you near the completion of each cut. Take your time to make one perfect key instead of rushing the job and making many incorrectly cut keys.

The cuts on the key should be as deep and as wide as necessary, and no more. Overly large cuts interfere with the action of the lock, and may force you to cut another blank.

Once the key is completed, it must be *dressed out*. As mentioned earlier, use emery paper to remove burrs before they break off and fall into the lock. Now polish the entire key with emery paper. You might also give it a light buffing on your wheel for appearance. It does you no good to give the customer a dirty and smudged key. Show pride in your work!

Pass keys

Skeleton-type pass keys (master keys) are sold in variety stores. These keys will fit many old locks and more than a few new ones. As such, they are a convenience. But no reputable locksmith stocks them. Why? The ethics of the profession forbid it. You certainly do not want to supply someone with a key that could open his neighbor's lock. Nor should you duplicate a pass key without authorization from the owner. Be leery of a customer who wants a key duplicated, but with additional cuts. Locksmiths have lost their licenses for less. Don't let it happen to you.

5

The lever lock

*L*ever tumbler locks, or lever locks, have many uses in light security roles. Available in a variety of sizes and shapes, these locks are found on desks, mailboxes, lockers, bank deposit boxes, and so on.

Figure 5-1 illustrates a popular example. The circular "window" on the back of the case is a locksmith's aid—it reveals the heights of the levers without the necessity of dismantling the lock. Thanks to the window, it is relatively easy to make a key "in the blind."

PARTS

A lever lock has six basic parts:

- ☐ Cover boss.
- ☐ Cover.
- ☐ Trunnion.
- ☐ Lever tumblers (top, middle, and bottom).
- ☐ Bolt (bolt stop, notch, and post).
- ☐ Base.

These parts are shown in FIG. 5-2.

OPERATION

This lock requires a standard flat key. When the key is turned, the various key cuts raise corresponding lever tumblers to the correct height. As the levers are raised, the gates (FIG. 5-3) of the levers align and release the bolt. The bolt stop (some call it the post) must pass through the gating from the rear to the front "trap" or vice versa, either unlocking or locking the lock.

LEVER TUMBLER

The number of lever tumblers varies. Most locks have no more than five, although deposit box locks have more. The lever tumbler consists of six parts:

□ Saddle (or belly).
□ Pivot hole.
□ Spring.
□ Gating slot.
□ Front trap.
□ Rear trap.

Each lever is a flat plate which is held in place by a pivot pin and a flat spring. Each lever has a gate cut into it. The gates are located at various heights either with the saddle aligned for all levers, or staggered. The latter approach has long since been antiquated. When the levers are raised to the proper position, the gates are open and the bolt post can be shifted from one trap to another, thus locking or unlocking the lock.

Since the bolt post meets no resistance at the gating, the lock will work properly. On some designs the edge of the lever has serrated notches. The bolt post has corresponding notches. The notches on the lever and the bolt jam together if an improperly cut key is used. This effectively stops the bolt from passing through the gating and keeps the lock secure. Only a perfectly and properly cut key will open this type of lock. This feature adds immensely to the security of the lock.

Manufacturers have, over the years, developed a variety of lever types (FIG. 5-4). The operating principle is the same for all of them.

The width of the gate is also a critical factor in the operation of the lock. Some gates are just wide enough for the pin to pass through. A duplicate key, even slightly off on a single cut, will not work on the lock.

The saddle of the lever is also important. Recall from Chapter 1 that staggered saddles make it possible to cut a key by observation. In the case of modern lever tumblers, the gate traps have different heights, leaving the saddles in perfect alignment.

Gating changes are made by two methods. The most common is to substitute a lever tumbler with a different gate dimension. Locksmith supply houses stock a variety of levers, so all you are required to do is change the original for one with a higher or lower gate. The second method is to alter the tumbler by filing the saddle. This approach is used with levers whose movement is restricted at the gate. Unless the tumbler gating varies greatly, the curvature of the saddle must vary with the shape of the key.

When working on the lever lock, as on other types, it is best to have two available: one for disassembly to determine internal working parts, and the second for actual problem-solving use.

As I mentioned earlier, the typical lever lock contains two, three, five, or possible six levers. Bank deposit box locks may have as many as 14. Lever locks can be keyed individually,

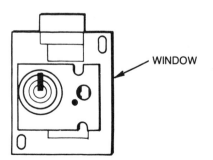

WINDOW

5-1 Note the "window" on the cover of this lever lock.

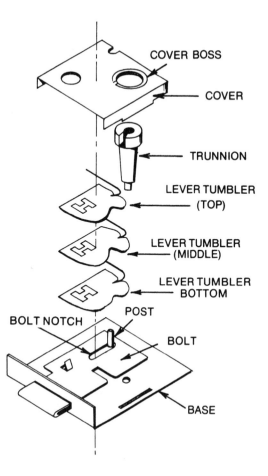

5-2   A lever lock in exploded view. This particular lock has three
tumblers; others may have a dozen or more.

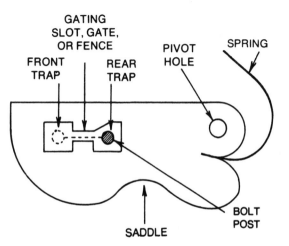

5-3   Lever-tumbler nomenclature. The key bears against the
saddle.

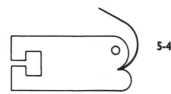

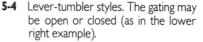

5-4   Lever-tumbler styles. The gating may be open or closed (as in the lower right example).

alike (two or more with the same key), or master keyed, depending upon the wishes of the buyer.

DISASSEMBLY

General-purpose lever locks come in three styles. In order of popularity these are the following:

- ☐ Solid case, usually spot-welded or riveted.
- ☐ Pressed form with the back and sides of the case one piece—small tabs from the sides bend to hold the cover in place.
- ☐ Cover plate secured by a screw.

Riveted or spot-welded locks should be discarded when they fail. It is cheaper to purchase a new lock; the time required to drill out the rivets or chisel through the spot welds costs the customer more than the lock is worth.

Locks secured by tabs can be disassembled and reassembled quite easily. To disassemble, insert a thin tool or small screwdriver under the flanges and pry upward. Remove the cover. Look for a small object jammed in the keyway; if something is found, remove it. At times like this you may wonder if being a locksmith is really worth it, but a locksmith must have patience with the small jobs as well as with the big ones.

Another common problem with lever locks is a broken spring. If it is not the top lever, then carefully remove each lever, in turn, placing them in a logical order so that they may be assembled as they were found. Remove the lever whose spring is broken. Select a piece of spring steel from your inventory. Cut and bend it to shape. If you have purchased an assortment of ready-made springs, select the proper one and replace the broken spring with a new one. Replace the levers and reassemble the lock.

SAFE DEPOSIT BOX LOCKS

Lever safe deposit box locks normally have a minimum of six and upwards of twelve or even fourteen levers. Two keys are required. One key is assigned to the individual who rents the box; the second or guard key is held by the bank. Both keys are needed to open the lock.

Disassembly is simple. Remove the screws and lift off the plate. You will notice that the lock is constructed differently than previously discussed lever locks. There are two sets of lever tumblers; two bolt pins must pass through the lever gates at the same time. Note the unique lever shape (FIG. 5-5).

Many safe deposit locks have a compression spring bearing against the upper lever that forces the lever stack down, allowing no play between them. This spring is an integral part of the lock's security mechanism. Without the spring, the levers would be able to move a fraction of an inch when a key or pick "irritates" them. This movement is enough to provide clues for the lockpick artist.

5-5 Notched tumblers are a security feature in some locks.

This lock, as stated earlier, requires two keys to open it. If one key could be turned to the *open* position by itself, it would mean that the key was faulty, the lock mechanism was in some way evaded, or the bolt post was bent or broken.

The levers have a V-shaped ridge that matches a similar V cut in the bolt post. Another key, even one that is 0.001 inch off on any cut, will mesh the notches in the post and lever. This is another built-in security feature of these locks.

Broken springs are the main difficulty. Or the levers may be at fault. The saddles may wear enough to affect the gate. In cases such as this, replace the lever with a new one. Do not file the gate cut wider to compensate. Sometimes a pivot post will work loose. Repair by rapping the post with a light hammer or by brazing. Never use solder. Slightly bent parts can be straightened out, but it is best to replace the entire damaged part.

Safe deposit locks are serviced by locksmiths who enjoy high standings within the community. Unless an emergency arises, a beginning locksmith or one who is a newcomer to the area would not be given the job. But this reluctance is only a matter of the conservatism of bankers. It does not reflect upon the skills of students and those working locksmiths who have had only a few months' experience. Many of these men and women could repair a safe deposit box lock without difficulty.

SUITCASE LOCKS

Suitcase locks are essentially simple but have been built to accept a staggering variety of keys. Ninety-nine locks out of a hundred are warded, with a primitive bolt mechanism to keep the case closed. Only a few, such as the Yale luggage lock, use a lever-type mechanism. These locks offer better security than warded locks.

The lock size, the depth the key is inserted, and the number of cuts are important clues to the lock type. A key for a lever lock will go ¼ inch or deeper into the lock before turning. Warded locks are shallower.

Opening suitcase locks is relatively simple. Many times one key will open several different suitcase and luggage locks. Many of these locks on the market are not designed for security. By cutting down almost any suitcase key, it is possible to make a skeleton key for emergencies.

LOCKER AND CABINET LOCKS

The Lock Corporation of America has come up with a new locking spring design accompanied by a free-turning keyway that combine to make the locks extremely jamproof and pick resistant in the flat key type of locks. Figure 5-6 shows such a lock.

These locks are made to last (and also carry an unconditional guarantee). They are constructed of a one-piece heavy duty steel case that will stand up to years of use, even under the most corrosive conditions. The hard wrought steel keys and the lock case have been electroplated and dichromated to ensure numerous years of trouble-free service.

Another advantage to these locks is that they have been designed to fit into current master key systems. The lock will fit left or right-hand steel locker or cabinet doors that have the standard piercings.

The keys can be removed from either the deadbolt or springbolt locks when the locker door is opened or closed. The deadbolt locks are available in models with the key removable in the locked or unlocked mode, or with the key removable in the locked position only. Also, with these locks, you have a self-locking springbolt with the key removable in the locked position only.

The series 4000 and 5000 flat key built-in lever tumbler (FIG. 5-7) specifications are as follows:

- Mounts in a standard three-hole piercing on steel lockers, cabinets and other shop equipment.
- Solid locking bolt moves horizontally with a ⁵⁄₁₆″ throw.
- Master keyed, keyed different, keyed alike, or group keyed.
- Key removable when locked, or when locked or unlocked (must be specified when ordered).
- .062″ gauge heavy-duty one-piece steel case.
- Free-turning keyway.
- Phosphor bronze locking springs.
- Five hard wrought steel levers.

The dimensions for the lock and the standard locker door punchings are shown in FIG. 5-8.

The model 5001 MK locker lock is a built-in flat key lock with a deadbolt locking mechanism; the key is removable in either the locked or unlocked position. A free-turning plug (with a minimum of five lever tumblers) provides for a minimum of 1400 different key changes. The locks (within a key change range specified) can be master keyed if desired. The lock will not be operated by keys of another lock within the specified key range, except a master key. The casing is of wrought steel and is 1⅜ inches × 1⅝ inches with top and bottom

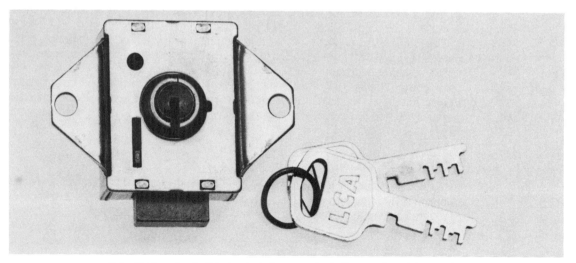

5-6   Flat key deadbolt locker lock. (courtesy Lock Corporation of America)

5-7 Flat key lever tumbler locker lock. (courtesy Lock Corporation of America)

attaching ears to fit the standard steel locker piercings. All parts are of phosphor bronze, zinc or rustproofed steel. The backset is 1 inch or less, and the solid locking bolt is ¾ inch × ¼ inch with a 5⁄16 inch throw.

Specifications are the same for the 5000 MK except for a key removable in the locked position only; the 4000 MK, differs only in having a solid spring bolt locking mechanism. The internal parts' relationship and positioning are illustrated in FIG. 5-9.

Looking at the inner workings of the LCA torsion lever flat key lock, we find that it contains a patented multi-movable lever flat key operated mechanism that has the latest improvements to the world's oldest and still most popular means of security—the warded lock.

Seven levers (including one master lever, if required) engage and restrain the locking bolt. The novel uni-spring construction is arranged in two banks of seven staggered interlocking arms constantly exerting a force on all the levers. By inserting and turning the precision cut key in the rotatable key guide, the latched levers thereby are moved from securing the deadbolt, enabling the lock to be opened.

Note: the model 4000 (basically the same) has a bolt split into two parts, including a separate spring which allows the bolt to retract without using a key to close or slam shut.

The grooved key built-in torsion tumbler cylinder lock (FIG. 5-10) has the following specifications:

□ Standard three-hole piercings for mounting.

□ Beveled locking bolt, moving horizontally.

□ Locking bolt is ¾″ × ⅜″ with a 5⁄16″ throw.

□ Over one million computerized key changes.

□ Eight nickel silver torsion tumblers (to provide the maximum security).

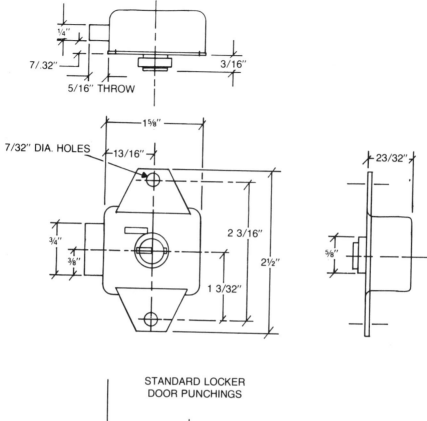

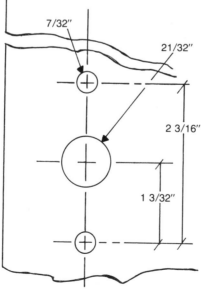

5-8   Specifications for the LCA series 4000 lever tumbler flat key lock. (courtesy Lock Corporation of America)

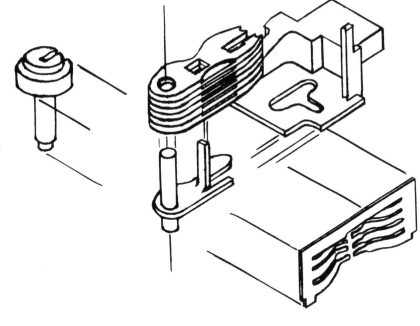

5-9 Internal parts positions for the 4000 and 5000 series flat key lever locker lock. (courtesy Lock Corporation of America)

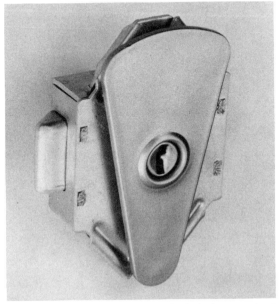

5-10 Built-in torsion tumbler cylinder lock. (courtesy Lock Corporation of America)

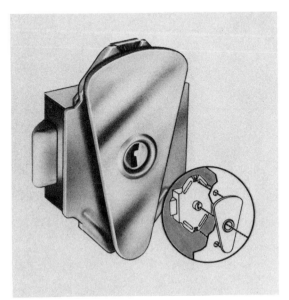

5-11 Series 6000 torsion tumbler lock with inset illustrating the snap-on face plate. (courtesy Lock Corporation of America)

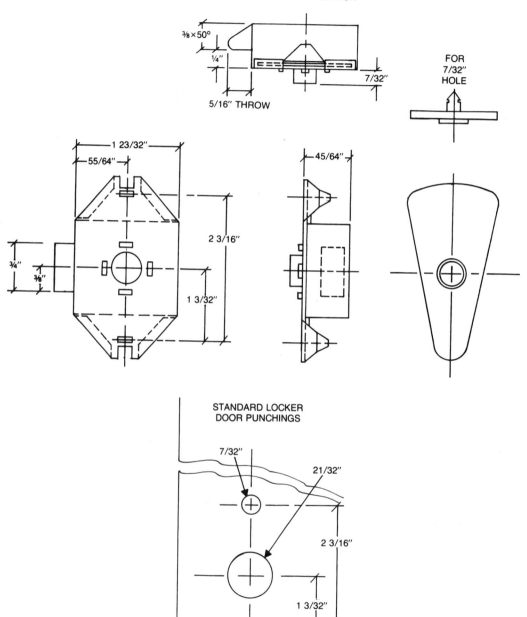

5-12 Specifications for series 6000/7000 "snap-on" installation lock (courtesy Lock Corporation of America)

- ☐ Regular, sub-master, master keyed, keyed different, keyed alike or group keyed.
- ☐ The master key design prevents the use of a regular key as a master key.
- ☐ The keys are embossed, nickel plated, and grooved.
- ☐ Cylinder plug of solid brass.
- ☐ .062″ gauge heavy-duty one-piece steel case construction.

The 6000 and 7000 series have a patented snap-on face plate for installation (FIG. 5-11). Technical drawings for the snap-on installation are at FIG. 5-12, along with the locker door hole punchings. Note that the series 6000 has a spring bolt, while the 7000 operates with a deadbolt.

The 7001 locker lock has a built-in high security type of ruggedness with a deadbolt locking mechanism and the cover plate; it is master keyed with the key removable in either the locked or unlocked position. Over 100,000 possible different key changes are available. The bolt is ¾ inch × ⅜ inch with a ⁵⁄₁₆ inch throw.

The LCA drawer and cabinet lock (FIG. 5-13) is constructed of high-pressure solid Zamak unicast casing and base and has a ⅞ inch in diameter cylinder case for use in up to 1⅜-inch thick material.

The torsion tumbler is based on a new and different design concept. Each of the eight nickel silver torsion tumblers is manufactured with its own integral spring; it is made strong and durable by an exclusive forging process. This tumbler-spring design will provide a much greater protection against unfavorable climatic conditions such as temperature and humidity changes because the torsion spring is heavier and sturdier than the finely coiled springs required by either pin or disc tumbler cylinders.

Variation in the eight tumblers (arranged four to each side of the cylinder) and in the double bitted key provide over one million key change possibilities. The unique open construction of a one-piece tumbler and spring eliminates clogging due to accumulated dirt

5-13 Eight torsion tumbler cylinder drawer lock. (courtesy Lock Corporation of America)

and small particles, and offers long and high performance reliability, smooth operation, and improved pick resistance.

Better master key control and protection are provided by an entirely different master key configuration. This special key design prevents the fashioning of a master key from regular keys or blanks, thus reducing the possibility of unauthorized entry. Practically unlimited master key changes are available without diminishing the security of the cylinder.

Figure 5-14 provides details of the maximum security eight torsion tumbler cylinder in an exploded view. Figure 5-15 shows the critical dimensions of the cylinder.

The LCA push lock for sliding door and showcases uses the same patented torsion tumblers and solid brass cylinder plug, and it has a stainless steel cap (FIG. 5-16). Three hundred key changes are standard, but unlimited changes are available.

The various series 3300 and 3400 locks can have a push bolt, automatically locked by pushing the cylinder inward or releasing the cylinder and catch by turning the key 90 degrees. The bolt is ⅜ inch in diameter with a ¹⁵⁄₃₂ inch throw and a ⅞-inch cylinder case for use in up to 1⅛-inch thick material mountings.

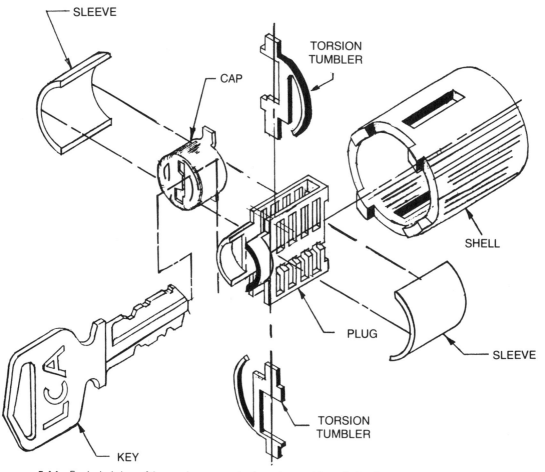

5-14   Exploded view of the maximum security 8-torsion tumbler cylinder. (courtesy Lock Corporation of America)

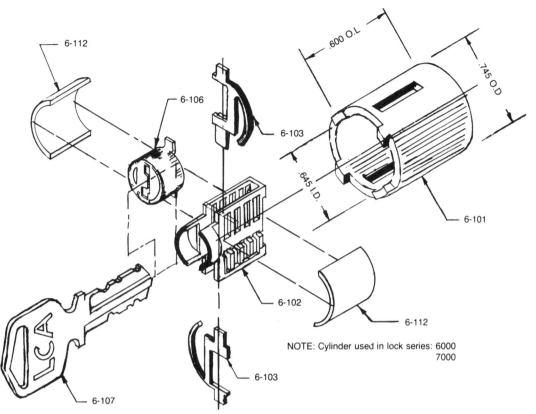

5-15 Critical cylinder dimensions of the 8-torsion tumbler cylinder lock. (courtesy Lock Corporation of America)

5-16 Sliding door and show case locks. (courtesy Lock Corporation of America)

The push-turn bayonet type lock in this LCA series has a different strike plate (FIG. 5-17). It is locked by turning the key 90 degrees, pushing the cylinder in and rotating the key again to lock the "T" type plunger behind the strike plate, thus preventing the doors from being pried apart. The cylinder locks in either the *in* or *out* positions for further protection. With the 3350 model lock, a 7/16 inch in diameter case hardened steel covered bolt is available. Figure 5-18 shows the 3300 model technical dimensions, while FIG. 5-19 provides the same information on the 3400 model.

For the locksmith concerned with businesses, these locks are a stock in trade, constantly proving themselves over and over.

LEVER LOCK KEYS

Unlike the warded or cylinder key, the lever lock key used by the average person almost never contains a keyway groove running along its side; it is a flat key. A number of different flat keys exist. Figures 5-20 and 5-21 show some of them. Figure 5-22 identifies the various parts of a typically cut lever lock key.

The lever lock key can be cut by hand or machine. In order to make the key you must have the proper key blank. The three critical dimensions of the lever lock key are the thickness, length, and height. If the key blank is slightly higher or wider than the original, the blank should be filed down to the proper size. If it is thicker, select another blank. Filing down the thickness of the blank weakens it structurally.

Laying the keys (original and blank) side by side, run your finger lightly across them. If the blank is thicker than the original, you will feel your finger catch as it passes from one to the other. When you insert the key into the keyway, it should not be tight fitting or bind.

The first cut that must be made is the *throat cut*, which enables the key to turn within the keyway. To do this, insert the blank in the keyway and scribe each side of the blank where it comes in contact with the cover boss. Determine the point where the trunnion or pin of the lock turns. Draw a vertical line there to indicate the depth of the throat cut. Remove the key

5-17 Grooved key push lock for sliding doors and display/show cases. (courtesy Lock Corporation of America)

LCA MODEL #3300

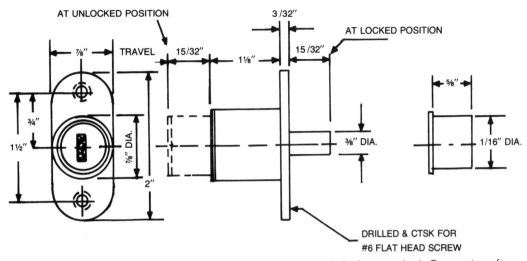

5-18 Technical dimensions for the sliding door push button lock. (courtesy Lock Corporation of America)

LCA MODEL #3400

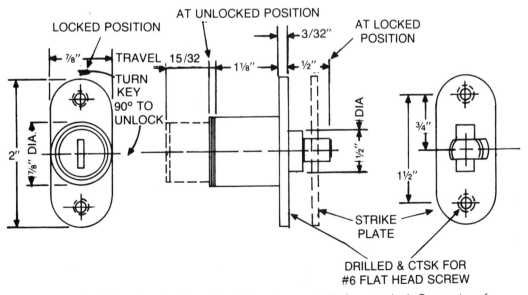

5-19 Technical dimensions for the push and turn showcase lock. (courtesy Lock Corporation of America)

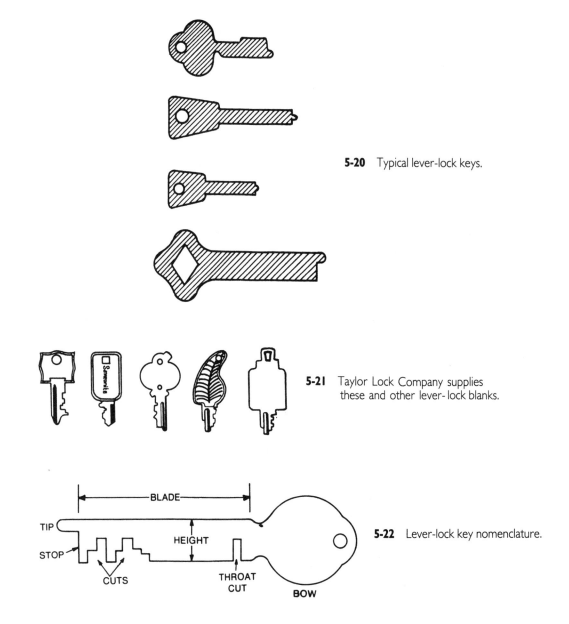

5-20   Typical lever-lock keys.

5-21   Taylor Lock Company supplies these and other lever-lock blanks.

5-22   Lever-lock key nomenclature.

from the lock, place it in a vise, and use a 4 inch warding file to cut alongside the vertical line to the proper depth. Cut on the side of the line towards the tip of the key.

There is a small round window in the back of most lever lock cases. The window is positioned, so you can see where the bolt pin meets the lever gates. You can get a general idea of the proper cuts to be made on a key blank by observing the lever action through the window.

After the throat cut is filed, the other cuts must be made. Follow these steps:

Smoke the blank, insert it in the keyway, and turn. Remove it. The lever locations will be marked on the blank.

File the marks slightly, starting with the marks closest to the throat cut.

Insert the key and turn it. Notice (through the window) the height that each lever comes up and the position of the pin in relation to the lever gates. The distance each gate is from the pin indicates the depth the key should be cut for each particular lever. File each key cut a little at a time, periodically inserting and turning the key to check the gate/pin relationship. Be sure to file *thin* cuts. Continue until you can insert the key and have the gate and pin line up exactly.

If the key in the keyway binds, observe the levers. One or more may not be correctly aligned. If the pin is too high, you have cut too deeply; if the pin is too low, you have not cut deeply enough. It may be necessary to resmoke the blank and reinsert it. The point where the key binds hardest will be indicated by the shiniest spot on the key. Just a touch with the file will usually alleviate the problem. Ensure that each filed cut is directly under its own lever.

At this point, you should have corrected any variations between the original key and the blank. The dimensions must be identical: the key height, thickness, and the length.

You will need a vise for holding the two keys, a small C-clamp, a warding file, a candle, and a pair of pliers. Follow this procedure:

Holding the original key over a candle flame, smoke it thoroughly.

Allow a few minutes for the key to cool; clamp the original and the blank at the bows. Most locksmiths use a C-clamp for this initial alignment.

Once aligned, place the key and blank in a vise. If you wish, you can leave the C-clamp in position.

Using the warding file, make the tip cut first. File in even and steady strokes, bearing down in the forward, or cutting stroke. Keep a careful eye on the original. Stop when the file just disturbs the blackening.

Once the tip cut is completed, move to the next cut.

Remove the keys from the vise and inspect the cuts. Each should be rectangular and flat.

Using emery paper, lightly sand away the burrs on the edges of the cuts. Wipe off the candle black from the original key.

Test the duplicate in the lock. Should it stick, blacken the duplicate and try it again in the lock. Breaks in the blackening will show where the key is binding. A light stroke with the file should correct this.

6

The disc tumbler
lock

*T*he *disc tumbler* (or *cam*) lock gets its name from the shape of the tumblers. These are about as secure as lever locks, and as such are superior to warded and other simple locks. However, disc tumbler locks are far less secure than pin tumbler locks.

Disc tumbler locks are found in automobiles, desks, and in a variety of coin-operated machines. Some padlocks are built on this principle. Because the cost of manufacture is very low, replacement is cheaper than repair.

While similar to the pin-tumbler lock in outside appearance and in the broad principle of operation, the internal design is unique.

The disc tumblers are steel stampings, arranged in slots in the cylinder core. Figure 6-1 shows a typical disc tumbler. The rectangular hold, or cutout, in the center of the disc matches a notch on the key bit. The protrusion on the side, known as the *hook*, locates the spring. The disc stack is arranged with alternating hooks, one on the right of the cylinder, one on the left. The complete lock is illustrated in FIG. 6-2.

OPERATION

The disc lock employs a rotating core, as does the more familiar pin-tumbler design. The disc tumbler core is cast, so that the tumblers protrude through the core and into slots on the inner diameter of the cylinder. The core is locked to the cylinder, so long as the tumblers are in place.

The key has cuts that align with the cutouts in each tumbler. The key should raise the tumblers high enough to clear the lower cylinder slot, but not so high as to enter the upper cylinder slot. In other words, the correct key will arrange the tumblers along the upper and lower shear lines (FIG. 6-2). The plug is free to rotate and, in the process, throw the bolt.

The key resembles a cylinder pin-tumbler key, except that it is generally smaller and always has five cuts. A cylinder pin-tumbler key might have six or seven cuts.

DISASSEMBLY

Good quality disc locks feature a small hole on the face of the plug, usually just to the right of the keyhole. Insert a length of piano wire into the hole and press the retainer clip. Turn the plug slightly to release. The key gives enough purchase to withdraw the plug. If a key is not

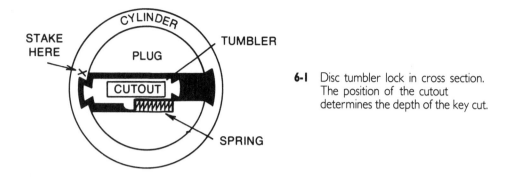

6-1 Disc tumbler lock in cross section. The position of the cutout determines the depth of the key cut.

available, you can extract the plug with the help of a second length of piano wire inserted into the keyhole. Bend the end of the wire into a small hook. Other locks attach the plug to the cylinder with the same screw that secures the bolt-actuating cam. Others (fortunately, a minority) have the plug and cylinder brazed together. File off the brass.

KEYING

Manufacturers have agreed upon five possible positions for the cutouts relative to the tumblers. Keying is a matter of arranging the tumblers in a sequence that matches the key cuts. Once the sequence has been identified, install the tumbler springs over their respective hooks and mount the tumblers in the plug. The tumblers are spring-loaded and, until the plug is installed in the cylinder, are free to pop out. Lightly stake them in place with a punch or the corner of a small screwdriver blade. One or two pips are enough, since the tumblers will have to be broken free once the assembly is inside the cylinder. Inserting the key is enough to release the tumblers.

SECURITY

These locks have no more than five tumblers, and each tumbler cutout has five possible positions. These variations allow 3125 or (5^5) key changes, at least in theory. In practice, the manufacturer will discard some combinations as inappropriate and may further simplify

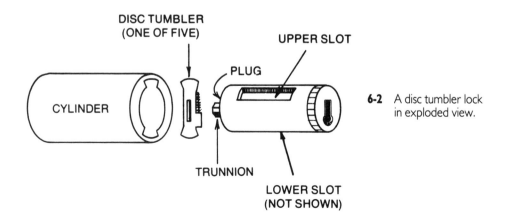

6-2 A disc tumbler lock in exploded view.

matters by limiting the key changes to 500 or less. Obviously, disc tumbler locks are not high-security devices.

CAM LOCKS

Cam locks are used for many general and special purposes for a variety of needs in business (FIG. 6-3). More than likely, you will see approximately 90 percent of all these units in office situations. From the face (front) these locks look pretty much the same after they have been installed. Beyond that point, however, they may differ in various regards. The illustrations on the next several pages provide exploded views of different models and the various parts associated with each model. These models are all five-disc tumbler locks, each with a possibility of 200 different key combinations. Because of the key variations possible, these locks are often master keyed prior to purchase by the customer. Only after disassembly or by viewing the tumblers through the keyway will you know for sure whether or not the lock is master keyed (FIGS. 6-4 through 6-9).

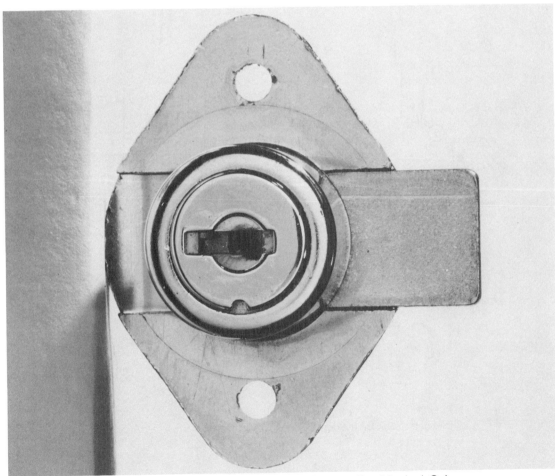

6-3   Five-tumbler wooden desk cam lock. (courtesy Dominion Lock Co.)

Parts
Desk Lock

1. Nut
2. Bolt
3. Shell
4. Retainer clip
5. Spring
6. Plug
7. Springs
8. Disc tumblers
9. Key

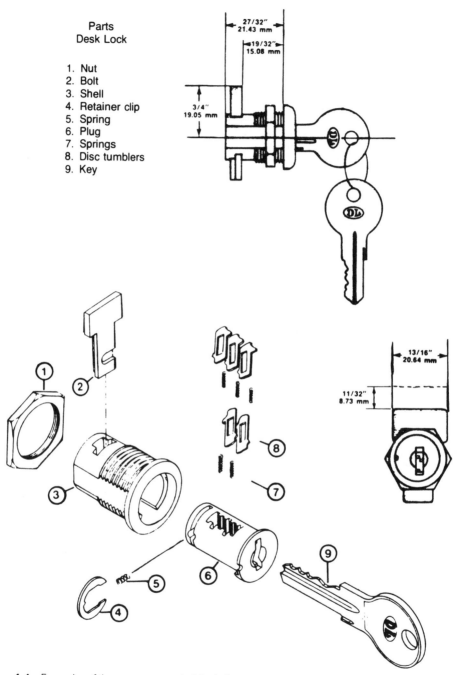

6-4   Examples of the common use desk locks in exploded view. (courtesy Dominion Lock Co.)

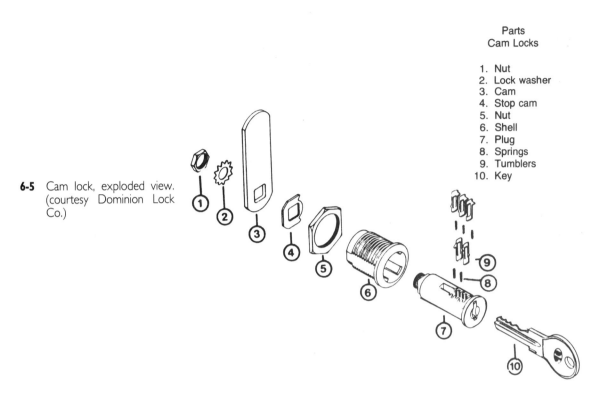

6-5 Cam lock, exploded view. (courtesy Dominion Lock Co.)

Parts
Cam Locks

1. Nut
2. Lock washer
3. Cam
4. Stop cam
5. Nut
6. Shell
7. Plug
8. Springs
9. Tumblers
10. Key

Two other core type cam locks are shown in FIGS. 6-10 and 6-11. These cam locks have a removable core, and are also of the seven-pin variety which uses the Ace-type circular key.

The variety of cams is used with these locks. Figure 6-12 is a chart showing the various types of cams available. From the types available you can find a cam for the lock that will meet every possible need. Included are the length and offset specifications also.

Figure 6-13 shows the hook, bent, double-ended and other miscellaneous cams that you may come across in the course of your work.

The standard thickness of the cam is ³⁄₃₂ inch (2.667 mm); they are of steel, with many being cadmium plated for durability and longevity.

READING DISC AND LEVER LOCKS

It's not unusual for a locksmith to be asked to make a key for a lock when the original key has been lost or misplaced. If the lock does not have a code number on its face or if the owner neglected to write down the number on the key, the locksmith has three choices. He can pick the lock, impression a key, or "read" the lock.

Like impressioning, reading a lock is a skill that must be developed through patient practice. You cannot expect to master this skill quickly, nor can you expect to remain proficient in it without constant practice. At first, practice daily, then weekly or twice-weekly to maintain your skill.

When called upon to fit a key, either by impressioning or reading of the lock, the locksmith invariably looks into the lock keyway. A quick glance determines whether it is a lever, disc, or pin-tumbler lock.

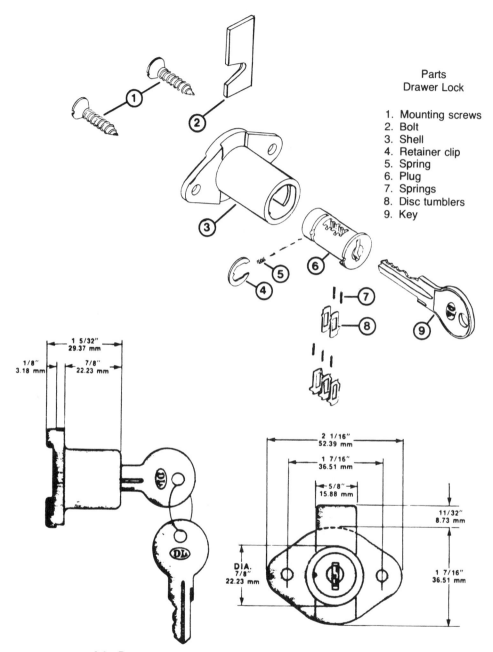

Parts
Drawer Lock

1. Mounting screws
2. Bolt
3. Shell
4. Retainer clip
5. Spring
6. Plug
7. Springs
8. Disc tumblers
9. Key

6-6   Drawer lock, exploded view. (courtesy Dominion Lock Co.)

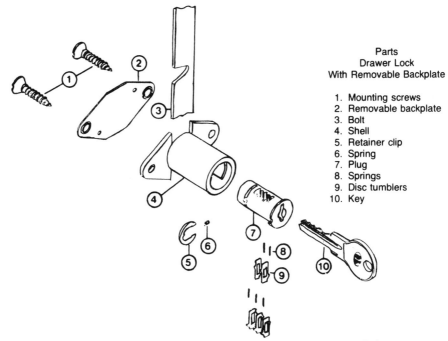

Parts
Drawer Lock
With Removable Backplate

1. Mounting screws
2. Removable backplate
3. Bolt
4. Shell
5. Retainer clip
6. Spring
7. Plug
8. Springs
9. Disc tumblers
10. Key

6-7 Drawer lock, exploded view. (courtesy Dominion Lock Co.)

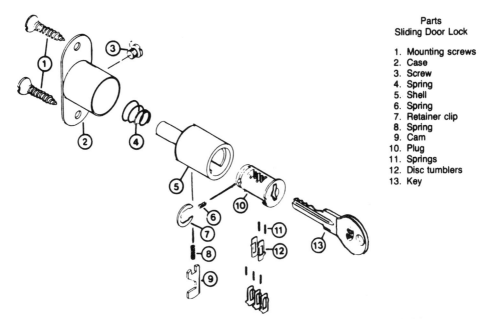

Parts
Sliding Door Lock

1. Mounting screws
2. Case
3. Screw
4. Spring
5. Shell
6. Spring
7. Retainer clip
8. Spring
9. Cam
10. Plug
11. Springs
12. Disc tumblers
13. Key

6-8 Sliding door lock, exploded view. (courtesy Dominion Lock Co.)

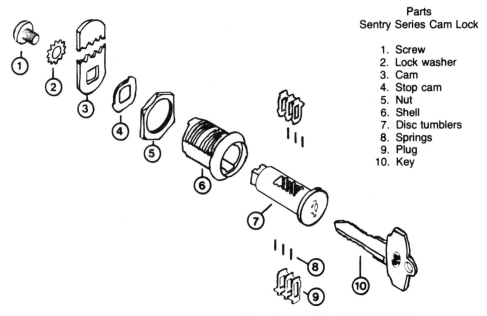

Parts
Sentry Series Cam Lock

1. Screw
2. Lock washer
3. Cam
4. Stop cam
5. Nut
6. Shell
7. Disc tumblers
8. Springs
9. Plug
10. Key

6-9   Sentry series cam lock, exploded view. (courtesy Dominion Lock Co.)

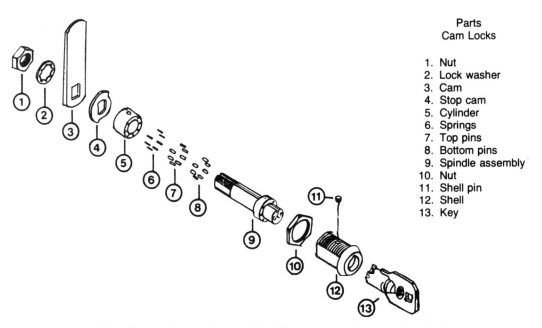

Parts
Cam Locks

1. Nut
2. Lock washer
3. Cam
4. Stop cam
5. Cylinder
6. Springs
7. Top pins
8. Bottom pins
9. Spindle assembly
10. Nut
11. Shell pin
12. Shell
13. Key

6-10   Ace type key cam lock, exploded view. (courtesy Dominion Lock Co.)

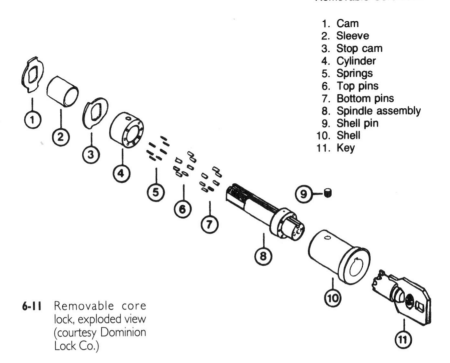

Parts
Removable Core Lock

1. Cam
2. Sleeve
3. Stop cam
4. Cylinder
5. Springs
6. Top pins
7. Bottom pins
8. Spindle assembly
9. Shell pin
10. Shell
11. Key

6-11 Removable core
lock, exploded view
(courtesy Dominion
Lock Co.)

Disc locks

The view down the keyway of a disc tumbler lock will show a row of discs with their centers
cut away and staggered. The cutaways are at different heights; the discs themselves are the
same diameter. The cutaway looks like a small staircase with a surrounding wall (formed by
the vertical edges) about the steps. Since each disc is the same size, only the height of the
"steps" varies, and this variation is predetermined. A No. 1 disc has its cutaway toward the top
of the tumbler; a No. 5 has its cutaway situated low on the tumbler (FIG. 6-14).

Here is where the skill of reading a disc lock is learned. Study the discs through the
keyway. Notice the relationship of the discs to each other and to the keyhole. Through
constant study and practice, which includes mixing the various discs, you will be able to
determine which disc is which (FIG. 6-15).

Lift each disc and compare it to the disc in front of or behind it. To do this, you will need
a reading tool. This is nothing more than a stiff length of wire about three inches long,
mounted in a small dowel handle. Think of the tool as a long hairpin attached to a short piece
of wood for convenience in holding. The wire should be bent slightly, so that you can see the
tumblers.

Insert the tool into the lock, holding it so you can see the interior and observe the discs.
By shifting the tool about, raising and lowering each disc, you can see the relationship of
each disc cutaway to the next disc and to the keyway. Using your knowledge about the
cutaway relationships, you will be able to decode the tumblers and cut a key for the lock.

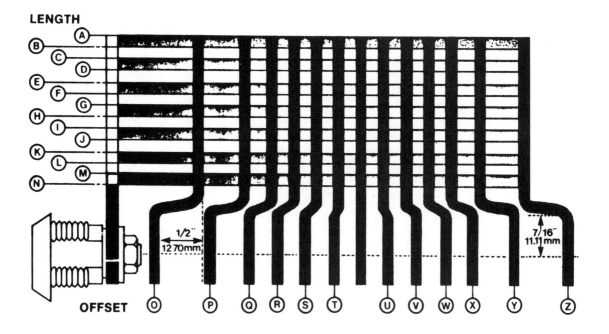

LENGTH

LENGTH			
	A	2 3/8"	60.325 mm
	B	2 1/4"	57.150 mm
	C	2 1/8"	53.975 mm
	D	2"	50.800 mm
	E	1 7/8"	47.625 mm
	F	1 3/4"	44.450 mm
	G	1 5/8"	41.275 mm
	H	1 1/2"	38.100 mm
	I	1 3/8"	34.925 mm
	J	1 1/4"	31.750 mm
	K	1 1/8"	28.575 mm
	L	1"	25.400 mm
	M	7/8"	22.225 mm
	N	3/4"	19.050 mm

OFFSET			
	O	1/2"	12.700 mm
	P	3/8"	9.525 mm
OUTSIDE	Q	1/4"	6.350 mm
	R	3/16"	4.775 mm
	S	1/8"	3.175 mm
	T	1/16"	1.600 mm
	U	1/16"	1.600 mm
	V	1/8"	3.175 mm
INSIDE	W	3/16"	4.775 mm
	X	1/4"	6.350 mm
	Y	3/8"	9.525 mm
	Z	1/2"	12.700 mm

6-12 Cam lengths and offsets. (courtesy Dominion Lock Co.)

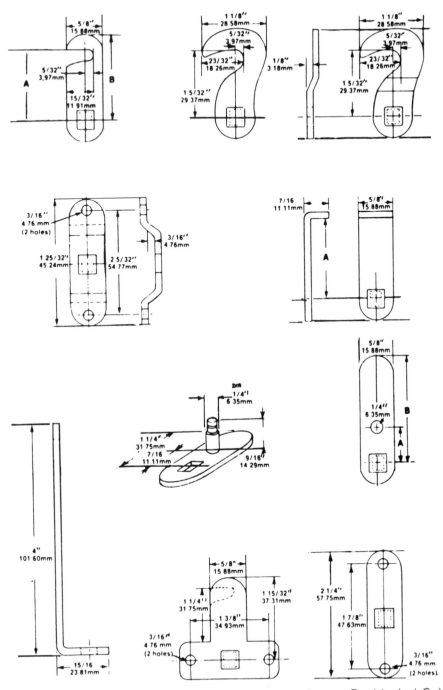

6-13 Hook, offset, double-ended and miscellaneous type cams. (courtesy Dominion Lock Co.)

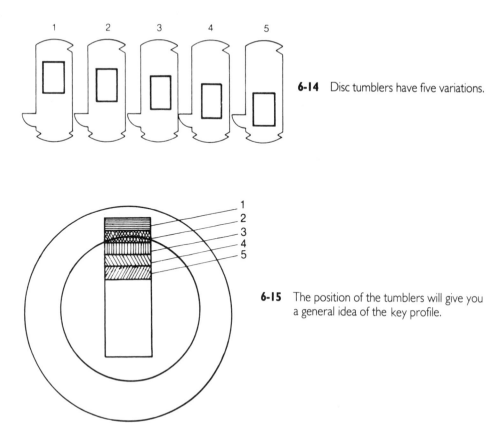

6-14   Disc tumblers have five variations.

6-15   The position of the tumblers will give you a general idea of the key profile.

You might ask yourself, "How do I know where to cut the key—and how deep?" Recall that when impressioning a key, you blackened the key and determined the cutting point by the pressure of the levers upon the key as you tried to turn it. The same technique applies here. Insert a blackened key and give it a slight turn. This brings the key in contact with the sides of each tumbler cutout. The cutout will leave a mark on the blank, indicating the portion of each cut.

Determining the depth of the cuts requires experience. You have already learned that cutting a key is slow, patient work. As you get closer to the proper depth, the file is moved with less pressure than before. The same care and precision is needed here. A disc with a No. 1 cut hole requires a deep key cut, as compared to a No. 4 or 5 disc, which requires a very shallow cut.

Reference aids include various extra keys that you have collected. Since the depths of the cuts are standardized in the industry, you can have a key with a 13354 cut and, by observing the differences in the depths of each cut, know exactly how deep a No. 2 cut should be. Obtain a depth key set from a locksmith supply house. This set has a different key for each depth, with the same cut in all five tumbler positions. Thus, a No. 2 key has five No. 2 cuts; a No. 3 key has five No. 3 cuts; and so on throughout the series. With the help of these keys and your trained eyesight, you can place a blank in a vise and cut the key by hand.

Without a depth key set, you can use a variety of disc keys as guides. Select a key with the proper cut and align your blank to it. Another approach is to make your own set of depth keys. Making the set teaches you how deep to make each cut at any given position on the blank.

Lever locks

As you already know, a lever-lock keyway is narrow, and the view of the tumblers is further restricted by the trunnion. While this is a handicap, it can, in part, be overcome with the help of an appropriate reading tool.

The positions of the lever saddles is one clue to reading the lock. The saddle of the lock can tell you quite a bit. The wider the saddle, the deeper the cut; the narrower the saddle, the shallower the cut. Using the reading tool, you can feel the different saddle widths to determine the cuts and develop some idea of the cuts and of the key shape.

In order to do this properly, you must have an appreciation of the internal workings of the lock. Depth key sets can be useful when you are ready to cut the keys. Lever cuts are usually in 0.015—0.025 inch increments.

Practice is required. Begin with a lock that has a window in it. If possible, obtain extra levers, so you can change the keying of the lock at will.

Raise each lever with the reading tool and try to determine the proper position for each one. Remember the general rule: a wide saddle requires more movement than a narrow saddle. Once you are satisfied that you have read the lock, disassemble it and examine the lever tumblers. You may have misread the tumblers, but do not be disheartened. The only way to achieve competence in this skill is practice and more practice.

Other tumblers are designed with uniform saddle widths. Keying is determined by the positions of the gates in the tumblers. These locks can be quite difficult to read.

Working with the reading tool, raise one of the levers as high as it will go. While this movement is not a direct indication of the depth of the key cut, it is important. The amount of upward movement establishes the minimum key cut; a shallower cut would jam the levers against the top of the lock case. In addition, the individual gates and traps are usually in some rough alignment. You will usually find that two are in almost perfect alignment.

A cut for a lever that has the post on the upper half of the trap will be shallow; a cut for a lever that has its post on the lower half will be deep. A lever that has its gate in the intermediate position will require a key cut between these extremes.

Once you have established the general topography of the cuts, refer to your set of depth keys for the exact dimensions. Established locksmiths have reference manuals that may simplify the work.

7

Pin tumbler cylinder locks and locksets

Pin tumbler cylinder locks come in a wide variety of shapes and styles. From the outside, they look similar to other types of locks.

A pin tumbler cylinder lock is simply one that uses a pin tumbler cylinder. When a lock is fully assembled you can usually see only the lock's *plug* or the face of its cylinder. Figures 7-1, 7-2, 7-3 show examples of common pin tumbler cylinder locks.

Pin tumbler cylinders are basically very simple mechanisms, but they can be built to provide a high degree of security.

CONSTRUCTION

Most pin tumbler cylinders are self-contained mechanisms and are designed to be used with a wide variety of locks and locksets. Figures 7-4, 7-5, and 7-6 show examples of various types of pin tumbler cylinders.

The basic parts of a pin tumbler cylinder include: a cylinder case (or shell); plug (or core); keyway; upper pin chambers; lower pin chambers; springs; drivers (or top pins); and bottom pins. It's easy to remember all those parts, once you understand their relationships to each other. (See FIGS. 7-7 and 7-8 for inside views of a pin tumbler cylinder.)

The cylinder case houses all the other parts of the cylinder. The part that rotates when the proper key is inserted is called the plug. The keyway is the opening in the plug that accepts the key. The drilled holes (usually five or six) across the length of the plug are called lower pin chambers; they each hold a bottom pin. The corresponding drilled holes in the cylinder case directly above the plug are called upper pin chambers; they each hold a spring and a driver.

How pin tumbler cylinders work

When a key isn't inserted into the cylinder, the downward pressure of the springs "drive" the drivers (top pins) partially down into the plug to prevent the plug from being rotated. Only the lower portions of drivers are pushed into the plug, because the plug holds bottom pins. There isn't enough room in a lower pin chamber to hold the entire length of a driver and a bottom pin.

7-1   Pin tumbler cylinder key-in-knob lock. (courtesy Schlage Lock Company)

There's a small amount of space between the plug and the cylinder case. That space is called the shear line. Without a shear line, the plug would fit too tightly in the cylinder case to rotate. When a properly cut key is inserted, it causes the top of all the bottom pins and the bottom of all the drivers to meet at the shear line. While the pins are in that position, the plug is free to rotate to the open position.

Plug

The plug contains a keyway, usually of the paracentric, or off-center, type. Although they usually contain five or six lower pin chambers, some plugs have four or seven. The lower pin chambers are spaced fairly evenly along the upper surface of the plug, and they are aligned as closely as modern production techniques allow.

7-2   Pin tumbler cylinder deadbolt lock. (courtesy Schlage Lock Company)

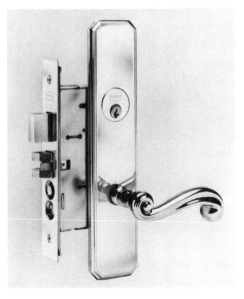

7-3 Pin tumbler cylinder mortise lock. (courtesy Omnia Industries, Inc.)

7-4 Pin tumbler cylinder for rim locks. (courtesy Ilco Unican Corp.)

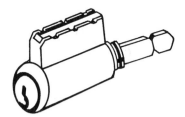

7-5 Pin tumbler cylinder for key-in-knob locks. (courtesy Ilco Unican Corp.)

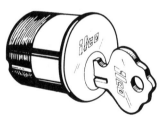

7-6 Pin tumbler cylinder for mortise locks. (courtesy Ilco Unican Corp.)

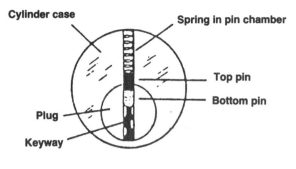

Parts of a pin tumbler lock—front view

7-7 Cutaway view of face of a pin tumbler cylinder. (courtesy The Locksmith Guild)

The plug may be machined with a shoulder at its forward surface; this shoulder mates with a recess in the cylinder and provides:

A reference point for regulating the alignment of pin chambers in the case and the plug.

A safeguard to prevent the plug from being driven through the cylinder, either deliberately or through resistance developed as the key enters the keyway.

A safeguard to discourage a thief from shimming the pins. Without this shoulder it would be possible to force the pins out of engagement with a strip of spring steel.

The plug is retained at the rear by a cam and screws, a retainer ring, or a driver that locks the plug into the cylinder.

Pins

Pins (including drivers) come in a variety of lengths, diameters, sizes and shapes. Figure 7-9 shows some of the most popular pin designs. Although they're small, pin tumblers are usually made of brass and are very strong.

The shape of the pins can resist attempts at picking a lock. A standard cylindrical driver can easily be manipulated to the shear line, as the plug is being turned slightly. A driver with a broken profile tends to "hang up" before it passes the shear line which makes a lock more difficult to pick. (See FIG. 7-10.)

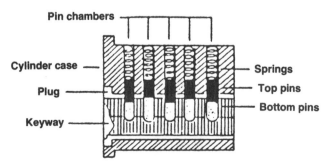

Parts of a pin tumbler lock—side view

7-8 Cutaway view of side of a pin tumbler cylinder. (courtesy The Locksmith Guild)

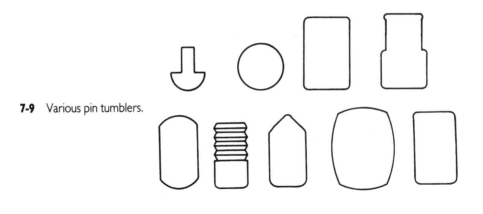

7-9 Various pin tumblers.

Tailpieces and cams

Most pin tumbler cylinders have either a tailpiece or a cam attached to the rear of the plug (FIG. 7-11A and FIG. 7-11B.) The tailpiece is loose to allow some flexibility in the location of the auxiliary lock on the other side of the door. Alignment should be as accurate as possible; under no circumstances should the tailpiece be more than ¼ inch off the axis of the plug.

On the other hand, pin tumbler cylinders for mortise locks do not have such tolerance. The best is driven by a cam on the back of the cylinder (FIG. 7-11C). If the lock is used on office equipment, the cam is probably a milled relief on the back of the plug, or else it is a yoke-like affair, secured to and turning with the plug. It is important that these locks are aligned with the bolt mechanism.

DISASSEMBLY

To disassemble a pin tumbler cylinder, you'll need a small screwdriver, a following tool of the correct diameter, and a pin tray. Proceed as follows:

Turn the plug about 30 degrees. That is done with the key, by picking the cylinder, or by shimming the cylinder.

7-10 High-security drivers "hanging-up" below the shear line. (courtesy DOM Security Locks)

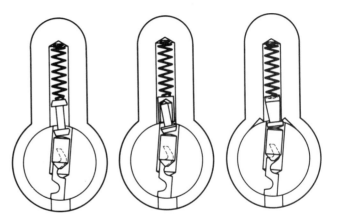

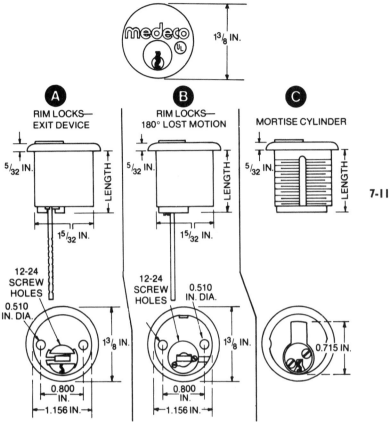

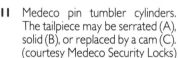

7-11 Medeco pin tumbler cylinders. The tailpiece may be serrated (A), solid (B), or replaced by a cam (C). (courtesy Medeco Security Locks)

Remove the cam or tailpiece, by removing two screws or by removing a retainer ring. (Although they're not essential, a small pair of snap ring pliers can be useful for removing retainer rings.)

While holding the cylinder so the pins are in a vertical position, slip the appropriate size plug follower tool into the cylinder directly behind the plug. Use the plug follower tool to slowly push the plug out of the face of the cylinder. Do not allow a gap to be created between the plug and the plug follower tool, or the springs and drivers will fall out.

Hold the plug as it leaves the cylinder; don't let the bottom pins fall out of the plug. Use your index finger to cover all pin chambers in the plug except the one closest to the face of the plug. Empty that chamber and put the pin into the first compartment of your pin tray.

Uncover the next pin chamber in the plug and empty the pin into the corresponding compartment of the pin tray. Continue this procedure until all the bottom pins are out of the plug, and they are in proper order in the pin tray. Then you can put the plug down for awhile.

Hold the cylinder case, and slowly pull the plug follower out the back of the cylinder. Stop each time a driver and spring fall from an upper pin chamber. Place them in the tray directly above each corresponding bottom pin.

ASSEMBLY

To assemble a pin tumbler cylinder, you will need a plug follower, a small screwdriver, and pin tweezers. A plug holder is optional; it holds the plug in an upright position, and leaves your hands free while you insert the bottom pins.

This procedure assumes that you have a key but do not know the pin combinations. (In normal reassembly, you already know which pins go into which chambers; in which case, you simply insert the springs and pins in the proper chambers.)

Place the plug into the plug holder and insert the key into the plug's keyway.

Taking one bottom pin at a time, insert it into the lower pin chamber nearest the face of the plug. If the pin stands above or below the shear line, try another size pin. When a pin appears to be just about at the shear line, turn the plug with the key. If the plug turns 360 degrees in the plug holder, the bottom pin is probably the right one.

Repeat the procedure for each lower pin chamber until all the bottom pins just reach the shear line. *Note: Do not remove the key or turn the plug over.*

Move to the cylinder case and insert the plug follower through the back of the cylinder until it covers the third upper pin chamber from the face of the cylinder. Then, turn the cylinder case upside down so the upper pin chambers are facing toward the floor.

Using your tweezers, drop a spring into the second pin chamber from the face of the cylinder. Then, use your tweezers to insert a driver on the spring. By pushing the driver into the pin chamber while applying slight forward pressure on the plug follower, the plug follower will bind the driver into place. That will allow you to release the driver with your tweezers, and use your tweezers to push the pin into the upper pin chamber while simultaneously pushing the plug follower over the driver. The plug follower will hold the driver and spring in place so they don't pop out of the cylinder. Repeat the procedure with the first upper pin chamber from the face of the cylinder.

At this point, you should have two upper pin chambers filled and covered by the plug follower. Now push the pin follower out of the face of the cylinder to the point that your plug follower is covering only the upper pin chambers you've already filled. The remainder of the upper pin chambers should be exposed.

Insert a spring and driver in the exposed upper pin chamber nearest the plug follower. Then, cover that pin chamber with your plug follower. Repeat this step until all the upper pin chambers are filled.

Turn the cylinder case upright, so the drivers and springs are in a vertical position.

Slowly remove the key from the plug, so the bottom pins don't pop out of the plug. Then, insert the plug into the cylinder by pushing the back of the plug against the plug follower from the face of the cylinder. After the plug has forced the plug follower out of the rear of the cylinder case, rotate the plug around until the upper pin chambers are aligned with the lower pin chambers. The pins should lock the plug into place. Note: when inserting the plug, be sure the lower pin chambers are at least 30 degrees out of alignment with the upper pin chambers.

Now make sure that your key works by inserting it into the plug and rotating the plug. But be careful while rotating the plug and while removing the key; only the pins are holding the plug in place. With the key in the plug, all the pins should be at the shear line, and the plug can easily be pulled out of the cylinder. If that happens, all the drivers and springs will fall out.

Attach the cam or tailpiece to the cylinder, and retest your key.

CHOOSING A PIN TUMBLER LOCKSET

Pin tumbler cylinder locksets are the most popular types of locking devices. These locksets are available in a wide variety of types and styles. Some of the general characteristics of such locksets include the following:

Security   The pin tumbler cylinder lockset provides better than average security by virtue of its internal design. Generally speaking, the more pins in a pin tumbler cylinder, the better security the lock offers. However, security also depends upon the quality of the specific lock and upon its application.

Quality   Quality depends upon the intended service of the lockset. Light, medium, and heavy duty locksets are available.

Type   Locksets are identified by their function lavatory, classroom, residential, etc.

Visual appeal   The lockset should match the decor of its surrounding; this is particularly important in new constructions.

Hand   The location and direction of swing of the hinges determines the *hand* of the door (FIG. 7-12). Taking the entrance side of the door as the reference point, there are four possible hands: left-hand, left-hand-reverse (the door opens outward), right-hand, and right-hand reverse.

It is important to match the lockset to the door hand. Failure to do so can cause bolt/striker misalignment and might require that the cylinder be rotated 180 degrees, so that the weight of the springs is on the pins. The pins weigh only a few grams, but that is enough to collapse the springs and disable the lock in time when a cylinder is installed upside down. Unless stipulated in the order, manufacturers supply right-hand mortise locksets; some of these can be modified in the field.

PIN TUMBLER CYLINDER MORTISE LOCKS

The pin tumbler cylinder mortise lock is frequently used in homes, apartments, businesses, and large institutions. It's extremely popular, providing excellent security. Never mistake the

LEFT HAND: HINGES ON
LEFT, OPENS INWARD.

RIGHT HAND: HINGES ON RIGHT,

7-12   The hand of a door is a term that describes the location of the hinges and the direction of swing. (courtesy Eaton Corp.)

LEFT HAND REVERSE:
HINGES ON LEFT, OPENS
OUTWARD.

RIGHT HAND REVERSE:
HINGES ON RIGHT, OPENS
OUTWARD.

pin tumbler cylinder mortise lock with its inferior cousin the bit key mortise lock. Figure 7-13 shows two basic pin tumbler cylinder locks without the cylinders in them.

Although the individual parts may vary slightly among different models, most basic pin tumbler cylinder mortise locks contain essentially the same parts. Practice disassembling and reassembling a few of these locks, and study the relationships of the various parts.

Specific pin tumbler cylinder locksets

The information presented in the following pages has been supplied by the various manufacturers. While it does not cover all pin tumbler cylinder locksets, it does give an overview of the current state of the art. If you need further information, contact one of the factory representatives.

Kwikset corporation's products

Kwikset's key-in-leversets are part of the company's Premium line. (See FIG. 7-14.) The leverset is constructed of solid brass and steel. It's engineered to meet most building code requirements including major handicapped codes and built to absorb the punishment of high-traffic areas. It's UL listed and backed by a 10-year limited warranty. The leverset is available in a wide range of finishes.

Kwikset offers several styles of entrance handlesets (see FIG. 7-15) to complement any entrance design architecture from traditional to contemporary.

Kwikset's Series 500 Premium Entrance Lockset was designed and built in response to specific needs of the building industry for an improved, heavier duty lockset for offices, apartments, townhouses, condominiums, and finer homes. (See FIG. 7-16.)

The Series 500 Locksets include features such as a panic-free operation on the interior knob; a solid steel reinforcing plate to protect the spindle assembly; a reversible deadlatch with a ⅔ inch throw solid brass beveled latchbolt; a solid steel strike plate with heavy duty screws; and a free-spinning exterior knob for protection against wrenching.

Kwikset's Protecto-Lok (see FIG. 7-17) combines an entry lockset with a 1 inch cylinder deadlock. Both deadbolt and deadlatch retract with a single turn of the interior knob or lever, assuring immediate exit in case of an emergency. Protecto-Lok is U.L. listed when ordered and, it is used with a special U.L. latch and strike.

Kwikset's Safe-T-Lok (see FIG. 7-18) combines deadbolt security with instant exit. When you lock it from the interior, the key remains in the cylinder and cannot be removed until the deadbolt is fully retracted in an unlocked position.

Safe-T-Lok has a 1 inch-steel deadbolt with a heat treated steel insert that turns with any attempted cutting.

Schlage Lock Company's products

Schlage's "B" Series deadbolt locks provide primary or auxiliary security locking for both commercial and residential applications. It includes the standard duty B100 Series for residential structures (see FIG. 7-19), the heavy duty B400 Series for commercial requirements, and the most secure B500 Series for maximum performance. All products in this series feature a full 1-inch throw deadbolt with Schlage's exclusive wood frame reinforcer to deter "kick-in" attack.

Schlage's commercial quality "D" Series locks (see FIG. 7-20) are specified when the highest quality mechanisms are required. Precision manufactured to exact tolerances, the "D" Series is best suited for commercial, institutional, and industrial use. Many specialized functions satisfy many unique locking applications.

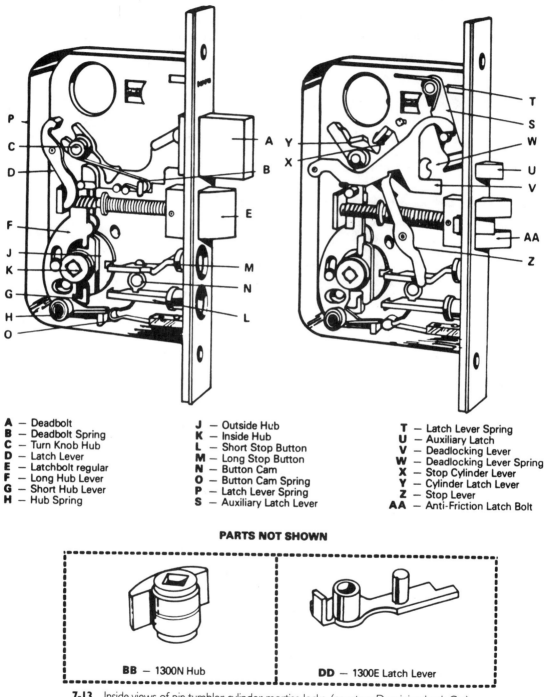

A — Deadbolt	**J** — Outside Hub	**T** — Latch Lever Spring
B — Deadbolt Spring	**K** — Inside Hub	**U** — Auxiliary Latch
C — Turn Knob Hub	**L** — Short Stop Button	**V** — Deadlocking Lever
D — Latch Lever	**M** — Long Stop Button	**W** — Deadlocking Lever Spring
E — Latchbolt regular	**N** — Button Cam	**X** — Stop Cylinder Lever
F — Long Hub Lever	**O** — Button Cam Spring	**Y** — Cylinder Latch Lever
G — Short Hub Lever	**P** — Latch Lever Spring	**Z** — Stop Lever
H — Hub Spring	**S** — Auxiliary Latch Lever	**AA** — Anti-Friction Latch Bolt

PARTS NOT SHOWN

BB — 1300N Hub   **DD** — 1300E Latch Lever

7-13   Inside views of pin tumbler cylinder mortise locks. (courtesy Dominion Lock Co.)

7-14 Kwikset's Premium Keyed Lever-set. (courtesy Kwikset Corporation)

kwikset ENTRANCE HANDLESETS: TOLEDO DESIGN

Heavy Duty

Heavy Duty with Protecto-Lok® Function

One Piece

One Piece with Matching Deadlock and Trim

7-15 Kwikset's Entrance Handlesets. (courtesy Kwikset Corporation)

7-16 Kwikset's Premium Entrance Lockset. (courtesy Kwikset Corporation)

Schlage's "L" Series mortise lock line is a heavy duty commercial mortise lock series containing a wide array of knob, lever, and grip handle designs. (See FIG. 7-21 for an example.) It's available in a variety of keyed and non-keyed functions, and in most decorative finishes. It is U.L. listed and can be specified for applications in offices, schools, hospitals, hotels, and commercial buildings, as well as residences. The "L" lock meets or exceeds the ANSI A156.13 specification, making it an excellent choice for any building where security, safety, and design compatibility are prime considerations.

The "L" lock has one common mortise lock case for knob, lever, and grip handle trim, providing tremendous versatility. In addition, it provides excellent design flexibility in trim combinations such as knob by lever, lever by knob, etc. All may be specified on the same "L" lock case.

7-17 Kwikset's Protecto-Lok. (courtesy Kwikset Corporation)

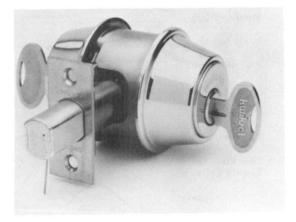

7-18 Kwikset's Safe-T-Lok. (courtesy Kwikset Corporation)

THE CYLINDER KEY

Figure 7-22 illustrates basic key nomenclature, and FIG. 7-23 shows an assortment of key blanks. Notice the differences in the bow, blade, length, and in the width, number, and spacing of the grooves.

The bow usually has a specific shape that identifies the lock manufacturer. The blade length is indicative of the number of pins within the lock that the key operates, e.g., a five-pin key is shorter than a six-pin key. The height of the blade sometimes indicates the depths of the cuts in a series; generally, the higher the blade the deeper the cuts.

Key blanks are made of brass, nickel-brass, nickel-steel, steel, aluminum, or a combination of these and other alloys. The metal determines the strength of the key and its resistance to wear.

Duplicating a cylinder key by hand

Duplicating a cylinder key by hand isn't much different than duplicating a lever or warded key by hand. The procedure is as follows:

Smoke the original and mount it together with the blank in a vise. The blank must be in perfect alignment with the original.

Using Swiss round and warding files, begin at the tip of the key and work towards the bow. Start in the center of each cut.

Go slowly near the end of each cut, but continue to cut with firm and steady strokes. Stop when the file touches the soot on the original key.

Once you have finished, remove the keys from the vise. Wire-brush the duplicate to remove any burrs, and polish the key.

Hold the duplicate up to the light and place the original in front of it and then behind it, to determine if the duplicate is accurately cut. Shallow cuts can be cut a little deeper with a file; if a cut is too deep, you'll need to start over again with a new key blank.

Impressioning

The technique for impressioning cylinder keys is different than for warded keys. The cylinder key is not smoked because the soot would wipe off the blade when the key was inserted into a pin tumbler cylinder.

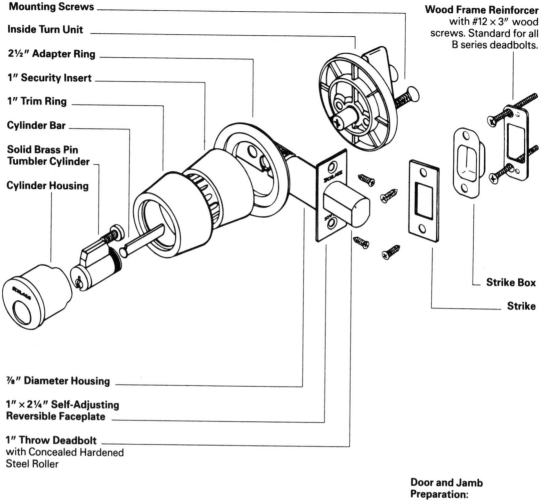

Mounting Screws

Inside Turn Unit

2½" Adapter Ring

1" Security Insert

1" Trim Ring

Cylinder Bar

**Solid Brass Pin
Tumbler Cylinder**

Cylinder Housing

Wood Frame Reinforcer
with #12 × 3" wood
screws. Standard for all
B series deadbolts.

Strike Box

Strike

⅜" Diameter Housing

**1" × 2¼" Self-Adjusting
Reversible Faceplate**

1" Throw Deadbolt
with Concealed Hardened
Steel Roller

**Door and Jamb
Preparation:**
See Templates

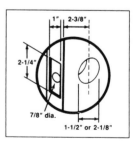

7-19   Schlage's B Series deadbolt lock. (courtesy Schlage Lock Company)

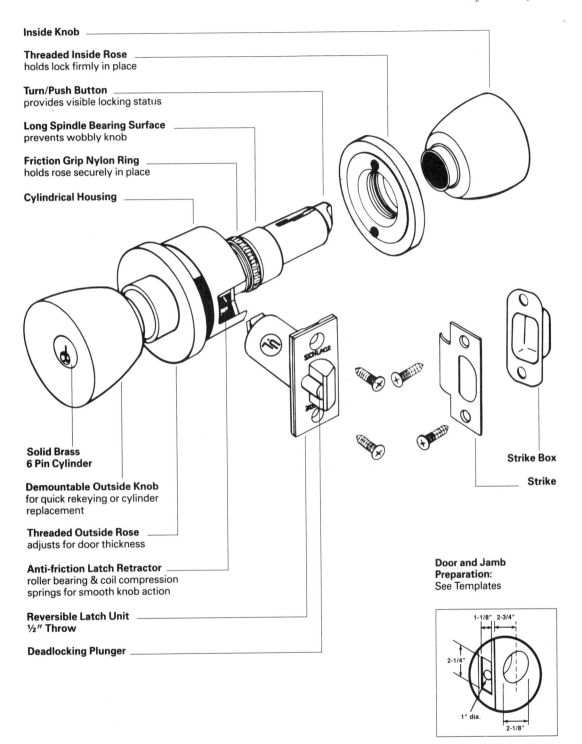

Inside Knob

Threaded Inside Rose
holds lock firmly in place

Turn/Push Button
provides visible locking status

Long Spindle Bearing Surface
prevents wobbly knob

Friction Grip Nylon Ring
holds rose securely in place

Cylindrical Housing

**Solid Brass
6 Pin Cylinder**

Demountable Outside Knob
for quick rekeying or cylinder
replacement

Threaded Outside Rose
adjusts for door thickness

Anti-friction Latch Retractor
roller bearing & coil compression
springs for smooth knob action

**Reversible Latch Unit
½″ Throw**

Deadlocking Plunger

Strike Box

Strike

**Door and Jamb
Preparation:**
See Templates

1-1/8″   2-3/4″

2-1/4″

1″ dia.

2-1/8″

7-20   Schlage's D Series key-in-knob lockset. (courtesy Schlage Lock Company)

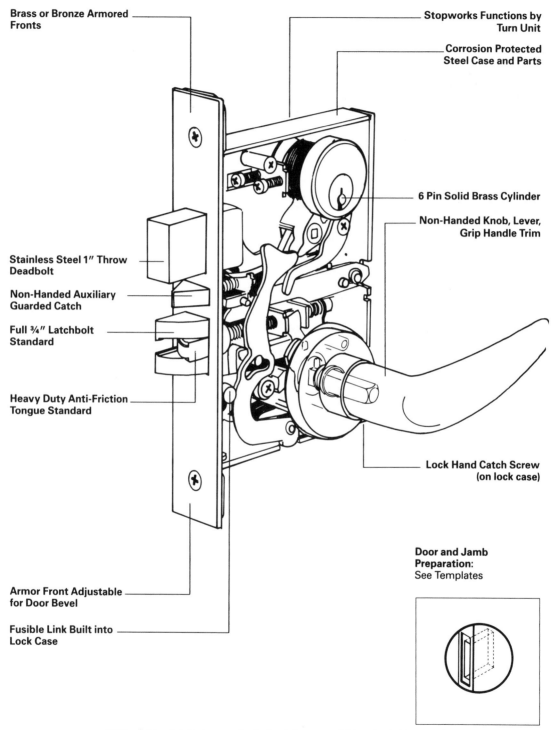

Brass or Bronze Armored Fronts

Stopworks Functions by Turn Unit

Corrosion Protected Steel Case and Parts

6 Pin Solid Brass Cylinder

Non-Handed Knob, Lever, Grip Handle Trim

Stainless Steel 1" Throw Deadbolt

Non-Handed Auxiliary Guarded Catch

Full ¾" Latchbolt Standard

Heavy Duty Anti-Friction Tongue Standard

Lock Hand Catch Screw (on lock case)

Armor Front Adjustable for Door Bevel

Fusible Link Built into Lock Case

Door and Jamb Preparation: See Templates

7-21 Schlage's L Series mortise lockset. (courtesy Schlage Lock Company)

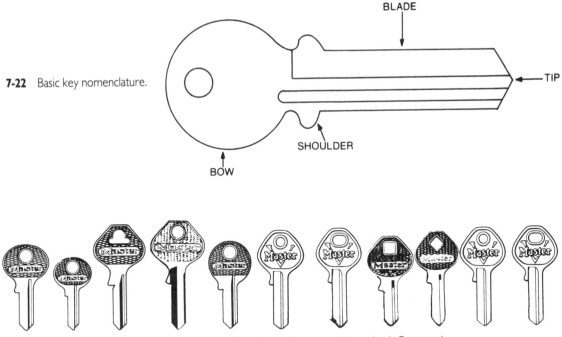

7-22 Basic key nomenclature.

7-23 An assortment of key blanks. (courtesy Master Lock Company)

Instead, depend upon small marks left on the key blank by the pin tumblers. These marks indicate the position of the tumblers relative to the blank and the depth of each cut. Proceed as follows:

Examine the blank. The marks left by the tumblers are difficult to see in the best of circumstances; scratches across the top of the blade will make them impossible to see. Polish the top of the blade with emery paper.

Place the bow of the key in a small C-clamp, and insert the blade into the keyway.

Twist the blank side to side and gently rock it up and down; there's no need to twist very hard. If you break the key blank, you've used far too much pressure.

The tumblers should leave their "footprints" on the blade. Examine the blank under a bright light.

Identify each mark with a light file cut.

Make a shallow cut at the mark closest to the tip of the blank with a Swiss No. 4 file.

Insert the blank into the cylinder, twist the blank from side to side, gently rock it up and down, and then remove it. Check the mark and file a little more. Continue until the tumbler no longer leaves a mark.

When you finish one cut, move to the next and repeat the procedure. Keep the slopes even and smooth.

When the last cut is finished, the newly cut key should operate the cylinder. At this point, smooth the various angles on the key and remove any burrs from it.

8

High-security mechanical locks

W ith respect to locking devices, the term "high-security" has no precise meaning. Some manufacturers take advantage of that fact by arbitrarily using the term to promote their locks.

Most locksmiths would agree that, at the very least, a locking device should have features that offer more than ordinary resistance against standard burglary attacks in order to be considered a high-security device. For our purposes, we'll consider a high-security lock as one that has a cylinder or other part that offers extra resistance against picking, impressioning, drilling, and wrenching.

HIGH-SECURITY MECHANICAL LOCKS

This chapter provides detailed information about some of the most popular high-security mechanical locks available. Much of the information was obtained by manufacturer's technical bulletins and service manuals.

Schlage's primus

The Schlage Primus is one of Schlage Lock Company's newest high-security mechanical locking systems. It features a specially designed "patent protected" key, which operates either the U. L. listed #20-500 High Security Series Cylinders or the Controlled Access #20-700 Series Cylinders (non-U.L. listed). Both series are available in Schlage "A," "B," "C/D," "E," "H," and "L" Series locks. (See FIGS. 8-1 and 8-2.)

Both series are easily retrofitted into existing Schlage locks, and they can be keyed into the same master key system and operated by a single Primus key. The Primus key is cut to operate all cylinders while those that operate the standard cylinders will not enter a Primus keyway.

The Primus security cylinder is machined to accept a side bar and a set of five "finger pins," which, in combination with Schlage's conventional 6 pin keying system, provides two independent locking principles operated by a Primus key. Hardened steel pins are incorporated in the cylinder plug and housing to resist against drilling attempts.

Mortise Cylinder

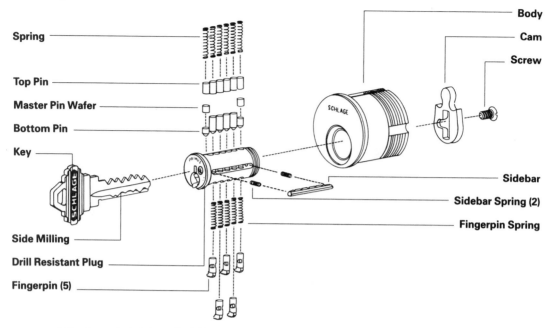

8-1   An exploded view of a Schlage Primus mortise cylinder. (courtesy Schlage Lock Company.)

Key-in-Knob Cylinder

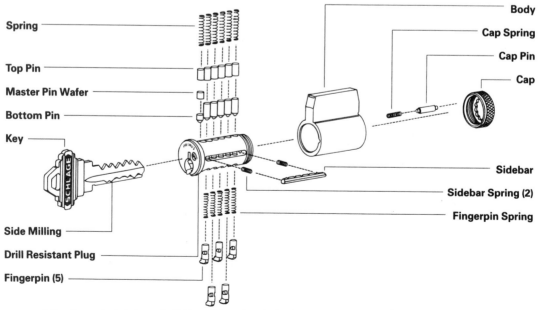

8-2   An exploded view of a Schlage Primus key-in-knob cylinder. (courtesy Schlage Lock Company.)

Primus security levels

The Primus system features four different levels of security. Each level requires an appropriate I.D. card for key duplication.

Security level one   Primus cylinders and keys have a standard side milling allocated to this particular level and are on a local "stock" basis. Level one also provides the flexibility of local keying into most existing Schlage key sections; it utilizes a Primus key to operate both systems. Key control is in the hands of the owner who holds an I.D. card for the purpose of acquiring additional keys. Level One is serviced through qualified locksmiths, fully trained at a Schlage "Primus I Center."

Security level two   At this level, service and inventory of Primus keys and cylinders are available, but on a more restricted basis. Level two has a restricted side milling which is allocated expressly to this level. Local or factory master keying for new or existing systems is available. Key duplication is controlled with an I.D. card and the authorized purchaser's signature. Level Two is serviced through Schlage Primus II Centers, who with proper I.D. or an authorized order form will process key blank orders from a locksmith.

Security level three   At this level, the factory maintains control of the Primus key system. Level three has a restricted side milling that is specifically selected for this level. Keying, key records, installation data and the owner's signature are controlled at the Schlage factory. To obtain duplicate keys, the owner must present an I.D. card, indicate the number of keys desired and sign a special order form, which is then verified for proper signature with the I.D. card. A copy of the authorization form is forwarded to the Schlage Lock Company for processing.

Security level four   The factory maintains control of the Primus key system. Level Four uses a single restricted side milling that is selected for each project, in conjunction with a restricted key section. All records, verified by the owner's signature and processing of orders, is controlled at the factory level. To obtain duplicate keys the owner must present an I.D. card, indicate the number of keys desired, and sign a special order form which is forwarded to the Schlage Lock Company for processing.

Kaba

Kaba locks are dimple key locks. There isn't just one, but an entire "family" of different Kaba cylinder designs. Each one is designed to fill specific security requirements. (See FIGS. 8-3, 8-4, and 8-5 for examples of Kaba locks.)

The "family" includes the following designs: Kaba 8, Kaba 14, Kaba 20, Kaba 20S, Saturn, Gemini, and Micro. Those which use numbers in their designations generally reflect not only the order of their development, but also the number of possible tumbler locations in a cylinder of that particular design. (See FIGS. 8-6 and 8-7 for examples of Kaba cylinders.)

Handing and key reading

Basic to the understanding of all Kaba cylinders is the concept of handing. With Kaba cylinders "handing" is not a functional installation limitation as you might expect. A left-hand cylinder will operate both clockwise and counter-clockwise, and function properly in a lock of any hand. The handing of Kaba cylinders refers only to the positions where the pin chambers are drilled.

8-3   Some Kaba locks fit Adams Rite narrow stile metal doors. (courtesy Lori Corp.)

Figure 8-8 shows the orientation of the two rows of side pins in a Kaba 14 cylinder. Notice that they are staggered much like the disc tumblers in some foreign car lock cylinders.

There are two possible orientations of these staggered rows of side pins. Either row could start closer to the front of the cylinder. The opposite row would then start farther from the front. These two orientations are referred to as right or left-hand. If the cylinder is viewed from the right side up, the hand of most Kaba designs is determined by the side whose row of pins begins farther from the front when viewing the face of the cylinder.

If a key is to do its job and operate a cylinder, obviously the cuts must be in the same positions as the pins in the cylinder; that means there are two ways to drill dimples in the keys as well.

To determine the hand of a Kaba key, view it as though it were hanging on your key board (FIG 8-9). You will notice that one row of cuts starts farther from the bow than the other. The row which starts farther from the bow determines the hand. If the row starting farther from the bow is on the left side, it is a left-hand key. If the row starting farther from the bow is on the right side, it is a right-hand key.

Occasionally, you may find a Kaba key with both right-hand and left-hand cuts (FIG. 8-10). This key is called *composite bitted*. It is primarily used in maison key systems.

8-4   Double cylinder deadbolt lock with "Inner Sanctum" core. (courtesy Lori Corp.)

8-5   Micro-Kaba switch lock and keys. On the right is the lock with the core removed. The switch size (compare with coin) provides for exceptionally wide varieties of uses with electrical and electronic circuitry. (courtesy Lori Corp.)

8-6   Kaba standard mortis cylinders can be used on Adams Rite doors. (courtesy Lori Corp.)

8-7   Cutaway view of a Kaba 20 cylinder. (courtesy Lori Corp.)

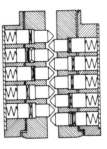
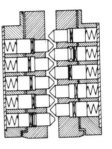

8-8   Kaba keys are handed. (courtesy Lori Corp.)

Right Hand   **Left Hand**

8-9   It's easy to determine the hand of a Kaba key.
(courtesy Lori Corp.)

8-10   Examples of Kaba composite bitted keys.
(courtesy Lori Corp.)

Composite Bitted Keys

Key reading

Determining the hand is the first step in key reading. The next thing to know is the order in which the various dimple positions are read.

The positions are always read bow to tip, beginning with the hand side. After the hand side comes the non-hand side and finally, the edge. Remember to go back to the bow to start each row (FIG. 8-11). For composite bitted keys, use a Kaba key gauge to help mark the cuts of both hands.

All Kaba designs except Micro have four depths on the sides; Micro has three. What is difficult at first is that Kaba's depths are numbered opposite from the way we normally think of them. #1 is the deepest and #4 is the shallowest. These depths can be read by eye with very little practice. (see FIG. 8-12.) The increment for side depths of Kaba 8 and 14 is .4mm (.0157″). For Gemini, it is .35 mm (.0138″).

For the edges, reading is a bit different for the various Kaba designs. The Kaba key gauge is very useful for reading the edges. Find the section of the gauge which corresponds to the design of the Kaba key you're attempting to read, e.g., Kaba 8. Then place the key under the gauge, so it shows up through the edge slot. When the shoulder of the key hits the stop on the gauge, you're ready to read the positions of the cuts.

Because of the non-standard cylinder drilling for Kaba 8, the codes will show two columns for the edge (FIG. 8-13). There are only two depth possibilities on the edge of Kaba 8 and Kaba 14: *#2 cut = cut* and *#4 = no-cut*. The combinations in FIG. 8-13 call for a key with a cut only in position 4, while the cylinder is drilled in positions 1-4-7. This means the

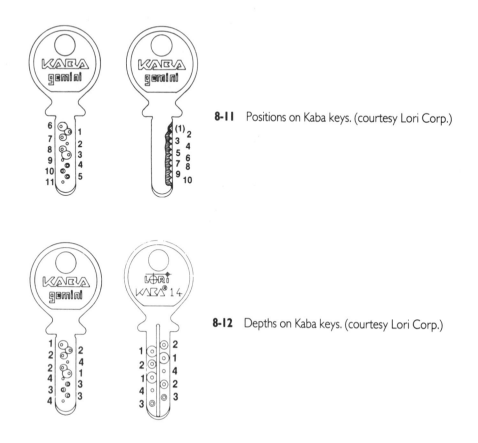

8-11   Positions on Kaba keys. (courtesy Lori Corp.)

8-12   Depths on Kaba keys. (courtesy Lori Corp.)

	SIDES		EDGE	
			key	cyl.
R2214	4233		4	147
R2242	4233		4	147
R2412	4233		4	147
R4212	4233		4	147
R2244	1233		4	147
R2414	1233		4	147

8-13 Kaba 8 combinations. (courtesy Lori Corp.)

cylinder receives a #2 pin in position 4 for the key cut. Positions 1 and 7 require a #4 pin because there is no cut on the key.

For the Kaba 14 and new Kaba 8 using ME series codes, the edge is read differently because you know automatically which positions are involved in every case. These cylinders are all drilled with odd edges (positions 1-3-5-7).

Knowing the positions involved, the edge combination of these keys is notated in terms of depths, rather than positions. A Kaba key gauge can be used to determine the positions of the #2 edge cuts. If there is a cut in position 5 only, the combination would be 4224. If there are cuts in positions 1, 3 and 7, the combination would be 2242. If there are cuts in positions 3 and 5, the combination would be 4224, etc. In other words, we know there are four chambers in the cylinder and they are in the odd numbered positions. Therefore, we have to come up with four bittings in the edge combination. *Remember:* cut = #2 and no-cut = #4.

For Kaba Gemini, there are three active depths on the edge, plus a high #4 cut used in master keying. Because there is no no-cut on Gemini, a key gauge should not be necessary to read the edge combination. Right-hand stock keys will always have cuts in positions 3-5-7-9 and left-hand keys for factory master key systems will almost always have cuts in positions 2-4-6-8-10. (See FIG. 8-14.)

Bitting notation

Before the bitting comes an indication for the hand, "R" or "L." As was mentioned, all bittings are read and notated, bow to tip. The key combination is broken up into separate parts

8-14 Kaba Gemini edge. (Courtesy Lori Corp.)

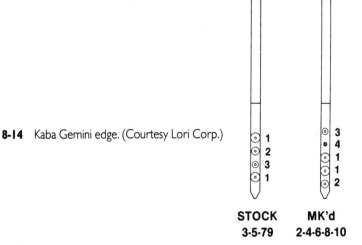

STOCK   MK'd
3-5-79   2-4-6-8-10

Kaba 8	R 1421-2414-17	
Kaba 14	L 34121-14123-2442	
Kaba Gemini (Stock)	R 14214-241414-1231	**8-15** Writing key combinations.
Kaba Gemini (MK'd)	L 11314-442121-14321	(courtesy Lori Corp.)
Kaba Saturn	L 142-313	

corresponding to each row of pins in the cylinder. The first group of bittings is the hand side, the second group is the opposite side, and the third is the edge (FIG. 8-15).

This holds true for all Kaba designs, but the guard pin cut on a Saturn key (a #3 depth) is not part of the key combination and should be ignored for all phases of key reading.

Notation of composite bitted key combinations isn't much different from that of regular keys. Composite bitted keys have both right and left-hand side cuts and often have both even and odd position edge cuts. Such keys are normally used only in selective key systems and maison key systems.

If most of the key system is made up of left-hand cylinders, that is the hand which is listed first. Conventionally, the edge bittings are all listed as part of the first line. Then the opposite hand bittings are written under those of the main hand, as illustrated in FIG. 8-16.

Identifying non-original keys

Because the dimensions of the key blanks for K-8, K-14, and Saturn are identical, persons in the field sometimes duplicate a key from one design onto another's key blank. This can lead to problems later on, especially when quoting prices to a customer. If a key says "Kaba 8" but it has really been cut for a Kaba 14 cylinder, you may quote a Kaba 8 price and order a Kaba 8 cylinder, only to find that you can't set it to the customer's key. A genuine key blank with a system designation such as "Kaba 8," "Kaba 14," or "Kaba Saturn" should only be used to make keys for that particular design in order to avoid confusion at a later date.

If you didn't sell the job originally, you should always check to verify the design by counting the dimples on the key.

If a new cylinder is needed to match a non-original key or if the key was made poorly and a code original must be made to operate the lock properly, you must also be able to determine which Kaba design the key was cut for.

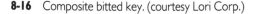

8-16   Composite bitted key. (courtesy Lori Corp.)

L 43121-124231-314213112
R 24133-122431

8-17 8 bittings equal Kaba 8. (courtesy Lori Corp.)

This is easily done by simply counting the dimples on the sides, being careful not to overlook any of the tiny #4 cuts:

Kaba 8 has two rows of 4 cuts (= 8 total) FIGURE 8-17. Kaba 14 has two rows of five cuts (= 10 total) FIGURE 8-18. Saturn has a row of 3 and a row of 4 (= 7 total). Kaba Gemini uses a different key blank which is thicker and narrower. The dimples are oblong rather than round (FIG. 8-19). It has a row of 5 and a row of 6 (= 11).

Composite bitted keys would have exactly twice as many dimples on a side.

Medeco locks

Perhaps locks manufactured by Medeco Security Locks, Inc. are the most well-known high-security mechanical locks in North America. For that reason, much of this remaining section is devoted to reviewing how the various types of Medeco locks operate.

General information

Before studying in detail the specifications of Medeco locks, it's first necessary to understand how these locks operate. Medeco's 10-through 50 series locks incorporate the basic principles of a standard pin tumbler cylinder lock mechanism a plug, rotating within a shell, that turns a tailpiece or cam when pins of various lengths are aligned at a shear line by a key. Figure 8-20 shows an exploded view of a Medeco cylinder.

8-18 10 bittings equal Kaba 14. (courtesy Lori Corp.)

8-19 11 oblong bittings equal Kaba Gemini. (courtesy Lori Corp.)

11 oblong = Kaba Gemini

The rotation of the plug in a Medeco lock is blocked by the secondary locking action of a sidebar protruding into the shell. Pins in a Medeco lock have a slot along one side, and the pins must be rotated so that this slot aligns with the legs of the sidebar. The tips of the bottom pins in a Medeco lock are chisel pointed, and they are rotated by the action of the tumbler spring seating them on the corresponding angle cuts on a Medeco key. (See FIG. 8-21.)

The pins must be elevated to the shear line and rotated to the correct angle simultaneously before the plug will turn within the shell (See FIG. 8-22). This dual-locking principle and the cylinder's exacting tolerances account for Medeco's extreme pick resistance.

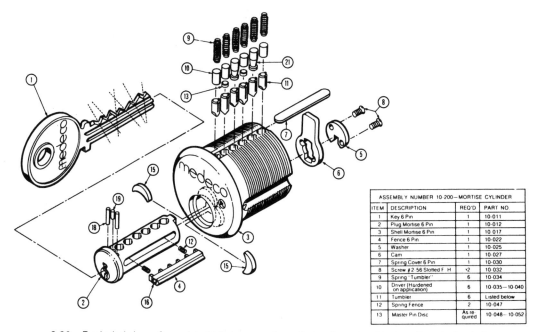

ASSEMBLY NUMBER 10-200—MORTISE CYLINDER			
ITEM	DESCRIPTION	REQ'D	PART NO.
1	Key 6 Pin	1	10-011
2	Plug Mortise 6 Pin	1	10-012
3	Shell Mortise 6 Pin	1	10-017
4	Fence 6 Pin	1	10-022
5	Washer	1	10-025
6	Cam	1	10-027
7	Spring Cover 6 Pin	1	10-030
8	Screw #2-56 Slotted F. H.	*2	10-032
9	Spring "Tumbler"	6	10-034
10	Driver (Hardened on application)	6	10-035—10-040
11	Tumbler	6	Listed below
12	Spring Fence	2	10-047
13	Master Pin Disc	As required	10-048— 10-052

8-20 Exploded view of an original Medeco mortise cylinder. (courtesy Medeco Security Locks.)

8-21   Angle cuts on a Medeco key cause the tumblers in a Medeco cylinder to be raised to the shearline while simultaneously rotating into position to allow the sidebar's legs to push into the pins. (courtesy Medeco Security Locks.)

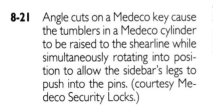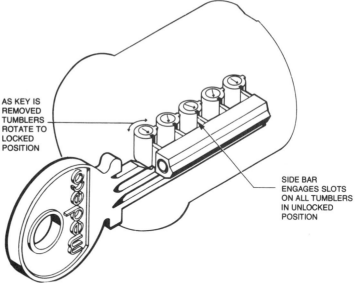

AS KEY IS REMOVED TUMBLERS ROTATE TO LOCKED POSITION

SIDE BAR ENGAGES SLOTS ON ALL TUMBLERS IN UNLOCKED POSITION

8-22   A cutaway view of an original Medeco mortise cylinder showing how both the pins and side bar obstruct the plug from rotating. (courtesy Medeco Security Locks.)

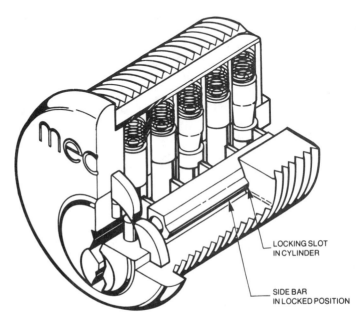

LOCKING SLOT IN CYLINDER

SIDE BAR IN LOCKED POSITION

Medeco cylinders are also protected against other forms of physical attack by hardened, drill resistant inserts. There are two hardened crescent shaped plates within the shell, protecting the shear line and the side bar from drilling attempts. There are also hardened rods within the face of the plug, and a ball bearing in the front of the sidebar. Figure 8-23 shows the shapes and positions of the inserts.

To fit within the smaller dimensions necessary in a cam lock, Medeco developed the principle of a driverless rotating pin tumbler. It is used in the 60- through 65-series locks. The tumbler pin and springs are completely contained within the plug diameter. (See FIG. 8-24.)

The rotation of the plug is blocked by the locking action of a sidebar protruding into the shell. The pins are chisel pointed and have a small hole drilled into the side of them. The pins must be rotated and elevated by corresponding angled cuts on the key, so that each hole aligns with a leg of the sidebar and allows the plug to rotate.

In addition, the cylinder is protected against other forms of physical attack by four hardened, drill resistant rods within the face of the plug.

In spite of the exacting tolerances and additional parts, Medeco's cylinders are less susceptible to wear problems than are conventional cylinders. As in all standard pin tumbler cylinder locks, the tips of the pins and the ridges formed by the adjacent cuts on the key wear from repeated key insertion and removal.

In a Medeco lock, this wear has little effect on its operation because, in contrast to a standard lock cylinder, the tips of the pins in a Medeco lock never contact the flat bottoms of the key cuts. Instead, they rest on the sides of the key profile; thus, the wear on the tips of the pins does not affect the cylinder's operation. (See FIG. 8-25.) Cycle tests in excess of one million operations have proven Medeco's superior wear resistance.

Medeco keys

There are four dimensional specifications for each cut on a Medeco key. They are as follows: the cut profile; the cut spacing; the cut depth; and the cut angular rotation. The profile of the cut on all Medeco keys must maintain an 86 degree angle.

This dimension is critical because the pins in a Medeco lock are chisel pointed and seat upon the sides of the cut profile rather than at the bottom of the cut. Prior to June, 1975, Medeco keys were manufactured with a perfect "V" shaped profile. After this date, the keys were manufactured with a .015 inch wide flat at the bottom.

Spacing of the cut on a Medeco key must be to manufacturers' specifications. For the 10-series stock keys, Medeco part KY-105600-0000 (old #10-010-0000) and KY-106600-0000 (old #10-011-0000), the distance from the upper and lower shoulder to the center of the first cut is .244 inch. Subsequent cuts are centered an additional .170 inch. For the 60-series stock keys, Medeco part KY-105400-60000 (old #60-010-6000) (5 pin) and KY-104400-6000 (old

8-23   Face of a Medeco cylinder showing the positions of the hardened steel drill resistant pins and crescents. (courtesy Medeco Security Locks.)

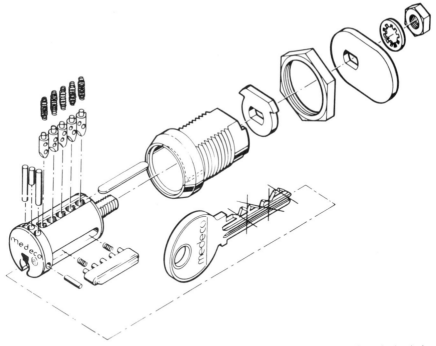

8-24   Exploded view of an original Medeco cam lock. (courtesy Medeco Security Locks.)

8-25   Angles of an original Medeco bottom pin and key cut. (courtesy Medeco Security Locks.)

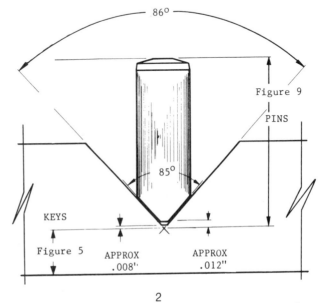

2

#60-011-6000) (4 pin), the distance from the upper shoulder to the center of the first cut is .216 inch. Subsequent cuts are centered an additional .170 inch. The distance from the bottom shoulder to the center of the first cut is .244 inch on this and all Medeco keys. For the 60-series thickhead keys, Medeco part KY-114400-6000 (old #60-611-6000) (5 pin) and KY-114400-6000 (old #60-611-6000) (4 pin), the distance from the shoulder to the center of the first cut is .244 inch. Subsequent cuts are centered an additional .170 inch.

Standard Medeco keys in the 10-through 50-series are cut to six levels with a full .030-inch increment in depths. Keys used in extensive master keyed systems and on restricted Omega keyways are cut to eleven levels with a half step .015-inch increment in depths.

Because of the size limitations, Medeco keys in the 60- through 65-series are cut to four levels with a .030-inch increment in depth. Keys used in extensive master keyed systems and Omega keyways are cut to seven levels with a .015-inch increment in depth.

In addition to the dimensions above, each cut in a Medeco key may be cut with any one of three angular rotations.

These rotations are designated as right (R), left (L), or center (C). When looking into the cuts of a Medeco key as illustrated in FIG. 8-26, concentrating on the flats of the cut, one will notice that flats which are positioned perpendicular to the blade of the key are designated as a center angle. Flats which point upward to the right are designated as right angles, and flats pointing upward to the left are designated as left angles. Right and left angles are cut on an axis 20 degrees from perpendicular to the blade of the key.

Keyways and key blanks

The entire line of Medeco locks is available in numerous keyways. The use and distribution of key blanks of various keyways is part of Medeco's systematic approach to key control.

10-series pins

Medeco bottom pins differ significantly from standard cylinder pins in four respects. The differences occur in the diameter; the chisel point, the locator tab; and the sidebar slot.

Medeco pins have a diameter of .135 inch, that is .020 inch larger than the .115" diameter pin in a standard pin tumbler cylinder.

All bottom pins are chisel pointed with an 85 degree angle. The tip is also blunted and beveled to allow for smooth key insertion.

The locator tab is a minute projection at the top end of the tumbler pin. The locator tab is confined in a broaching in the shell and the plug, it prevents the bottom pin from rotating

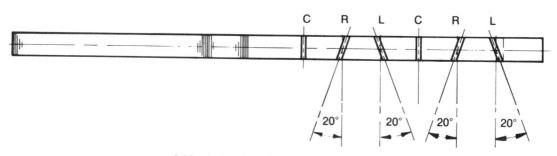

8-26   Angles of cuts in an original Medeco key.

a full 180 degrees. A 180 degree rotation would cause a lockout because the sidebar leg would not be able to enter the sidebar slot in the pin.

The sidebar slot is a longitudinal groove milled in the side of the bottom pins to receive the sidebar leg. The slot is milled at one of three locations in relationship to the axis of the chisel point.

In 1986, the patent for the original Medeco lock system expired. Now other key blank manufacturers can produce those key blanks. The original system, referred to by Medeco as "controlled," is still being used throughout the world. To allow for greater key control, however, Medeco produced a new lock system called *Medeco Biaxial system*.

Medeco's biaxial lock system

From the outside, Biaxial Medeco cylinders look similar to those of the original system. Internally, however, there are some significant differences. Figures 8-27 and 8-28 show exploded views of Biaxial Medeco Cylinders.

Biaxial Pins differ from original Medeco Pins in three respects: the chisel point; the locator tab; and the pin length. Biaxial pins are made of CDA340 hard brass and are electroless nickel plated. They have a diameter of .135 inch and are chisel pointed with an 85 degree angle. However, this chisel point is offset .031 inch in front of or to the rear of the true axis or center line of the pin. (See FIG. 8-29.)

Fore pins are available in three angles, B, K, and Q and Aft pins are available in three angles, D, M, and S. Pins B and D have a slot milled directly above the true centerline of the pins. Pins K and M have a slot milled 20 degrees to the left of the true centerline of the pin. Pins Q and S have slots milled 20 degrees to the right of the true centerline of the pin. (See FIG. 8-30.)

The locator tab, the minute projection limiting the pin rotation, has been moved to the side of the pin, roughly 90 degrees. It is now located along the centerline of the pin opposite the area for sidebar slot. (See FIG. 8-31.)

Biaxial Medeco key specifications

There are four dimensional specifications of a Biaxial Medeco key: the cut profile; the cut depth; the cut spacing; and the cut angular rotation. The cut profile of the Biaxial Medeco key remains at 86 degrees. However, keys in the 10 thru 50 series locks are cut to six levels with a full .025-inch increment in depths.

Spacing of the cut on Biaxial Medeco keys must be to manufacturers' specifications. Because Biaxial Medeco pins have the chisel point forward or aft of the pin centerline, the dimensional spacing on the Biaxial key blank can change from chamber to chamber.

From the shoulder to the center of the first cut, using a fore pin (either B, K, or Q), the dimension will be .213 inch. From the shoulder to the center of the first cut, using an aft pin (either a D, M, or S), the dimension will be a .275 inch. Subsequent fore cuts and subsequent aft cuts are spaced at .170 inch. (See FIG. 8-32.)

While there are only three angular rotations on a key, each rotation can be used with either a fore pin or aft pin. Angular cuts B and D are perpendicular to the blade of the key; "B" will be a fore cut and "D" an aft cut. Angular cuts "K" and "M" have flats pointing upwards to the left; "K" will be a fore cut and "M" will be an aft cut. Angular cuts "Q" and "S" have flats pointing upwards to the right; "Q" will be a fore cut and "S" will be an aft cut. K, M, Q, and S angles are cut on an axis, 20 degrees from perpendicular to the blade of the key. (See FIG. 8-33.)

Mortise and Rim Cylinders

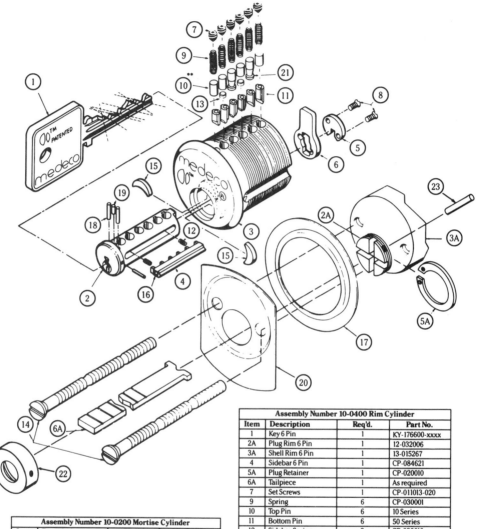

Assembly Number 10-0400 Rim Cylinder			
Item	Description	Req'd.	Part No.
1	Key 6 Pin	1	KY-176600-xxxx
2A	Plug Rim 6 Pin	1	12-032006
3A	Shell Rim 6 Pin	1	13-015267
4	Sidebar 6 Pin	1	CP-084621
5A	Plug Retainer	1	CP-020010
6A	Tailpiece	1	As required
7	Set Screws	1	CP-011013-020
9	Spring	6	CP-030001
10	Top Pin	6	10 Series
11	Bottom Pin	6	50 Series
12	Sidebar Spring	2	CP-030018
13	Master Pin (Wafer)	As required	
14	Mounting Screws	2	CP-011417-000
*15	Steel Insert (Hardened)	2	CP-180011
*16	Steel Ball (Hardened) pre-assy. w/item #4	1	CP-250060
*17	Cylinder Collar	As required	CP-180021
*18	Security Pin pre-assy. w/item #2	2	CP-060103
*19	Security Pin pre-assy. w/item #2	1	CP-060100
20	Mounting Plate	1	CP-180031
*21	Mushroom Drivers	As required	10 Series
22	Tailpiece Retainer	1	CP-021121
23	Roll Pin	1	CP-060203

Assembly Number 10-0200 Mortise Cylinder			
Item	Description	Req'd.	Part No.
1	Key 6 Pin	1	KY-176600-xxxx
2	Plug Mortise 6 Pin	1	12-005006-xx
3	Shell Mortise 6 Pin	1	13-009267
4	Sidebar 6 Pin	1	CP-084621
5	Cam Washer	1	CP-021011
6	Cam	as required	
7	Set Screws	6	CP-011013-020
8	Screw #2-56 Slotted FH	2	CP-010103-050
9	Spring	6	CP-030001
10	Top Pin	6	10 Series
11	Bottom Pin	6	50 Series
12	Sidebar Spring	2	CP-030018
13	Master Pin (Wafer)	As required	

**Certain mortise and rim top pins are hardened steel to prevent drilling of shear line.

*These parts included on 10-0200 MORTISE CYLINDER
Patent No. 3499302

8-27   Exploded view of a Medeco Biaxial mortise and rim cylinder. (courtesy Medeco Security Locks.)

Knob Cylinder

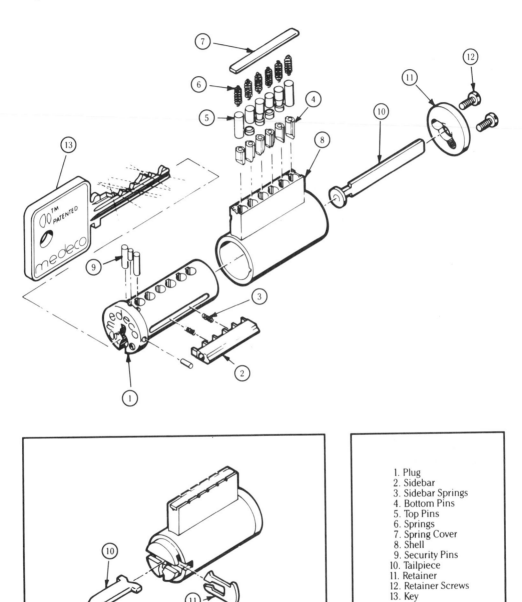

8-28 Exploded view of a Medeco Biaxial key-in-knob cylinder. (courtesy Medeco Biaxial key-in-knob cylinder. (courtesy Medeco Security Locks.)

1. Plug
2. Sidebar
3. Sidebar Springs
4. Bottom Pins
5. Top Pins
6. Springs
7. Spring Cover
8. Shell
9. Security Pins
10. Tailpiece
11. Retainer
12. Retainer Screws
13. Key

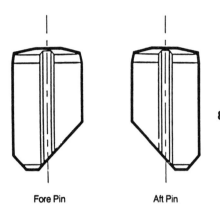

Fore Pin   Aft Pin

8-29 Medeco Biaxial pins. (courtesy Medeco Security Locks.)

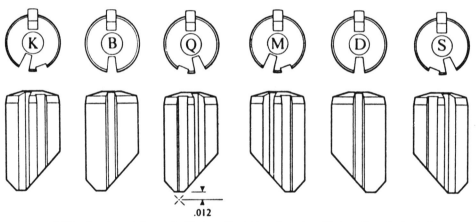

.012

8-30 Angular rotations of a Medeco Biaxial key. (courtesy Medeco Security Locks.)

8-31 The locator tab is located along the centerline of the Medeco Biaxial pin opposite the area for sidebar slot. (courtesy Medeco Security Locks.)

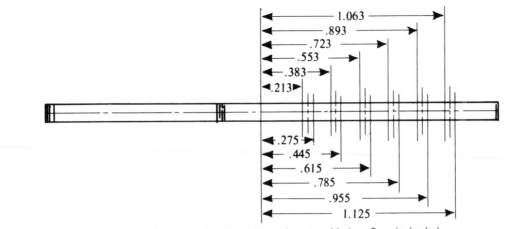

8-32 Spacing for Medeco Biaxial keys. (courtesy Medeco Security Locks.)

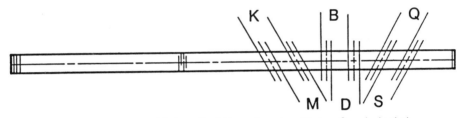

8-33 Angles for cuts in Medeco Biaxial keys. (courtesy Medeco Security Locks.)

9

Master keying

*M*aster keying can provide immediate and long range benefits that a beginning locksmith can find most desirable. This chapter covers the principles of master keying, as well as techniques for developing master key systems.

CODING SYSTEMS

Coding systems help the locksmith to distinguish various key cuts and tumbler arrangements. Without coding systems, master keying would be nearly impossible.

Most coding systems (those for disc-, pin-, and lever-tumbler locks) are based upon depth differentiation. Each key cut is coded according to its depth; likewise, each matching tumbler receives the same code. Depths for key cuts and tumblers are standardized for two reasons: (1) It is more economical to standardize these depths; mass production would be impossible without some kind of standardization. (2) Depths, to some extent, are determined by production.

MASTER KEY SYSTEMS

In most key coding systems, tumblers can be set to any of five possible depths. These depths are usually numbered consecutively 1 through 5. Since most locks have five tumblers, each one having five possible settings, there can be thousands of combinations. Master keys are possible because a single key can be cut to match several lock combinations.

In developing codes, there are certain undesirable combinations which cannot be used. The variation in depths between adjoining tumblers cannot be too great. For example, a pin-tumbler key cannot be cut to the combination 21919 because the cuts for the 9s would rule out the cuts for the 1s. Likewise a pin-tumbler lock with the combination 99999 would be too easy to pick. The undesirable code combinations vary depending upon the type of tumblers, the coding system, and the number of possible key variations. The more complex the system, the greater the possibility of undesirable combinations.

MASTER KEYING WARDED LOCKS

Since a ward is an obstruction within a lock that keeps out certain keys not designed for the lock, a master key for warded locks must be capable of bypassing the wards. Figure 9-1 shows a variety of side ward cuts that are possible on warded keys. The master key (marked *M*) is cut to bypass all the wards in a lock admitting the other six keys.

As explained earlier, cuts are also made along the length of the bit of a warded key. These cuts correspond to wards in the lock. To bypass such wards, a master key must be narrowed.

Because of the limited spaces on a warded key, master keying is limited in the warded lock. The warded lock, because it offers only a very limited degree of security, uses only the simplest of master key systems. Figure 9-2 shows some of the standard master keys that are available from factories.

MASTER KEYING LEVER LOCKS

Individual lever locks may be master keyed locally, but any system that requires a wide division of keys would have to be set up at the factory. A large selection of tumblers is required. The time involved in assembling a large number would make the job prohibitive for the average locksmith.

There are occasions when you will be asked to master key small lever locks. There are two systems. The first is the double-gate system (FIG. 9-3); the other is the wide-gate system (FIG. 9-4). Double gating is insecure. As the number of gates in the system increases, care must be taken to prevent cross-operation between the change keys. For example, you may find a change key for one lock acting as the master key.

With either system, begin by determining the tumbler variations for the lock series in question. If the keys to all the locks are available, read the numbers stamped on the keys.

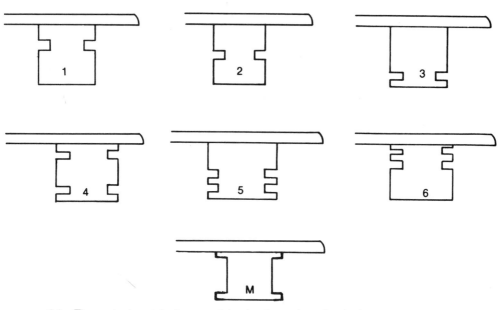

9-1   The master key at the bottom of the drawing replaces the six change keys above it.

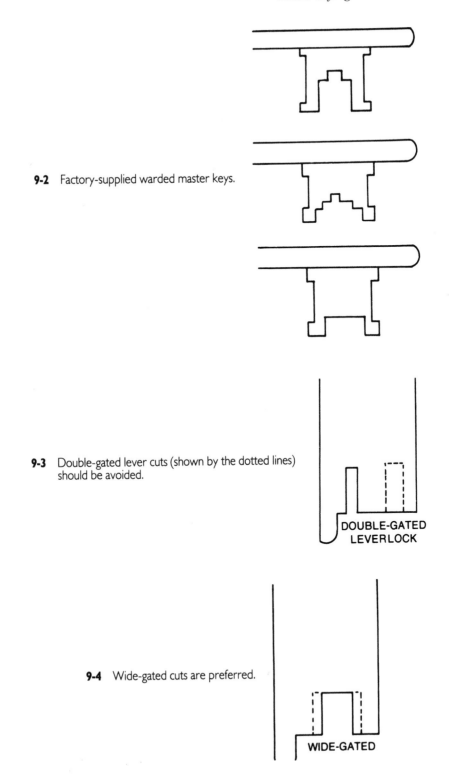

9-2 Factory-supplied warded master keys.

9-3 Double-gated lever cuts (shown by the dotted lines) should be avoided.

DOUBLE-GATED
LEVERLOCK

9-4 Wide-gated cuts are preferred.

WIDE-GATED

Otherwise, the locks must be disassembled so that you can note the tumbler depths for each one. Next, make a chart listing the tumbler variations (FIG. 9-5).

The master key combination can be set up fairly easily now. The chart in FIG. 9-5 is for ten lever locks, each having five levers with five possible key depths per lever. The master key for these locks will have a 21244 cutting code.

Suppose the tumblers in the first position have depths of 1, 2, 3, and 5. Depths 1 and 3 must be filed wider to allow the depth 2 cut of the master key to enter. The tumbler with the depth 5 cut requires a separate cut, or double gating. It will have a cut that will align it properly at two positions.

Moving to position 2 on the chart, we see that four levers will have to be cut; all of these will require a double gating cut. In position 3, four will require a double gating and one will need widening. In position 5, only one will require another gating to be cut, while four will require widening at the current gate.

Another master keying method is to have what is known as a master-tumbler lever in each lever lock. The master tumbler has a small peg fixed to it that passes through a slot in the series tumblers. The master key raises the master tumbler. The peg, in turn, raises the individual change tumblers to the proper height, so that the bolt post passes through the gate in each lever. The lock is open.

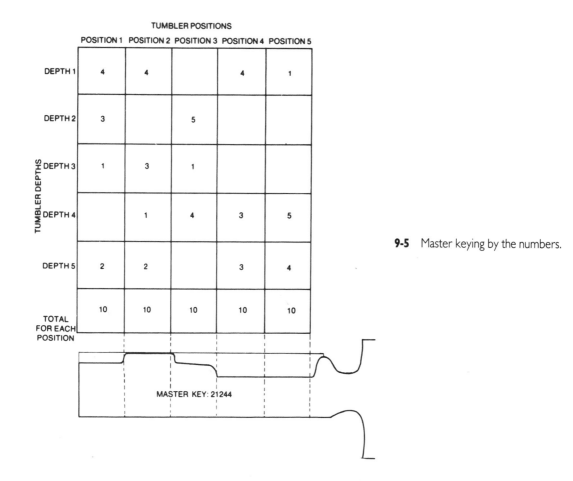

9-5 Master keying by the numbers.

TUMBLER POSITIONS

TUMBLER DEPTHS	POSITION 1	POSITION 2	POSITION 3	POSITION 4	POSITION 5
DEPTH 1	4	4		4	1
DEPTH 2	3		5		
DEPTH 3	1	3	1		
DEPTH 4		1	4	3	5
DEPTH 5	2	2		3	4
TOTAL FOR EACH POSITION	10	10	10	10	10

MASTER KEY: 21244

This system should be ordered from the manufacturer. The complexities building one yourself require superhuman skill and patience.

Master keying disc tumbler locks

Disc tumbler locks have as few as three discs and as many as twelve. The most popular locks have five.

Figure 9-6 shows how the tumbler is modified for master keying. The left side of the tumbler is cut out for the master; the right side responds to the change key. The key used for the master is distinct from the change key in that its design configuration is reversed. The cuts are, of course, different. The keyway in the plug face is patterned to accept both keys. Since the individual disc tumblers are numbered from 1 to 5 according to their depths, it is easy to think of the master key and disc cuts on a 1 to 5 scale, but on different planes for both the key and the tumblers.

Uniform cuts are taboo. The series 11111 or 22222 would give very little security since a piece of wire could serve as the key. Other uniform cuts are out because they are susceptible to shimming. To keep the system secure, it is best to keep a two-depth interval between any two change keys. For example, 11134 is only one depth away from 11133, so 11134 should be used and 11133 omitted. The rationale is that 11134 is the more complex of the two.

The next step is to select a master key combination composed of odd numbers. At the top of your worksheet, mark the combination you have selected for the master key. Below it add a random list of possible change-key numbers. If you choose a systematic approach in developing change-key numbers, you compromise security. On the other hand, the systematic approach ensures a complete list of possible combinations. Begin systematically; then, randomly select the change-keys combinations.

A single code could be used for all customers. The main point to remember is to *use different keyways.*

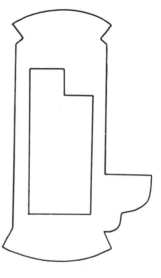

9-6　The master key operates on the left side of the tumbler.

Master keying pin-tumbler locks

Master keying is more involved than modifying the cylinder. It requires the addition of another pin sandwiched between a top and bottom pin. This pin is, logically enough, called the *master pin*.

The master system is limited only by the cuts allowed on a key, the number of pins, and the number of pin depths available. Since this book is for beginning and advanced students, the subject will be covered on two levels: the simple master key system for no more than 40 locks in a series, and the more complex system involving more than 200 individual locks.

It is important to remember that a master key system should be designed in such a way as to prevent accidental cross keying.

Pins are selected on the basis of their diameters and lengths. Master pin lengths are built around the differences between the individual pin lengths. Consider, for a moment, the Yale five-pin cylinder, with pin lengths ranging from 0 to 9 cuts. Each pin is 0.115 inches in diameter. Lower pin lengths are as follows:

$$0 = 0.184'' \quad 5 = 0.276''$$
$$1 = 0.203'' \quad 6 = 0.296''$$
$$2 = 0.221'' \quad 7 = 0.315''$$
$$3 = 0.240'' \quad 8 = 0.334''$$
$$4 = 0.258'' \quad 9 = 0.393''$$

As an illustration, let's master key ten locks with five tumblers each. Each tumbler can have any of ten different individual depths in the chamber.

Determine the lengths of each pin in each cylinder. Mark these down on your worksheet. The master key selected for this system may have one or more cuts identical to the change keys.

Cut a master key to the required depths. In this instance, each cylinder plug is loaded by hand.

Using the known master key depth, subtract the depth of the change key from it. The difference is the length of the master key pin. If the change key is 46794 and the master key is 68495, the master pin combination will be as follows:

Chamber Position	Bottom	Master
1	4	2
2	6	2
3	7	3
4	9	0
5	4	1

This procedure is followed for each plug. Notice that not all the chambers have a master pin. Such complexity is not necessary and makes the lock more vulnerable to picking. Each master pin represents another opportunity to align the pins with the shear line. Figure 9-7 is an extreme instance, with five master pins and three master keys.

In practice, locksmiths avoid most of this arithmetic by compensating as they go along. For example, an unmaster keyed cylinder has double pin sets. Master keying means that an additional pin is added in (usually) position 1.

This drives the bottom pin lower into the keyway. The bottom pin must be shortened to compensate.

A grand master key or a great grand master key adds complexity to the system. You have two choices: add master pins in adjacent chambers (the Yale approach shown in FIG. 9-8), or stack pins in the first chamber. Suppose you have a No. 4 bottom pin and the appropriate

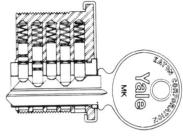

9-7 Master keying pin-tumbler locks means a combination other than the change key will raise the pins to the shear line.

MASTER KEY INSERTED

master pin (FIG. 9-9A). In order to use grand and great grand master keys, you must add two No. 2 pins so that all four pins will operate the lock (FIG. 9-9B).

DEVELOPING THE MASTER KEY SYSTEM

A master key system should be planned to give the customer the best security that the hardware is capable of. Begin by asking the customer these questions:

- ☐ Do you want a straight master key system with one master to open all locks? Or do you want a system that will have submasters? That is, do you want a system with several submasters of limited utility and a grand or great grand master?
- ☐ What type of organizational structure is within the business? Who should have access to the various levels?
- ☐ How many locks will be in each submaster grouping?
- ☐ Is the system to be integrated into an existing system, or will the system be developed from scratch?
- ☐ What type of locks do you have?

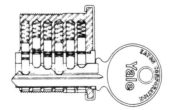

CHANGE KEY INSERTED

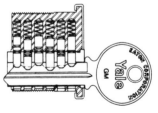

GRAND MASTER INSERTED

9-8 The Yale great grand master system requires five master pins.

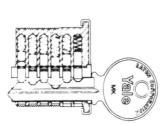

MASTER KEY INSERTED

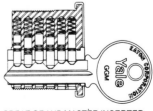

GREAT GRAND MASTER INSERTED

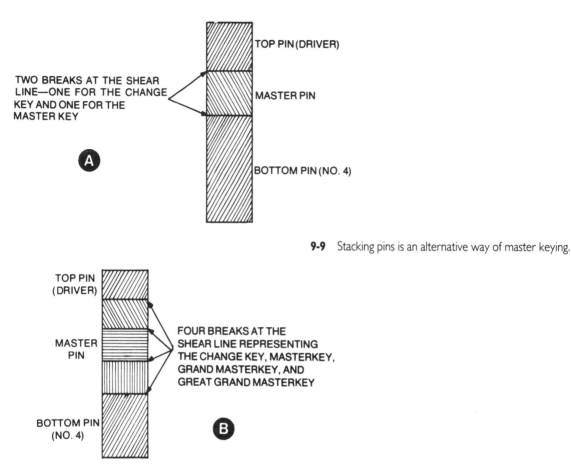

9-9 Stacking pins is an alternative way of master keying.

Once you have (with the help of your catalogs, specification sheets, technical bulletins, and experience) considered these answers, you are ready to begin development of a master key system.

The purpose of master keying is to *control access*. A key may open one lock, a series of locks, a group composed of two or more series, or every lock in the system.

Master key terms

Before we go much further into the subject, it is wise to spend a few minutes defining the terms. This glossary has been prepared by Eaton Corporation (Yale locks) and is reprinted with their permission.

bicentric cylinder—A pin-tumbler cylinder with two plugs, which effectively make it two cylinders in one. The bicentric cylinder is recommended for large, multilevel master key systems, where maximum security and expansion is required.

building master key—A master key which opens all or most of the locks in an entire building.

change key (or individual key)—A key which will usually operate only one lock in a series, as distinguished from a master key which will operate all locks in a series. Change keys are the lowest level in a master key system.

changes (key)—See **key changes**.

construction breakout key (CBOK)—A key used by the owner to make all construction master keys permanently inoperative.

construction master key (CMK)—A key normally used by the builder's personnel for a temporary period during construction. It operates all cylinders designated for its use. The key is permanently voided by the owner when he accepts the building or buildings from the contractor.

control key—A key used to remove the central core from a removable core cylinder.

controlled cross keying—See **cross keying**.

cross keying—When two or more different change keys (usually in a master key system) intentionally operates the same lock.

cross keying controlled—When two or more change keys under the same master key operate one cylinder.

cross keying uncontrolled—When two or more change keys under *different* master keys operate one cylinder.

department master key—A master key that gives access to all areas under the jurisdiction of a particular department in an organization, regardless of where these areas are in a building or group of buildings.

display room key—A special hotel change key that will allow access to only designated, even if the lock is in the shutout mode. With many types of hotel locks, this key will also act as a shutout key, making all other change and master keys inoperative, except the appropriate individual display room key and the emergency key.

dummy cylinder—One without an operating mechanism; used to improve the appearance of certain types of installations.

emergency key (EMK)—A special, usually top level, hotel master key that will operate all the locks in the hotel at all times. An emergency key will open a guestroom lock even if it is in the shutout mode. With many types of hotel locks, this key will act as a shutout key, making all other change and master keys inoperative, except the appropriate individual display room key and the emergency key.

engineer's key (ENG)—A selective master key which is used by various maintenance personnel to gain access through many doors under different master and grand master keys. The key can be set to operate any lock in a master key system and, typically, fits building entrances, corridors, and mechanical spaces. Establishing such a key avoids issuing high level master keys to maintenance personnel. See also **selective master key**.

floor master key—A master key that opens all or most of the locks on a particular floor of a building.

grand master key (GM or GMK)—A key that operates a large number of keyed-different or keyed-alike locks. Each lock is usually provided with its own change key. The locks are divided into two or more groups, each operated by a different master key. Each group can be operated by a master key only, or by the grand only, or by a master and the grand.

great grand master key (GGM or GGMK)—A key that operates a large number of keyed-different and/or keyed-alike locks. Each lock is usually provided with its own change key. The locks are divided into two or more groups, each operated by a different grand master key. Each of these groups is further subdivided into two or more groups, each operated by a different master key. A group can be operated by the great grand only, a grand only, a master only, or any combination of the three.

guestroom change key—The hotel room key that is normally issued to open only the one room for which it was intended. The guestroom key cannot be used to set a hotel lock in the shutout mode from the outside of the room, nor will it open a hotel lock from the outside if the lock is in the shutout mode.

hotel-function shutout—When a hotel-function lock is in the shutout mode, regular master keys of all levels and the guestroom key will not open the lock from the outside of the room. Most hotel-function locks can be set in the shutout mode with the thumbturn or pushbutton from the inside, or with the emergency or display room key from the outside.

hotel keying—Keying for hotel-function locks.

hotel great grand master keys, grand master keys, or master keys—Depending on the level of the system, these keys function as they would in a normal system except they cannot be used to set a hotel lock in the shutout mode from the outside of the room, nor can they open a hotel lock from the outside if the lock is in the shutout mode.

housekeeper's key—A grand master key in a great grand level hotel system that normally operates all the guestrooms and linen closets in the hotel.

interchange—See **key interchange**.

key bitting—A number that represents the depth of a cut on a tumbler-type key. A bitting is often expressed as a series of numbers or letters that designate the cuts on a key. The bittings on a key are the cuts which actually mate with the tumblers in the lock.

key bitting depth—The depth of a cut that is made in the blade of a key. See also **root depth**.

key bitting list—A list originated and updated by the lock manufacturer for every master key system established. This list contains the key bittings of every master key and change key used in the system. Each time an addition is made to the system, all new bittings used are added to the list. It is essential that a complete copy of this list be furnished to any personnel servicing a master key system locally. The lock manufacturer should be informed of any changes made locally to a keying system.

key bitting or cut position (also called spacing)—The location of each cut along the length of a key blade. It is determined by the location of each tumbler in the lock. Bitting position is measured from a reference point to the center of each cut on the key. The most common reference point is the key stop, but the tip of the key is sometimes used.

key change number—A recorded number, usually stamped on the key for identification. A key change number can be either the direct bitting on the key or a code number.

key changes (chges)—The total possible number of different keys available for a given type of tumbler mechanism. In master key work, the number of different change keys available in a given master key system.

key interchange—An undesirable situation, usually in a master key system, whereby the change key for one lock unintentionally fits other locks in the system.

key section (KS)—The cross-sectional shape of a key blade that can restrict its insertion into the lock mechanism through the keyway. Each key section is assigned a designation or code by the manufacturer. A key section is usually shown as a cross section viewed from the bow towards the tip of the key. See also **keyway**.

key set (or set)—A group of locks keyed exactly the same way. A key set is usually identified with a key symbol. See also **standard key symbols**.

keyed alike (KA)—A group of locks operated by the same change key. Not to be confused with master keying.

keyed different (KD)—A group of locks each operated by a different change key.

keying—A term used in the hardware industry that refers to the arrangement of locks and keys into groups in order to limit access.

keying levels—The stratification of a master key system into hierarchies of access. Keying systems are available with one or more levels. The degree of complexity of the system depends on the number of levels used. Generally, the top level master key can open all locks in the system. Each successive intermediate level of master key can open fewer locks but can open more locks than a change key.

keying system chart—A chart indicating the structure and expansion of a master key system, showing the key symbol and function of every master key of every level.

keyway (Kwy)—The shape of the hole in the lock mechanism that allows only a key with the proper key section to enter. See also **key section**.

levels—See **keying levels**.

maid's key—A hotel master key given to the maid which will give access only to the guestrooms and linen closets in her designated area of responsibility. A hotel is normally divided into floors or sections with a different maid's key for each floor or section. A maid's key will not open a guestroom if the lock is in the shutout mode.

multiple key section system—(also called **sectional key sections**)—Used to expand a master key system by repeating the same or similar key bittings on different key sections. Keys of one section will not enter locks with a different section, yet there is a master key section milled so it will enter some or all of the different keyways in the system. See also: **simplex key section**.

paracentric keyway—A keyway in a cylinder lock with one or more side wards on each side projecting beyond the vertical centerline of the keyway to hinder picking. See also **simplex key section**.

pin tumblers—Small sliding pins in a lock cylinder, working against drivers and springs and preventing the cylinder plug from rotating until raised to the exact height by the bitting of a key.

plug (of a lock cylinder)—The round part containing the keyway and rotated by the key to transmit motion to the bolt or other locking mechanism.

plug retainer—The part of a lock cylinder which holds the plug in the shell.

privacy key—A change key set up as part of a master key system but not operated by any master keys or grand masters of any level. This key is set up for such areas as liquor-storage rooms in hotels, narcotic cabinets in hospitals, and food storage closets where valuables are kept.

removable core cylinder—A cylinder containing an easily removable assembly which holds the entire tumbler mechanism, including the plug, tumblers, and separate shell. The cores are removable and interchangeable with other types of locks of a given manufacturer by use of a special key called the control key. See also **control key**.

root depth—Refers to the distance from the bottom of a cut on a key down to the base or bottom of the key blade. Root depth is easy to determine since it measures the amount of blade remaining, rather than the amount which was cut away (bitting depth).

selective master key—A special top level master key in a grand or great grand system that can be set to operate any lock in the entire system, in addition to the regular floor or section master key, without cross keying. Typical selective master keys include an engineer's key (ENG), nurse's key (NUR), and attendant's key (ATT). The number of selective master keys is normally limited to one or two and should be setup when the original system is established. See **engineer's key**.

simplex key section—A single independent key section that cannot be expanded into a multiple key section system. Simplex key sections such as the Yale "Para" are used for stock locks and small master key systems.

spacing—See **bitting position**.

standard key symbols—A uniform way of designating all keys and cylinders in a master key system. The symbol automatically indicates the exact function of each key or cylinder in the system, without further explanation.

tailpiece—The connecting link attached to the end of a rim cylinder which transmits the rotary motion of the key through the door into the locking mechanism.

top level master key—The highest level master key in a multilevel keying system that fits most of the locks in the system.

tumbler—One or more movable obstructions in a lock mechanism which dog or prevent the motion of the bolt or rotation or the plug and that are aligned by the key to remove the obstruction during locking or unlocking.

uncontrolled cross keying—See **cross keying**.

visual key control—A system of stamping all keys and the plug face of all lock cylinders with standard key symbols for identification purposes. Other key and cylinder stamping arrangements are available but are not considered visual key control.

Standard key symbol code

Figure 9-10 illustrates keying levels of control and the rudiments of the standard key symbol code. Great grand master keys are identified by the letters GGM. Grand masters carry a single letter, beginning with the first in the alphabet and identifying the hierarchy of locks that the individual grand master keys open. Master keys carry two letters; the first identifies its grand master, the second identifies the series of locks under it. Thus, a master key labeled AA is in grand master series A and opens locks in master key series A. Master key AB is under the same grand master, but opens locks in series B. Master BA is under grand master B and opens locks in series BA. Change keys are identified by their master key and carry numerical suffixes to show the particular lock that they open. Any key in the series can be traced up and down in the hierarchy. Thus, if you misplace change key AB4, you know that master key AB, grand master A, or the great grand master will open the lock.

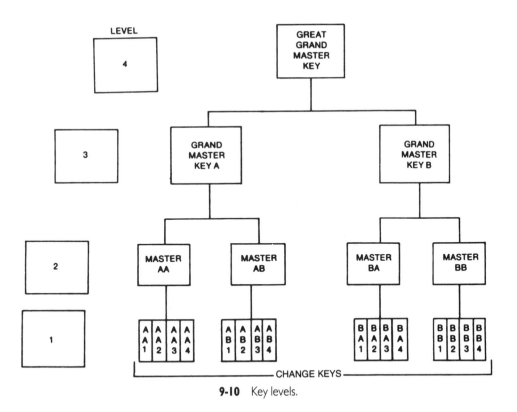

9-10 Key levels.

There are special keys that are out of series. Some of these keys are mentioned in the glossary, together with the appropriate key symbols for them.

If cross keying is introduced into the system—that is, if a key can open other locks on its level, the cylinder symbol should be prefixed with an X. If the cylinder has its own key, it is identified with the standard suffix. For example, XAA4 is a change-key cylinder that is fourth on this level. It may be cross keyed with AA3 or any other cylinder or cylinders on this level. By the same token, master key AA, grand master A, and the great grand master will open it. Elevator cylinders are often cross keyed without having an individual change key. It is no advantage to have a key that will operate the elevator cylinder and none other in the system. These cylinders are identified as X1X, X2X, and so on.

The symbols that involve cross keying apply to the cylinder only; all other symbols apply to the cylinder *and* the key. This point may seem esoteric, but ignoring it causes the factory and everybody else grief. There is, for example, no such thing as an X1X key. Nor is there an XAA4 key. Change key AA4 fits cylinder XAA4 that happens to be cross keyed with another cylinder. Key AA4 does not fit any cylinder except XAA4.

There are certain advantages to using the standard key symbol code:

☐ It is a standardized method for setting up the keying systems.

☐ It maintains continuity from one order to the next.

☐ It indicates the position of each key and each cylinder in the hierarchy.

☐ It helps to control cross keying, since each cross keyed cylinder is clearly marked.

☐ It offers a method of projecting future keying requirements.

- ☐ It can be easily rendered on a chart.
- ☐ It allows better control of the individual keys within the system.
- ☐ It is a simple method of selling and explaining the keying system to the architect or building owner.

Selling the system

It is important to be able to communicate the advantages of the key symbol code and the implications it has for setting up an ordered, coherent, and secure keying system. This may take some selling on your part, since architects and building owners tend to think of keys and locks as individual entities and not part of a larger system. Selling a comprehensive master keying system involves the following:

- ☐ Being able to explain the subject of master keying to the architect or owner.
- ☐ Reviewing the plans of the building(s).
- ☐ Choosing the proper level of control required.
- ☐ Selecting a keying system that will cover present and future expansion requirements.
- ☐ Presenting the system to the architect or owner at a meeting dealing *specifically* with keying.
- ☐ Recording all changes to the system as agreed upon at the meeting.
- ☐ Marking plans with the proper key symbol on the side of each door where the key is to operate.
- ☐ Presenting the owner with a schematic layout of the entire system, showing him the layout of the masters, grand masters, etc.

MASTER KEY SYSTEM VARIATIONS

So far the structure of the master key system has been discussed. Within this broad structure there are many opportunities for variations. Some of these variations involve the possible range of keyways; others involve special hardware, such as removable cylinders, master-ring cylinders, and rotating tumblers. Each of these variations can extend the range, flexibility, and security of the system. A locksmith must be conversant with all of them.

Keyway variations

Figure 9-11 illustrates a system of control based on keyway design. General Lock's series 800 key section passes all cylinders in the system; 29, 30, 31, 32, and 33 are submasters, each passing four cylinders; the change keys are restricted to their own individual cylinders. In its fullest expansion, the system includes 35 different keyways on the change-key level and 4 master key levels.

High security pin-tumbler cylinders

Figure 9-12 illustrates features of the General lock:

- ☐ A 100 degrees key bitting for long wear (A).
- ☐ Master pins have a minimum length of 0.040 inch, (B). Shorter pins tend to wedge in the chamber.
- ☐ Only two pins are standard (C).

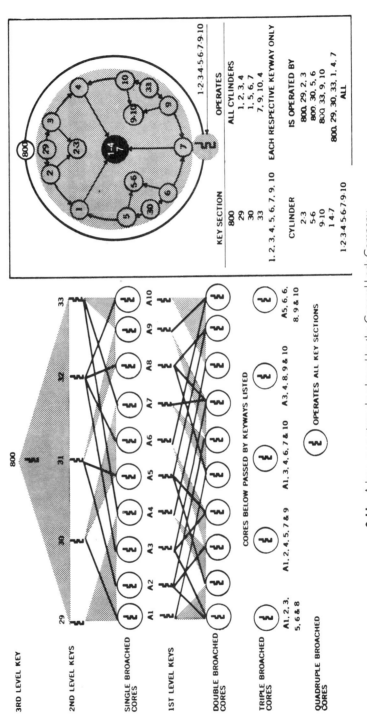

9-11 A keyway system developed by the General Lock Company.

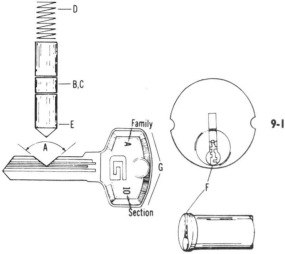

9-12 General locks have special features described in the text.

- ☐ Pins and springs are made of corrosion-resistant alloy (D and E).
- ☐ The keyway is part of the security system (F).
- ☐ If requested, the factory supplies special identification for keys, cores, and cylinders (G).

The Emhart (Corbin) High Security Locking System uses rotating and interlocking pins (FIG. 9-13). The pins must be raised to the shear line and, at the same time, rotated 20 degrees so the coupling can disengage (FIGS. 9-14 and 9-15). Rotation is by virtue of the skew-cut bitting on the key (FIG. 9-16). Figure 9-17 illustrates the way the cylinder is armored. The pins are protected by hardened rods and a crescent-shaped shield.

Removable-core cylinders

Removable-core cylinders are increasingly popular. Figure 9-18 illustrates the Corbin cylinder, a type that is typical of most. To rekey the change key, follow this procedure:

Obtain a Corbin rekeying kit. The kit includes the necessary pins, gauges, and tools.

Mount the cylinder in a vise.

Remove the plug retainer.

Select the key with the deepest bitting as the plug extractor. Normally the grand master key meets this specification; however, there are instances where the engineer's key will have the deepest bitting. A shallow-cut key complicates matters by forcing the control pins, drivers, and buildup pins into the cylinder.

Withdraw the plug and remove all pins from their chambers. Figure 9-19 illustrates this procedure.

Determine the bitting of the change key.

Write down the new combination. As an example, suppose the original change-key bitting is 513525 and we wish to reverse it to 525315 (FIG. 9-20).

Install the tumbler pins, ball end down.

Use a depth gauge to determine the master key bitting.

Calculate the master pins by subtracting the change-key bitting combination from the master key combination. If the master key combination were 525763, the difference between it and the new change-key combination would be 448.

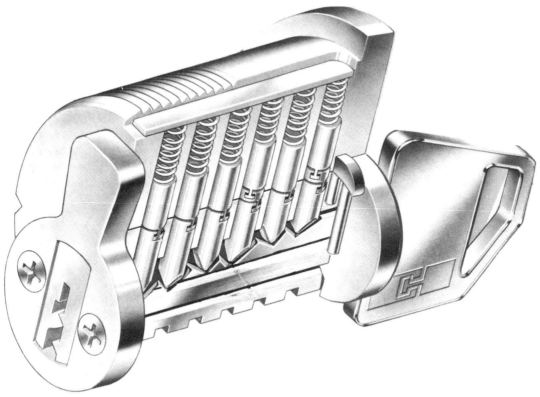

9-13 Corbin's High Security Locking System depends upon split and rotating pins.

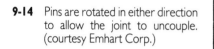

9-14 Pins are rotated in either direction to allow the joint to uncouple. (courtesy Emhart Corp.)

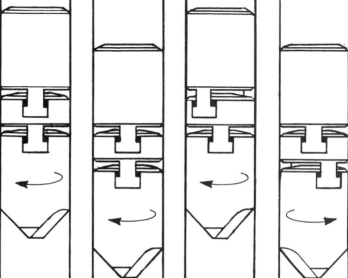

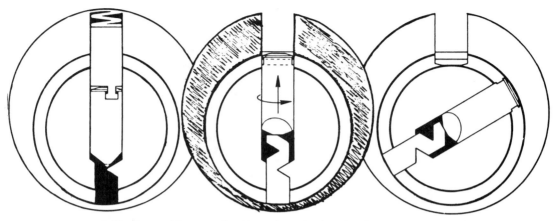

9-15   Pins must be rotated and brought to the shear line for the lock to open.

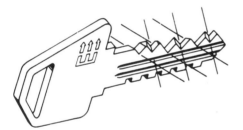

9-16   The corbin key has its bitting cut at 20 degree angles.

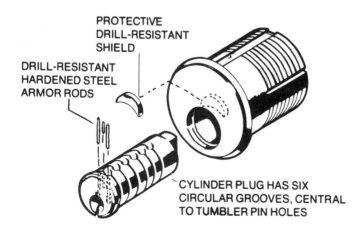

PROTECTIVE
DRILL-RESISTANT
SHIELD

DRILL-RESISTANT
HARDENED STEEL
ARMOR RODS

CYLINDER PLUG HAS SIX
CIRCULAR GROOVES, CENTRAL
TO TUMBLER PIN HOLES

9-17   Passive defense measures include hardened steel pins and armor plate. (courtesy Emhart Corp.)

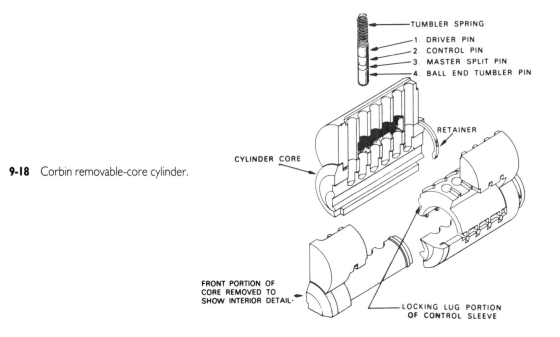

TUMBLER SPRING

1. DRIVER PIN
2. CONTROL PIN
3. MASTER SPLIT PIN
4. BALL END TUMBLER PIN

RETAINER

CYLINDER CORE

FRONT PORTION OF
CORE REMOVED TO
SHOW INTERIOR DETAIL·

LOCKING LUG PORTION
OF CONTROL SLEEVE

9-18 Corbin removable-core cylinder.

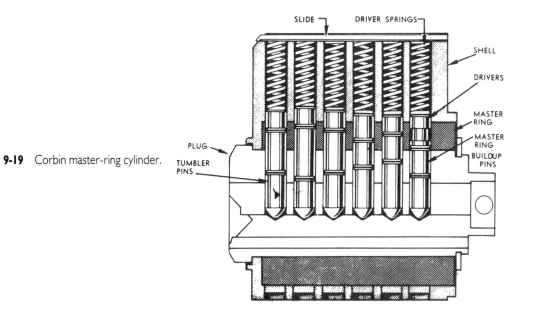

SLIDE   DRIVER SPRINGS

SHELL

DRIVERS

MASTER RING

MASTER RING

BUILDUP PINS

PLUG

TUMBLER PINS

9-19 Corbin master-ring cylinder.

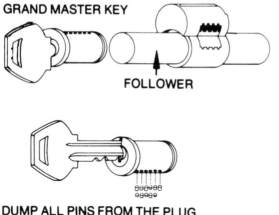

GRAND MASTER KEY

FOLLOWER

DUMP ALL PINS FROM THE PLUG

9-20 Using the appropriate follower, remove the plug. Dump the pins. (courtesy Emhart Corp.)

Insert the master key into the plug.

Install the appropriate master pins (FIG. 9-21).

Remove the master key carefully and insert the grand master key. Select the master split pins by subtracting the master key bitting from the grand master key bitting. All pins should be flush with the surface of the plug.

Assemble the plug and cylinder.

Test all keys.

Lubricate the keyway with a pinch of powdered graphite.

Master-ring cylinders

The Corbin master-ring cylinder is shown in FIG. 9-22. The change key operates the plug plungers, and the master key operates the plunger in the master ring. Sometimes called "two-in-one" cylinders, these cylinders increase the range of key combinations for any given system. To rekey the change key, follow this procedure:

Mount the plug in a vise, and remove the cylinder slide with a pair of pliers or a small chisel.

Remove the springs, drivers, and pins (FIG. 9-23).

Ream pin holes through the shell, master ring, and plug (FIG. 9-24).

Assuming that the original combination was 414472, reversing the combination gives 274414. This will be the combination of the new change key.

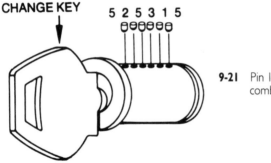

CHANGE KEY   5 2 5 3 1 5

9-21 Pin length corresponds to the change-key combination. (courtesy Emhart Corp.)

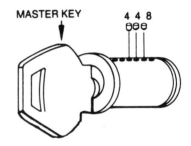

9-22 Subtract the change-key combination from the master key combination. The difference represents the length of the master pins. (courtesy Emhart Corp.)

9-23 After the slide is withdrawn, remove springs, pins, and drivers. (courtesy Emhart Corp.)

9-24 Ream the pin holes through the shell, master ring, and plug. (courtesy Emhart Corp.)

Reverse the pins to conform with the new combination. That is, the pin that was first goes into the last chamber; the pin that was second goes in to the fifth chamber, and so on.

As each pin is installed, tamp it home with a drill bit and turn the key. If the key will not turn, you have confused the pin sequence.

Assemble the lock and test.

To rekey the master key, follow this sequence:

Mount the cylinder in a vise and remove the cylinder slide with a pair of pliers or a small chisel.

Remove the springs, drivers, and pins.

Determine the master key bitting with a gauge. The combination runs from the shoulder to the tip of the key, the reverse of the usual sequence. As an example, let it be 678572.

Write down the change-key combination and reverse it. Suppose the combination is 275414. Reversed, it is 414572.

Subtract the reversed change-key combination from the master key combination (678572—414572). The difference is 264000.

Insert the master key into the cylinder and select the appropriate buildup pins. In this case the pins are 264.

Note: If any number in the change-key combination is greater than the master key number above it, you must use a negative number buildup pin in the chamber. For example, a 678572 master key combination with an 814572 change-key combination requires a −2 buildup pin, together with a 6 and 4 pin.

As each buildup pin is installed, seat it with a drill bit and turn the key to determine that the correct pin has been installed.

Insert the drivers into the cylinder chambers. Either of two drivers are used, depending upon the lock style. Spool drivers (No. J-172) are furnished with mortise cylinders; straight drivers (No. M-099) are used on other cylinders.

Insert the springs into their chambers.

Holding the springs down with your thumb, try the change key. Do the same for the master and grand master key.

Mount the cylinder in the slide, hammering the slide down for a secure fit. Be careful not to damage the threads on mortise cylinders.

Try all the keys.

A SIMPLE MASTER KEY SYSTEM

When it becomes necessary to install a large number of locks for any given concern, be it a business or a large residence, wherein the customer wants the locks master keyed, then it is time that your expertise in the realm of the master key systems is put to the test.

Probably the hardest part of master keying development is the creation of a working master key system that takes into consideration any variables that may be required by the customer. For this reason, many locksmiths do not perform their own master keying of any systems; rather, they rely upon the many years of experience, expertise, and professionalism from a factory-developed system. The use of a factory system ensures that the system will meet all the criteria set forth by the customer, the cylinders will be the right ones, and the possibility of extending the system in the future can be assured. For the locksmith, time, personnel, and money are saved through this process.

On the preceding pages, you have been introduced to various aspects of master keying, the types of locks that can be used, the differences in the systems, and the potential for system expansion at a later date. We will not be detailing the varied intricacies of such a system, but will view a small system that you may want to consider and keep in the back of your mind for those small jobs that come up. These are jobs wherein an existing system may be master keyed, using the same cylinders, or a job where a residence is to be master keyed for a homeowner.

Figure 9-25 is a sample master key system developed for a three-floor office building. In addition to the individual office keys on each floor, you have the master key, extra cylinders for additional inner offices on each floor (or for use as replacement cylinders, a main door key, a key for the building maintenance shop, lavatory keys for both the men's and women's rooms on the three floors, and also the start of an additional key code system you can

MASTER: 431472

First Floor	Second Floor	Third Floor
522534	652534	752534
522544	652544	752544
522554	652554	792554
522564	652564	792564
522574	652574	792574
522594	652584	792584
522504	652594	792594
522524	652504	792504
522634	652624	792624
522644	652634	792634
522654	652644	792644
522664	652654	792654
522674	652664	792664
522684	652674	792674
522694	652684	792684
522604	652694	792694
522624	652734	792604
522724	652744	792734
522734	652754	792744
522744	652764	792754

Extra cylinders for each floor:

552784	652514	752614
552514	652524	752601
552614	604714	752602

Main Door Key: 346572
Bldg Maintenance Shop Key: 434672
Construction Key: 466662

NOTE: If building were to have four or more floors, then the last three digits of the various individual keys could be as above, but the first three digits would be:

Fourth floor: 514___
Fifth floor: 614___
Sixth floor: 714___

If each floor is to have two lavatories, they could be keyed as follows:

First floor:	731573	331577
Second floor:	831573	331578
Third floor:	931573	331579

9-25 Sample master key system for a three-floor building.

develop should the building or other facility require more individual keys than the system was designed for.

Within this system, depending upon the size of the installation you will be developing, select the number of keys for individual locks to be used. In this case we are talking about 20 individual door locks on each floor. The master key has already been designated. The progression in the development of the master key system is that the depths of the various individual keys are such that any given key can be recut to become a master key.

In actually rekeying the various cylinders, you will need the standard pins for each cylinder, and then a complete set of master pins. In this case, though, you may choose to use the lower ranking pins of the current system. Why? Because the cylinder pins are worn in addition to the interior of the cylinder itself. Also, in this case, it will save you time and money in having to order the pins.

What you are doing is determining for each key the depths that will be required for each pinning. As an example, let's take two cylinders from the first floor for rekeying. The first cylinder, 522534 will have the following pinning:

Regular Pin: 4 − 2 − 1 − 4 − 3 − 2
Master Pin: 1 − 1 − 1 − 1 − 4 − 2

With these pins in the cylinder, with the driver pins and tension spring included, either the individual or the master key will open the lock.

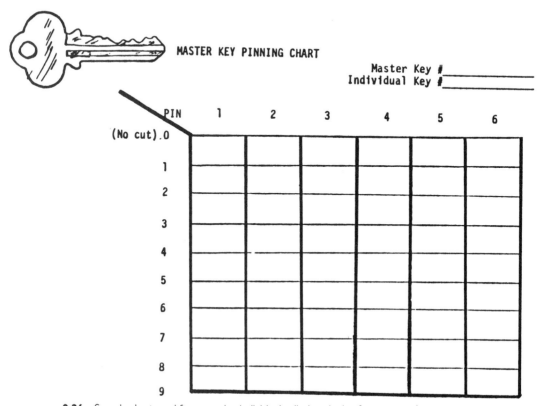

9-26   Sample chart used for arranging individual cylinder pinning for a master key system.

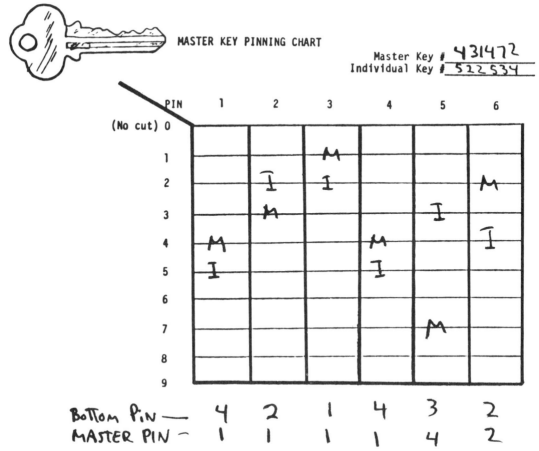

MASTER KEY PINNING CHART

Master Key # 431472
Individual Key # 522534

9-27 Layout chart with master key and individual key cylinder pinning indicated.

For the second cylinder in the series, 522534, the pinning is as follows:

Regular Pin: 4 – 2 – 1 – 4 – 4 – 2
Master Pin: 1 – 1 – 1 – 1 – 3 – 2

Again, with these pins, both the individual and the master key will operate the lock.

Now, how were these particular pins determined? The best way to figure this out is to lay out a simple chart, as shown in FIG. 9-26. As you progress in selecting pins for the individual cylinders, first mark the appropriate points on the chart that indicate the master key number. (You can preprint this on a form and save a lot of time.) Then, as you select each cylinder to work with, put down on the chart the code number of that particular key (FIG. 9-27). The difference between the two points, if any, is the difference between the code number variations. This means that the pinning difference must be the same also.

In FIG. 9-27, for column six, notice that the individual key requires a 4 pin, but the master key requires a 2 pin. The difference of 2 is obvious. The maximum allowable height for the overall height of both pins combined must be 4, so you will require a 2 pin for the master key to operate this particular pin, but also a 2 pin for the individual key. If you only had the 2 pin in the lock, only the master key would be able to open the cylinder at this point; by adding the second 2 pin, you make it possible for the individual key to also open the lock.

If you continue onward, you can readily see how each individual pinning is to be accomplished, what pins will be required, and the insertion sequence into the lock cylinder.

10

Double-bitted locks and keys

The Junkunc Brothers American double-bitted cylinder lock is usually found in padlocks and office and utility locks (FIG. 10-1).

OPERATION

When the double-bitted key is inserted, it passes through the center of the tumblers (as in a disc lock) and aligns them to the shear line, allowing the plug to be rotated. But the key and the tumbler arrangement is different from that of the regular disc lock.

The key cuts are wavy in appearance; thus the tumblers have to align in a wavelike configuration for the lock to open. Further, there are no definitive tumbler cuts on the key. This is because the key holds ten or more tumblers compressed together and held in a locked position by means of a Z-shaped wire within the tumblers.

All the tumblers are uncoded, meaning they are all of a standard cut. So in order for them to turn within the cylinder, the tumblers have to be cut down. A special keying tool is used for this purpose.

CUTTING DOWN THE TUMBLERS

Once the tumblers have been inserted into the plug (with the tumbler spring in place), insert the precut key into the plug.

Mount the plug in your vise firmly.

Attach the keying tool to a ¼ inch drill. Drill the back of the plug. Since the inside diameter of the drill is the same as the outside diameter of the plug, the individual tumblers will cut down to what will be the shear line.

Trim the tumblers with a light wire brush to take off any burrs.

Insert the plug into the cylinder and test it. Attach the retainer screw and withdraw the key.

The tumblers, since they are uncoded, can be used within any plug. The tumbler spring, because of the shape, holds the various tumblers in the locked position. Only with the insertion of a key, which forces the tumblers into another position, can the lock be opened.

10-1   A double-bitted lock and key. (courtesy American Lock Company)

SECTION 1   SECTION 2

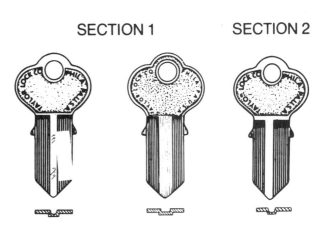

SECTION 3

10-2 The four basic double-bitted key sections. (courtesy Taylor Lock Company)

SECTION 4

KEYS AND KEYWAYS

The double-bitted lock takes four basic key sections (FIG. 10-2). These sections, of course, match the shapes of the keyways. Keyway one is referred to as a K4 and the center point is at the center of the tumbler. Keyway two is referred to as a K4L; the center point is just left of center. Keyway three is called a K4R; the center point is right of center. Keyway four, called a K4W, is shaped like a *W*. The keyway shape does not reflect the tumbler types that are within any given plug.

11

Vending machine locks

*T*he Ace lock, manufactured by the Chicago Lock Company, has become standard on vending machines. The arrangement of the pin tumblers increases security and requires a different type of key—one that is tubular. A typical Ace lock and key is shown in FIG. 11-1. The cam works directly off the end of the plug.

The key has its bitting disposed radially on its end. The depth and spacing of each cut must match the pin arrangement. In addition there are two notches, one on the inside, the other on the outer edge of the key. These notches align the key to the lockface. Otherwise, the key could enter at any position.

The key bittings push the pins back, bringing them to the shear line. Once the pins are in alignment, the key is free to turn the plug and attached cam.

The pins within the Ace lock are entirely conventional in construction, with the exception of the bottom one. A ball bearing is sandwiched between the pin and its driver. The bearing reduces friction and increases pin life. The pin in question is not interchangeable with others in the lock. Pin tolerance is extremely critical. There is no room for sloppy key cutting.

DISASSEMBLY

To disassemble the Ace lock, follow this procedure:

Place the lock into its holder. Ace makes a special vise for these cylinders.

Drill out the retainer pin at the top of the assembly. Use a No. 29 drill bit and stop before the bit bottoms in the hole (FIG. 11-2).

Remove what is left of the pin with a screw extractor.

Insert the appropriate plug follower into the cylinder. Apply light pressure. Plug follower dimensions are: length 1.50 inch; outside diameter 0.373 inch; inside diameter 0.312 inch.

Lift off the outer casing from the bushing assembly.

Scribe reference marks on the plug sections as an assembly guide.

Remove the follower from the cylinder. *Note:* Perform this operation carefully. The pins are under spring tension and must be kept in order. If not, you will have a monumental job sorting the pins. There are 823,543 possible combinations!

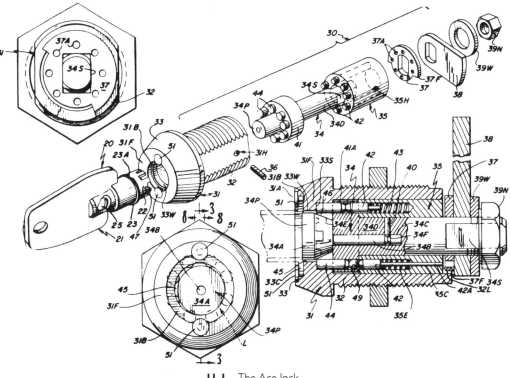

11-1   The Ace lock.

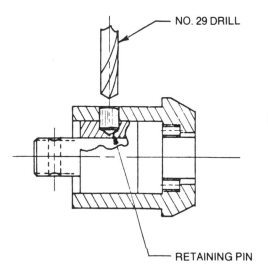

NO. 29 DRILL

RETAINING PIN

11-2   Use a No. 29 drill bit to remove the retaining pin. Alternately, use a No. 42 bit, thread a small metal screw into the hole, and extract the pin and screw by prying upward on the screwhead. (courtesy Desert Publications)

ASSEMBLY

Assembly is the reverse order of disassembly. Replace the retaining pin with an Ace part, available from locksmith supply houses.

REKEYING

Rekeying is not difficult if you approach the job in an orderly manner.

Cut the key. There are seven bit depths, ranging from 0.020 inch to 0.110 inch in 0.015 inch increments.

Remove the pins with tweezers. *Note:* Pin lengths range from 0.025 inch to 0.295 inch in increments of 0.015 inch. Drivers are available in 0.125 inch, 0.140 inch, and 0.180 inch lengths.

Select new pins, using the key combination as a guide.

Install the pins. The flat ends of the core pins are toward the key (See FIG. 11-1). This pattern must be followed:

Core pins 1, 2, and 3 require 0.180″ drivers.
Core pins 4 and 5 require 0.140″ drivers.
Core pins 6 and 7 require 0.125″ drivers.

Pins are numbered clockwise from the top as you face the lock.
Insert the key.
Install the plug in the cylinder with the scribe marks aligned.
Insert a new retaining pin and give it a sharp tap with a punch.
Test the key. It may be necessary to rap the cam end of the plug with a mallet.

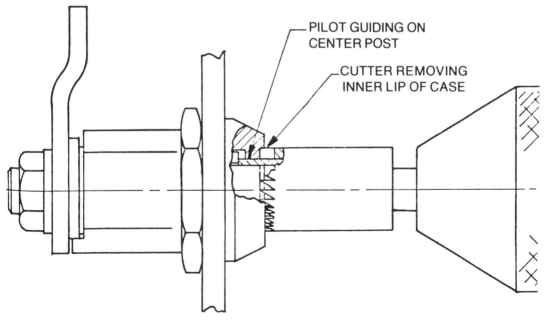

PILOT GUIDING ON CENTER POST

CUTTER REMOVING INNER LIP OF CASE

11-3 Drilling the lock requires a piloted hole saw, available from locksmith supply houses. (courtesy Desert Publications)

LOCKOUT

A lockout can be a real headache. Ordinary Ace locks—those without ball bearings—can be drilled out with the tool shown in FIG. 11-3. Hole saws are available for standard and oversized keyways.

12

Keyed padlocks

*P*adlocks have many uses. They are used to secure outbuildings, bicycles, buildings under construction, tool boxes, paint lockers, and even automobile hoods. Because of this wide use, it pays a locksmith to have a good knowledge of these locks.

While padlock exteriors vary, the functional and operational differences are few and are similar to other locks. Padlocks may use pin tumblers, wards, wafers, levers, or a spring bar. Some must be shackled closed before the key can be removed; this feature is made possible by a spring-loaded coupling. Key security is improved and there is less likelihood of leaving the lock open.

CHOOSING A PADLOCK

Ask the customer if he has a brand preference, and then ask the following questions:

- If width, case length, and shackle clearance are critical (FIG. 12-1).
- Where the lock will be used.
- How often the lock will be opened. The price of the lock has a direct relationship to its wearing qualities.
- If the lock is intended to secure valuable property. An inexpensive lever lock would be adequate to keep children from straying into the backyard, but it would be inappropriate for a boat trailer.
- If the lock will be used indoors or outdoors. If it is an outdoor lock, will it be protected from the elements?

WARDED LOCKS

Warded locks have limited life spans, particularly when used outdoors. The cheapest locks of this type are only good for a few thousand openings and, when locked, give minimal security.

Most of these locks have three wards, although the cheaper ones have only two. Figure 12-2 illustrates the principle. The key must negotiate the wards before it can disengage the

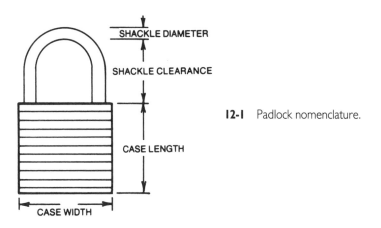

SHACKLE DIAMETER

SHACKLE CLEARANCE

12-1   Padlock nomenclature.

CASE LENGTH

CASE WIDTH

spring bar from the slot in the shackle end. Keys are flat or corrugated. The latter is a mark of the Master Lock Co.

Pass keys

Figure 12-3 shows how a pass key is cut to defeat the wards. The broad tip of the key opens the spring bar; locks with two spring bars require a key cut as shown in FIG. 12-4.

Some corrugated keys can be reversed to fit other locks by filing it on the back of the blades. Alternately, you can file all the unnecessary metal off, converting the key into a pass key.

In many localities pass keys are illegal and, unless you are a locksmithing student from an accredited school, a locksmith trainee, or a licensed locksmith, possession of such a key is a criminal offense. If you have the need for a pass key, keep it in a safe place in your home or office.

Key cutting

A warded padlock key is simple to duplicate. Follow this procedure:

Using the original key as a guide, select the appropriate key blank.

Smoke the original key and mount the original and the blank in your vise.

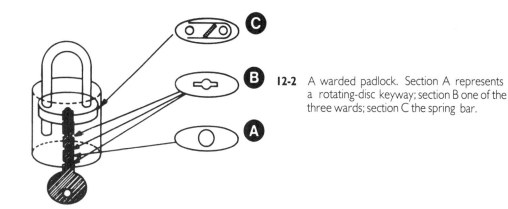

12-2   A warded padlock. Section A represents a rotating-disc keyway; section B one of the three wards; section C the spring bar.

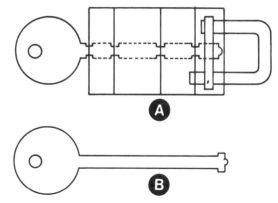

12-3 Wards and their limitations. View A shows the ward arrangement and the necessary key bitting; view B illustrates a pass key.

Using a 4 inch warding file, cut away the excess metal until the blank is an exact copy of the original.

Turn the keys over and repeat the operation.

Remove the burrs from the duplicate and test it in the lock. Should it stick, the cuts are not deep or wide enough. Make the appropriate alterations and you will have a perfect duplicate key.

Key impressioning

Impressioning a warded key should take less than five minutes. Follow this procedure:

Select the appropriate blank and thoroughly smoke it.

Insert the key and twist it against the wards. Do this several times to get a clear impression.

Mount the key in the vise.

Make shallow cuts where indicated.

Smoke the key again and try it in the lock.

Remove the key and file the cuts as indicated.

Continue to smoke, test, and file until the key turns without protest. Do not go overboard with the file. If the bits are too deep, the key will work but may break off in the lock.

Repairs to the lock itself are out of the question, since it would be cheaper to buy a new one.

12-4 A pass key for a lock with two spring bars.

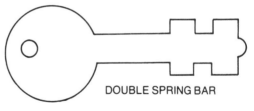

DOUBLE SPRING BAR

American (Junkunc Brothers)

All American (Junkunc Brothers) padlocks share the same patented locking device—two hardened steel balls fitting into grooves in the shackles. This arrangement is the best ever devised. Applying force on the shackle wedges the balls tighter. The H10 model, for example, requires more than 5000 lb. to force; test locks have been stressed to 6000 lb. and still worked. All these locks use a ten-blade tumbler. The exceptions to these statements are those locks with a deadlock feature. The key must be turned to lock the shackle. These padlocks are made on an entirely different principle and are not discussed here.

American padlocks have a removable cylinder that simplifies servicing. If a customer wants keyed-alike locks, the modification takes only a few minutes. One may also key alike different models.

These locks have three basic subassemblies: cylinder assembly, locking mechanism assembly, and the shackle assembly.

Cylinder removal and installation.   Follow this procedure.

Open the lock, exposing the retaining screw at the base of the shackle hole.

Remove the retaining screw (I in FIG. 12-5). If the screw is stubborn, use penetrating oil, letting it set a few moments before attempting to turn the retaining screw.

Strike the side of the lock with a leather mallet. The purpose is to force the retaining pin (C) into the space vacated by the retaining screw. Referring to the drawing, note how the pin retains the cylinder in the slot shown.

Withdraw the key together with the cylinder.

Place a new cylinder in the case.

Assemble in the reverse order of disassembly. *Note:* The H10 series uses a plated silver cap dimensionally identical to the brass caps on the other models.

The locking mechanism.   The brass retainer is the heart of the assembly (E in FIG. 12-5). Retainers vary in size according to the lock model. A30 and AC20 retainers have a groove milled in the body. L50, K60, and KC40 retainers are plain. The H10 is much larger than the others.

The retainer assembly has three parts—a washer, coil spring, and retainer body. The washer (D) fits over the retainer. One end is hooked to accept the coil spring. The other end of the spring is moored in a hole on the retainer.

To remove the retainer, follow this procedure:

Grasp the retainer body (E) with needle-nosed pliers.

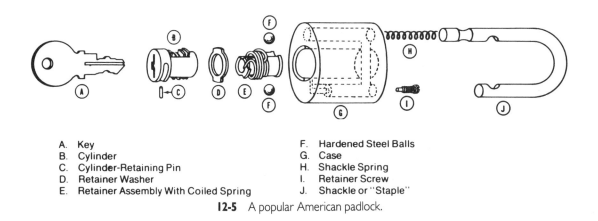

A. Key	F. Hardened Steel Balls
B. Cylinder	G. Case
C. Cylinder-Retaining Pin	H. Shackle Spring
D. Retainer Washer	I. Retainer Screw
E. Retainer Assembly With Coiled Spring	J. Shackle or ''Staple''

12-5 A popular American padlock.

Rotate the retainer 45° and pull.

Remove the steel balls.

To replace, follow these steps:

Grease the balls so they will stay put.

Replace the balls and spread them apart.

Exert pressure against the shackle to hold the balls.

Determine that the washer is correctly aligned with the spring. The free end of the spring has to be in line with the retaining-pin hole in the case.

Place the retainer assembly about halfway in the case, with the washer riding in the grooves provided. The end of the spring also rides in the groove.

Twist the retainer about a quarter turn to the right and down.

Shackle.   The various models have different shackle lengths and diameters. The spring (H in FIG. 12-5) must match. For example, an L-shaped spring is used on the L50 lock. An extra pin is used to guide the long spring on the K60. The shackle is secured in the H10 with a hardened steel pin; other models secure the shackle with the retainer mechanism.

Pin-tumbler padlocks.   The American five pin-tumbler padlock has been designed so the cylinders may be quickly changed (FIG. 12-6). Late production 100 and 200 series use the same cylinder assembly and keyway, reducing the inventory load.

Assembly and disassembly of the pin-tumbler padlock is only slightly different from that of the other American padlocks. To change the cylinder, follow this procedure:

Open the padlock, exposing the retaining screw at the base of the shackle.

Remove the retaining screw with a small screwdriver.

Pull the cylinder out of the case. *Note:* Leave the lock unlocked; do not depress the shackle.

Insert a new cylinder in the case. Replace the screws, bringing the cylinder almost flush with the case.

To assemble the locking mechanism:

Assemble the retainer (D of FIG. 12-6) and the retainer washer (C) with the free end of the spring tightly against the left side projection on the washer.

Place the shackle spring (H) in the hole on the end of the shackle (I).

Insert the shackle spring into the deep well of the lock body.

Drop the balls (E) into the case bottom and move them into the pockets with a small screwdriver. There must be room for the retainer assembly.

Depress the shackle so the balls cannot slide back into the center of the case.

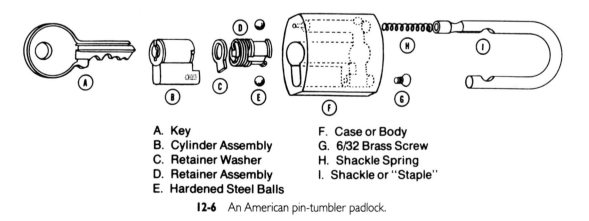

A. Key
B. Cylinder Assembly
C. Retainer Washer
D. Retainer Assembly
E. Hardened Steel Balls
F. Case or Body
G. 6/32 Brass Screw
H. Shackle Spring
I. Shackle or "Staple"

12-6   An American pin-tumbler padlock.

Insert the retainer assembly (C and D). The retainer washer (C) fits into the elongated hole in the case.

Make sure the assembly is bottomed in the hole.

Insert a retainer-assembly tool into the cylinder hole so the step on the tool engages the retainer step. Turn the tool clockwise until the lock opens.

Install the cylinder assembly, holding it flush with the case bottom. Insert the 6×32 brass screw (G) into the shackle hole, and tighten it down snugly.

The American Lock Company will provide the retainer-assembly tool free for the asking.

Because of the close tolerances of American padlocks, they will not operate properly in extreme cold weather unless Kerns ML3849 lubricant is used. Graphite is acceptable in milder climates.

To remove the cylinder in the 600 series pin-tumbler padlocks, follow this procedure:

Unlock the padlock and turn the shackle as shown in FIG. 12-7.

Depress the spring-loaded plunger (A) with a small screwdriver until the plunger is flush with the wall of the shackle hole. At the same time, pull on the key; withdraw the cylinder.

Master

The Master Lock Company makes a variety of lock types, ranging from simple warded locks to sophisticated pin-tumbler types.

An improved version of the familiar Master padlock has recently been introduced—the Master Super Security Padlock (FIG. 12-8). It was designed to give greater security than standard locks and is recommended for warehouses, storage depots, and industrial plants, as

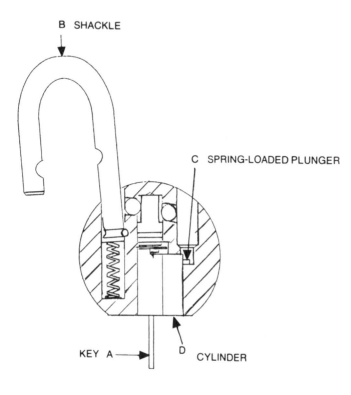

12-7   The American 600 series padlock.

12-8 The Master Super Security Padlock.

well as around the home and yard. Figure 12-9 is a cutaway view of the mechanism. Salient features include:

□ A patented dual-lever system to secure the shackle legs. Each lever works independently of the other and is made of hardened steel.

□ The long shackle leg is tapered to align the locking levers.

□ The case laminations are made of hardened steel and are chrome-plated for weather protection.

□ The case is larger than that of standard locks for better protection.

□ A rubberoid bumper prevents the lock body from scratching adjacent surfaces. This feature is not found in other locks.

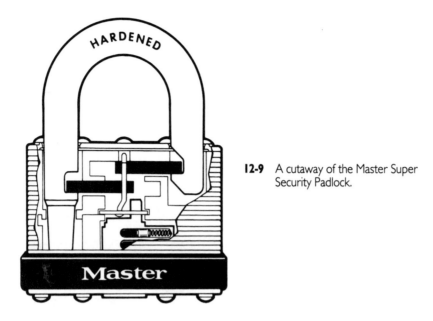

12-9 A cutaway of the Master Super Security Padlock.

□ Tension tests show that this high security lock can tolerate a force of 6000 lbs. on the shackle without damage.

Servicing.   Follow this procedure:
Drill out the bottom rivets.
Remove the plates, one at a time, and keep them in order.
Remove the lock plug.
Assemble the lock using new rivets.

ILCO

ILCO secures the cylinder with what the trade calls the "loose rivet" method. A single brass rivet extends through the case and into the cylinder. The rivet is headless (hence the term "loose") and is secure by virtue of its near invisibility (FIG. 12-10).

To remove the pin, follow these steps:

Hold the case up to the light and look for the shadow created by the pin. The pin is located ⅛ inch from the bottom of the case and is centered on the side.

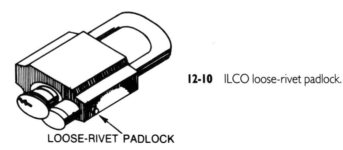

12-10   ILCO loose-rivet padlock.

LOOSE-RIVET PADLOCK

Using a No. 48 bit, drill a hole in the end of the pin.

Thread a small screw into the hole.

Gently pry up on the screw and remove it together with the pin.

To insert a new pin:

Select a piece of brass for stock that exactly matches the diameter of the pin hole.

Use the next size smaller rod (to make removal easier if it is ever necessary) and cut it so that ⅟₃₂″ stands out.

Peen over the end of the rod.

Burnish the case to camouflage the pin.

Another ILCO variant is a cylinder retaining plate on the bottom of the case. The rivet that holds the assembly together is cunningly disguised in the maker's name stamped on the plate.

To remove the retaining plate, follow this procedure:

Draw a line under the ILCO name.

At a right angle to the first line, draw a second line (on the outside edge of the O in ILCO).

Drill a shallow hole to accept the tip of a 6×32 machine screw.

Fit the screw with a washer, and thread it into the hole.

Using the screwhead for purchase, pry the retainer out of the case.

Withdraw the cylinder.

When the cylinder has been serviced and replaced into the case, mount the retainer in its original position, tapping it home with a flat-ended punch. Fill the hole with brass stock.

Kaba security padlocks

The Kaba KP 2008 security padlock (FIG. 12-11) is a good quality lock—heavy-duty, with a stainless steel body for a wide variety of applications in the home and industry. It is fitted with an interior dust cover which, when coupled with the stainless steel body, means it can be used in any weather conditions. The dimensions are in FIG. 12-12.

The KP 9 cable padlock has a plastic-coated steel cable. It is very well suited for securing bulky items such as equipment on building sites, bicycles, etc. (FIG. 12-13).

The KP 10 padlock (FIG. 12-14) is designed for strength, the shackle load being over two tons. Dimensions are shown at FIG. 12-15.

The "Epilok" security hasp and staple (FIG. 12-16) is a unique unit. A strong hasp and staple with hardened steel construction utilizes the unique "Epilok" spigot lock with its advanced Mini-Kaba locking cylinder. This unit will stand up to the most strenuous attack.

Figure 12-17 breaks the unit down a little more. You actually have the hasp unit and the locking spigot lock; separate is the rawbolt staple (on the right). Once the hole is drilled and the rawbolt inserted, the stapel is screwed on tight, making removal exceptionally difficult. The security provided in this manner is much greater than that of several screws that one would expect with a security hasp. The dimensions are provided at FIG. 12-18.

The 45201 security window/patio door lock has a universal locking mechanism that can be used for any application where a sliding door or window requires additional locking on the top and/or bottom part of the frame. It is also very suitable for hinged windows or pivoting windows (FIG. 12-19).

The lock is operated by a Kaba-20 cylinder mechanism. The same key can be linked to an unlimited number of locks with the same key code, including rim locks, padlocks, or mortise locks. All windows, patio doors, front and back doors can be quickly locked, every time. There is no need to search for individual keys on a large key ring. The dimensions for the window/patio door lock are in FIG. 12-20.

12-11   Heavy duty padlock with the Kaba 20 cylinder. (courtesy of Kaba Locks, Ltd.)

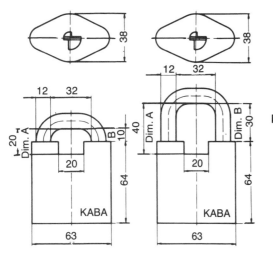

12-12   Specification data for the heavy-duty padlock. Notice the two different shackle sizes available. (courtesy Kaba Locks, Ltd.)

12-13 Cable padlock suitable for securing small pieces of equipment and bicycles. (courtesy Kaba Locks, Ltd.)

12-14   The KP 10 has a large shackle into a solid locking body for increased strength. (courtesy Kaba Locks, Ltd.)

These last two locks are excellent counter displays for the locksmith that practically sell themselves to the customers. There is no need to provide any form of sales pitch or "hype" for the product. They are top-of-the-line, with proven quality, reliability, and durability, and will give your customer years of valuable protective service.

WAFER-DISC PADLOCKS

Wafer-disc padlocks are recognized by their double-bitted keys and can be a headache to service. If you do not have a key, the lock can be opened by either of two methods:

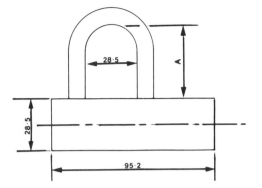

12-15 KP 10 specification data. (courtesy Kaba Locks Ltd.)

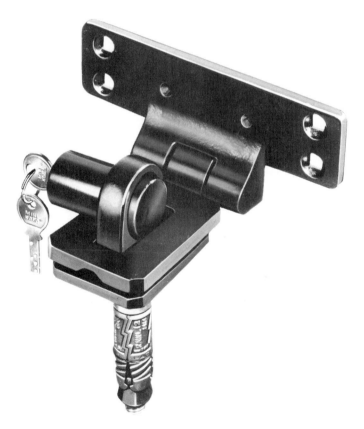

12-16 The Epilock high security hasp and staple unit used the Mini-Kaba cylinder. (courtesy Kaba Locks, Ltd.)

12-17   Epilok unit broken down into separate sections for installation. (courtesy Kaba Locks, Ltd.)

☐ Picking is possible, but it takes practice. Purchase a set of lockpicks for these locks and spend a few hours learning the skill.

☐ You can try your lock with a set of test keys, available from locksmith supply houses.

Disassembly

If the lock is already opened or a key is available, your job is almost half done.

Release the retaining ring clip with a length of stiff wire inserted into the toe shackle hole.

Remove the cylinder plug and lay out the parts.

Keys

Duplicate keys can be cut by hand or with a machine designed for this purpose. These machines are expensive. Locksmiths get around the problem by stocking cylinder inserts and precut keys. The insert replaces the original wafer mechanism. A precut key requires that the discs be realigned to fit. You will need some blank discs and should have access to the cutting tool described in the chapter on double-bitted locks and keys.

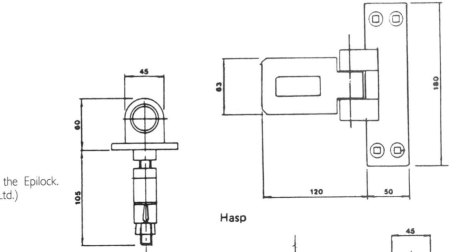

12-18 Specification data for the Epilock. (courtesy Kaba Locks, Ltd.)

Hasp

Rawbolt staple

Staple

12-19 Window or patio door security is assured with this lock unit. (courtesy Kaba Locks, Ltd.)

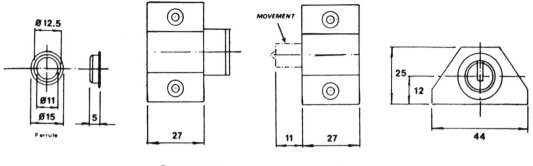

Open position   **Closed position**

12-20 Specification data for the window/patio lock. (courtesy Kaba Locks, Ltd.)

In extreme cases, you can insert the key and file the disc ends for shell clearance. This is not the sort of thing that a real professional would do, but it works.

Do not attempt to cut double-bitted keys by the impressioning method. A new plug and key is the better choice.

PIN-TUMBLER PADLOCKS

There are hundreds of pin-tumbler locks on the market, but most fall into two categories based on the case construction—laminated or extruded.

Laminated padlocks

Slaymaker pin-tumbler padlocks are made up of a series of steel plates held together by four rivets at the casing corners. Follow these service procedures:

Use a hollow mill drill to shear off the rivet heads on the case bottom; this technique allows the rivets to be reused.

Remove the bottom plate.

Remove the entire cylinder section in one piece.

Two cylinder-housing types are used. The most popular requires a follower tool to keep the pins intact.

Make the necessary repairs.

Insert the cylinder into the casing and replace the bottom plate over the four rivet ends.

After checking the action, use a ball-peen hammer and repeen the rivet ends.

Extruded padlocks

Extruded locks are made from a single piece of metal—usually brass. You will need these tools to service these locks:

□ A small nail with the point cut off.

□ Key blanks.

□ Pliers.

□ A small punch.

□ Hammer.

□ A small light (overhead lights are too bright; a flashlight is acceptable).

To disassemble the extruded lock, first locate the plug pins or cover on the bottom of the case. Since the case is highly polished, this is sometimes hard to do. Tilting the lock under the light will show a faint outline of the cover cap or plug pins.

To assemble extruded locks held together by exposed pins, follow this procedure:

Determine the diameter of the plug pins.

Select a drill bit smaller than the diameter of the pin.

Drill slowly and only a fraction of an inch deep—the pin does not extend into the lock very far. If you drill through it you can damage the spring.

Extract the plug.

Remove each spring and pin.

Remove the cylinder. The cylinder is held by a pin that may be covered by a small pin plug. Rapping the lock against a hard surface may shock the plug loose. If not, dislodge the pin plug with a small screwdriver.

Make the necessary repairs.

Assemble the lock.

Fill the pin holes with brass wire.

File the wire flush with the case and polish.

Extruded locks with a retaining plate over the pins require a slightly different procedure. If the lock is open, drive the plate out with a punch inserted at the shackle hole. If locked, follow this procedure:

Drill a small hole off-center in the plate.

Using an icepick or other sharp instrument, drive the plate down, toward the shackle. This will buckle the plate and cause the edges to rise.

Pry the plate loose by working a sharp instrument around the edges.

If the plate is not severly damaged, it can be reused. Bow the plate slightly so that the center bulges outward when the plate is installed.

Peen the edges of the plate to form a tight seam.

Other locks mount the plug and cylinder by means of horizontal pins running across the width of the lock. These pins can be seen under strong light.

Key fitting for pin-tumbler padlocks

Follow this procedure:

Insert the blank into the plug.

Place the No. 1 pin into its chamber.

Using a blunted nail as a punch, drive the pin into the blank. A single light hammer tap is enough to impression the blank.

Remove the blank.

File the blank at the impression.

Insert the key and try to turn the cylinder. File as necessary.

Move to the next pin and repeat the process.

HELPFUL HINTS

It's easier to work with padlocks that are open. If the lock is not open, pick it open. Sometimes the shackle bolt can be disengaged with the help of a hatpin inserted through the keyway. However, locks are getting better and this technique doesn't work as well as it used to.

☐ Polish the lock before returning it to the customer. Use emery paper to remove the deep scratches; then burnish with a wire wheel. Finish by buffing.

□ The best security in a padlock comes when the shackle is locked at both the heel and toe. The double bolt action (or balls) is the ultimate in padlock security, making it nearly impossible to force the shackle.

□ When picking fails (and even the best of locksmiths may occasionally have this problem), use penetrating oil. Yale, Corbin, and ILCO pin-tumbler locks are especially susceptible to penetrating oil.

13

Home and business services

Over the years a locksmith acquires a great deal of knowledge; much of this information is learned from correcting his own mistakes. This is a hard school. The best way to solve problems is to never let them arise in the first place.

This chapter includes some basic techniques. Some of this material may seem obvious, and you might wonder why it's in here. The answer is that these techniques have been tested and approved by the experts. They will save you time, money, and the embarrassment of callbacks.

The following few pages are supplied courtesy of Corbin (Emhart Corporation). The first section applies to all locksets, regardless of make. Other sections detail service and troubleshooting procedures for specific Corbin locksets, as well as other makes of locks.

COMMON PROBLEMS AND TROUBLESHOOTING

Should you encounter difficulties in the operation of a lockset, first review this checklist of common problems and solutions to see if you can clear up the difficulty.

- ☐ Is the door locked? (FIG. 13-1)
- ☐ Are you using the right key? (FIG. 13-2)
- ☐ Latchbolt or deadbolt does not engage or disengage the strike, or binds in the strike—usually due to bolt-strike misalignment.
- ☐ Has the door warped? (FIG. 13-3)
- ☐ Is the door binding? Frames which are out of plumb are frequently the cause of faulty operation of locksets and binding of bolts in the strike (FIG. 13-4).
- ☐ Are the hinges loose? Tighten the screws, filling holes if necessary, or rehang the door if the screws will not hold (FIG. 13-5).
- ☐ Are the hinges worn? If excessive wear has occurred on the hinge knuckles, the door will not be held tightly. Replace the hinges.
- ☐ Is the frame sagging? If sag cannot be corrected and the door and frame returned to plumb relationship, planing or shaving of the door and repositioning or shimming the strike may relieve this condition.

13-1   Is the door locked?

13-2   Are you using the right key?

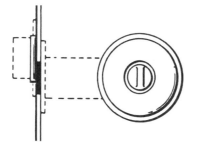

13-3   Has the door warped?

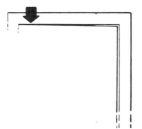

13-4   Is the door binding?

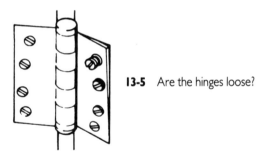

13-5   Are the hinges loose?

□ Key operates the latchbolt or deadbolt with difficulty. Usually due to bolt-strike misalignment (FIG. 13-6).

Corbin cylindrical locksets

Problem 1—Latchbolt will not deadlock. Caused by deadlocking latch going into strike. Either the strike is out of line or the gap between door and jamb is too greater. Realign the strike or shim the strike out towards the flat area of the latchbolt (FIG. 13-7).

   Problem 2—Latchbolt cannot be retracted or extended properly. Caused by latchbolt tail and latchbolt retractor not being properly positioned (FIG. 13-8). Remove the lockset from the door. Reinsert the latchbolt in the door. Looking through the hole in the door, the tail should be centered between the top and bottom of the hole. Remove the latchbolt and insert the lockcase. Looking through the latchbolt hole in the lock face of the door, the latchbolt retractor should be centered in the hole. Adjust the outside rose for proper position. Rebore the holes, if necessary, to line up the retractor and tail.

   Problem 3—Latchbolt will not project from the lock face (FIG. 13-9). Latchbolt tail and retractor may be misaligned. See *Problem 2*. If this is not the cause, the spring is probably broken.

   Problem 4—Key works with difficulty. Lubricate the keyway (FIG. 13-10). Do *not* use petroleum products. Spray powdered graphite into the cylinder or place powdered graphite or lead pencil shavings on the key. Move the key slowly back and forth in the keyway. Bitting (notches) on the key may be worn.

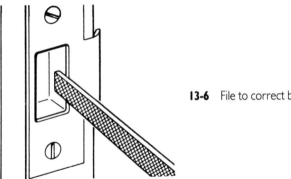

13-6   File to correct bolt-strick misalignment.

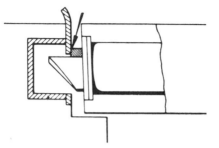

13-7   Latchbolt will not deadlock.

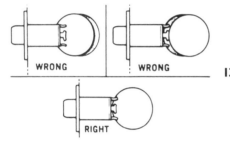

13-8   Latchbolt cannot be retracted or extended properly.

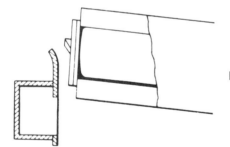

13-9   Latch bolt will not project from lock face.

13-10   Lubricating keyway.

Corbin heavy-duty cylindrical locksets: service procedures

Servicing these locks is not difficult if the task is approached methodically.

To tighten the locksets:

Tighten the inside rose thimble with a wrench. If the thimble needs to be taken up a great deal, tighten the *outside* rose at the same time to prevent possible misalignment and binding of both (FIG. 13-11).

If the lockset is still not tight, back off the thimble; using a screwdriver, push down the knob retainer and remove the knob and rose. If the spurs on the back of the rose are bent, straighten and reposition the rose so the spurs are embedded in the door. If the spurs are broken off, insert a rubber band or nonmetallic washer under the rose. Tighten the thimble.

To remove and install the locksets:

Remove the key from the knob.

Loosen the inside rose thimble with a thimble wrench. Pull the lock slightly to release the rose spurs from the door.

Disengage the inside knob retainer with a thimble wrench and pull out the inside knob. Slide the rest of the lockset from the outside of the door.

Remove the bolt faceplate screws and slide the latch unit from the lockface of the door.

To reinstall, reverse procedures above. When inserting the case and keyed knob from the outside of the door, be sure the bolt retractor in its case properly engages the latchbolt tail.

To remove and replace the cylinder in the locksets

Follow the procedure in the preceding section.

Remove one case screw and slightly loosen the other.

Swing out the outside knob retainer (it pivots on the case screw slightly loosened above).

Using a screwdriver, pry the knob filler cover off the outside knob.

Using special Waldes ring pliers No. 3, remove the large Waldes ring from the groove and withdraw the shank and cylinder (FIG. 13-12).

To reinstall the cylinder, reverse the above procedure. Be sure the knurled side of the Waldes ring is face up.

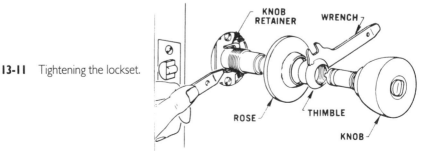

13-11   Tightening the lockset.

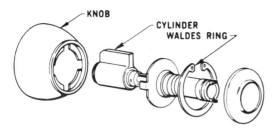

13-12   Removing cylinder.

To change the hand of the locksets:

Remember that the lock cylinder should always be in position to receive the key with the bitting (notches) facing upward.

Follow the procedure in the preceding section.

Slightly loosen *one* case screw and back off the other (FIG. 13-13).

Swing out the outside knob retainer (it pivots on case screw slightly loosened above).

Lift the knob out and rotate it 180 degrees. Replace it in case. Swing the knob retainer back into place.

Insert the case screw and tighten both screws.

Reinstall (FIG. 13-14).

Corbin standard-duty cylindrical locksets: service procedures

To tighten the locksets:

Depress the knob retainer and remove the inside knob.

Unsnap the rose from the rose liner (FIG. 13-15).

Place the wrench into the slot in the rose liner and rotate clockwise until tight on the door. If the lockset is extremely loose, tighten the outside rose and inside liner *equally* (FIG. 13-16).

Replace the inside rose; depress the knob retainer and slide the knob onto the spindle until the retainer engages the hole in the knob shank.

To remove locksets:

To remove, follow this procedure:

Depress the knob retainer and remove the inside knob.

Unsnap the rose from the rose liner.

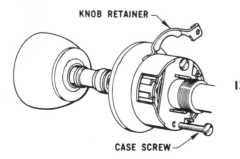

13-13   Changing lockset hand.

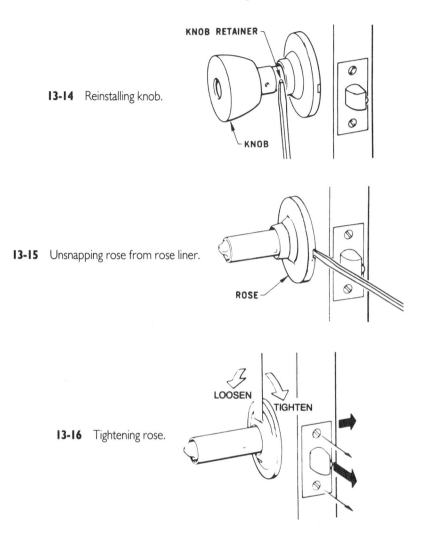

13-14   Reinstalling knob.

13-15   Unsnapping rose from rose liner.

13-16   Tightening rose.

Place the wrench into the slot in the rose liner and rotate counterclockwise until disengaged from the spindle.

Remove the lock by pulling the outside knob.

Remove the bolt after removing the screws in the lockface of the door.

To reinstall locksets:

Adjust the lock for the door thickness by turning the outside rose until the edge of the rose matches one of the lines marked on the shank of the knob. (First line is for 1⅜ inch door, the second for 1¾ inch door). When using trim rosettes, increase the adjustment to compensate for the thickness of the metal.

Install the latchbolt in the face of the door.

Install the lock from the outside of the door, so the case engages the slot in the latchbolt and the tail interlocks the retractor.

Replace the rose liner, rose, and knob on the inside of the door.

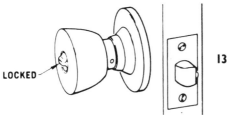

13-17   Locking exterior knob.

To reverse the knobs:

Lock the exterior knob (FIG. 13-17).

Hold the latchbolt in the retracted position with the key (FIG. 13-18). Depress the knob retainer with a screwdriver through the slot in the knob shank.

Pull the knob off the spindle (FIG. 13-19). Rotate the knob so the bitting on the key will be *up*.

Replace the knob by lining up the lance in neck of knob with the slot in the spindle.

Push the knob on the spindle until it hits the retainer button. Depress the retainer button, and push the knob until it snaps into position.

To remove the cylinders:

1. Remove the outside knob. Turn the key in either direction until it can be partially extended from the plug (FIG. 13-20).

2. Hold the knob and turn the key to the left, pulling slightly on the key until the cylinder disengages.

Troubleshooting corbin unit locksets (300 and 900 series)

Problem 1—Latchbolt binds or rattles in strike. Adjust the strike. Because the lockset is preassembled as one unit, there are no internal adjustments to be made. Check to be sure that the cutout for the lockset is square and at the right depth, so the face of the lockset is flush with the face of the door. Once this has been done, adjust the nylon adjusting screw in the strike (FIG. 13-21).

Problem 2—Key does not activate the knob or latchbolt. Check for worn key. If bitting (notches) in the key is worn down or the key is bent, the locking mechanism will not operate properly. If the key is not worn, spray powdered graphite into the keyway or put graphite or lead pencil filings on the key and move it back and forth slowly in the keyway (FIG. 13-22). Never use petroleum products. Check to be sure that a binding latch is not the cause.

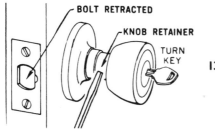

13-18   Retract latchbolt.

13-19   Pull knob off spindle.

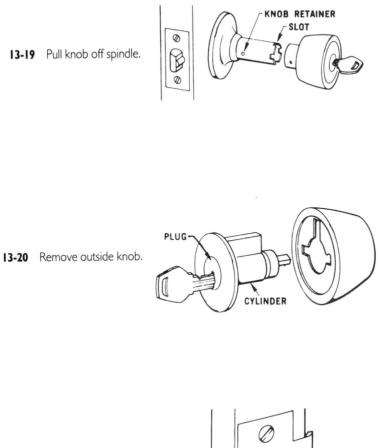

13-20   Remove outside knob.

13-21   Adjusting screw in strike.

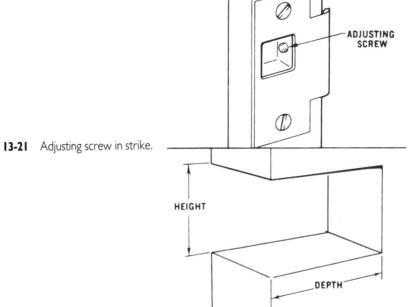

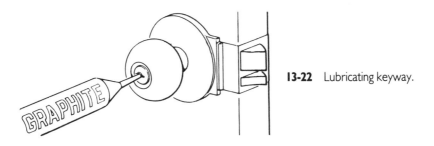

13-22   Lubricating keyway.

Problem 3—Lockset is loose in the door. Tighten the escutcheon screws (FIG. 13-23). Be sure to tighten evenly.

Problem 4—Lockset has the wrong bevel for the door. Reverse the lockset. Unit locksets with horizontal keyways may be changed from right to left hand regular bevel, or vice-versa—or from right hand reverse bevel to left hand reverse bevel, or vice-versa—by merely turning the lock upside down (FIG. 13-24).

Corbin unit locksets (300 and 900 series): service procedures

Usually the unit locksets are easier to service than the cylindrical locksets. To service this series of locksets, proceed as follows:

To remove the locksets from a door:

Remove the through-bolts on the inside of the door (FIG. 13-25).

Push the outside escutcheon away from the door so the lugs clear the holes.

Slide the assembly out of the door.

To remove 900 series knobs:

Remove the attaching screws (FIG. 13-26).

Snap off the dust cover.

Pry the wire retaining ring from the knob retaining key located in a slot in the frame tube. Remove the retaining key. Remove the inside knob, which is fastened to the knob shank.

Remove the inside escutcheon.

Loosen the escutcheon on the outside of the lock by inserting a screwdriver through the access hole from the inside of the lock frame; remove the screw and escutcheon fastener.

Pry the wire retaining ring from the knob retaining key located in a slot in the frame tube. Remove the retaining key. Remove the outside knob, which is fastened to the knob shank.

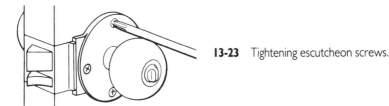

13-23   Tightening escutcheon screws.

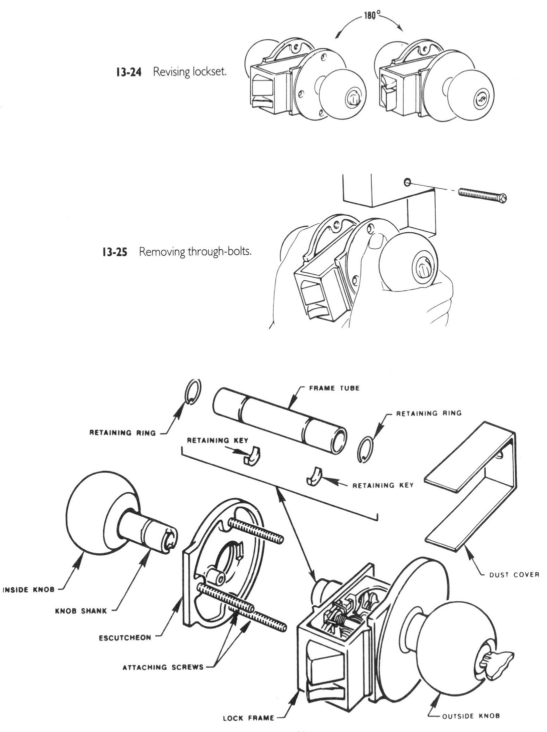

13-24   Revising lockset.

13-25   Removing through-bolts.

13-26   Removing attaching screws.

To remove cylinders:

Using a screwdriver, pry the knob filler cover off the outside knob.

Use No. 103F92 Waldes retaining pliers to remove the Waldes ring.

Remove the shank and lock cylinder. Rekey the lock.

Reverse the procedure to reinstall. Be sure that the beveled edge of Waldes ring faces away from the knob, and that the ring is properly seated in the groove (FIG. 13-27).

To remove 300 series knobs:

Remove the key from the lock (lock should be unlocked). Remove the three attaching screws (FIG. 13-26).

Snap off the dust cover.

On the inside knob side, pry the wire retaining ring away from the knob retaining key located in the slot in the frame tube. Remove the retaining key. Remove the knob that has the shank assembled to the knob.

Remove the inside escutcheon.

Loosen the escutcheon on the outside of the lock by inserting a screwdriver through the access hole from the inside of the lock chassis; remove the screw and washer.

Repeat step 3 on the outside knob and shank.

Troubleshooting corbin mortise locksets (7000, 7500, 8500 series)

Problem 1—With door open, latchbolt doesn't extend or retract freely. Check for binding against the rose. Adjust the knob. Loosen the roses or trim on the door. If the bolt now operates freely, the roses or trim must be realigned. A knob aligning tool is recommended. Check installation templates for proper position (FIG. 13-28). If the bolt does not operate properly with trim and roses loosened, remove the lockset from the door. If the lockset operates, properly when removed from the door, use a chisel to make the mortise larger so that the lockset enters freely.

Problem 2—With door closed, latchbolt doesn't extend or retract freely, or door won't latch at all. Open the door. If the latchbolt still doesn't operate properly, see *Problem 1.*

Problem 3—Latchbolt "stubs" on the strike lip. Bend the strike slightly back toward the jamb. Wax or paraffin makes an excellent lubricant, as does silicone spray (FIG. 13-29).

Problem 4—Deadbolt doesn't enter the strike. This is probably due to misalignment of strike and bolt, particularly in cases where door sag has taken place. Both latchbolt and deadbolt holes in the strike must be filed or the strike repositioned. Do *not* force the thumbpiece if the deadbolt doesn't extend and retract in the strike freely (FIG. 13-30).

Problem 5—To remove the cylinder. Loosen the cylinder locking screws in the face of the lockset. If the scalp covers the set screws, remove the scalp. Unscrew the cylinder. When replacing, be sure the locking screws are firmly seated (FIG. 13-31).

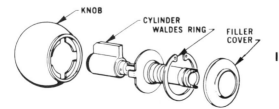

13-27 Removing cylinder.

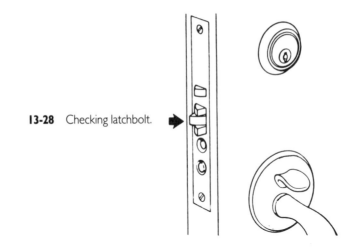

13-28   Checking latchbolt.

13-29   Bending strike.

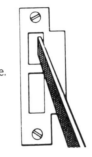

13-30   Filing strike.

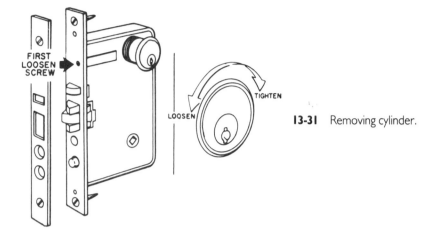

13-31   Removing cylinder.

Problem 6—Key does not operate the latchbolt or deadbolt. Loosen the cylinder locking screws. The cylinder is in the wrong position in the door, so the cylinder cam does not engage the locking mechanism properly. Turn the cylinder a whole turn to the left or right until it works properly. The keyway must always be in position to receive the key with bitting (notches) up. Tighten the cylinder locking screws.

Problem 7—Key turns hard when retracting the deadbolt. If the bolt operates freely with the door open, check the bolt-strike alignment. Check the scalp to be sure it is not binding the bolt, or that paint over the bolt is not causing the bind.

Problem 8—Key works hard in the cylinder. Lubricate the cylinder with powdered graphite, or place graphite on the key and move it back and forth slowly in the keyway. Never use any petroleum lubricants (FIG. 13-32).

Problem 9—Key breaks in the lock. Remove the cylinder from the lock. Insert a long pin or wire into the back end (cam end of the cylinder). Move it back and forth until the broken key stub is forced out through the front of the cylinder. Clean the cylinder with ethyl acetate and lubricate it with graphite before reinstalling (FIG. 13-33).

Problem 10—Thumbpiece trim doesn't retract latchbolt completely or doesn't extend bolt completely. Check for binding at inside trim; or if the outside thumbpiece trim is used in conjunction with the inside panic device, check to see if it is operating properly and is not dogged down. Remove the thumbpiece, and check the position of thumbpiece in relation to the latch trip at the bottom of the mortise lock case. When properly installed, the top of the thumbpiece should be up against the bottom of the latch trip, but not lifting it (FIG. 13-34). If

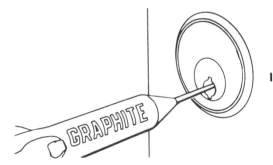

13-32   Lubricating keyway.

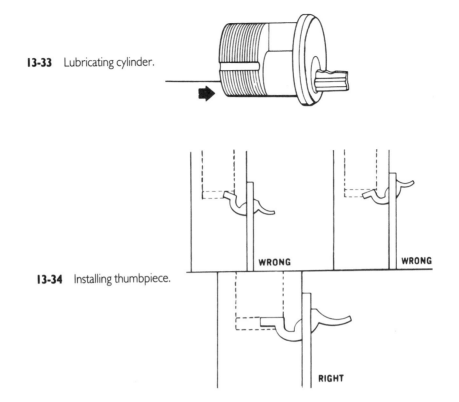

13-33   Lubricating cylinder.

13-34   Installing thumbpiece.

the bolt doesn't retract fully when the thumbpiece is pushed down, the thumbpiece is too low on the door. Move the trim up as needed. If the bolt doesn't extend completely when the thumbpiece is released, the thumbpiece is too high on the door. Move the trim down as needed. If the trim is fixed and cannot be moved, carefully bend the thumbpiece tail up or down.

Corbin mortise locksets (7000, 7500, 8500 series): service procedures

Mortise locksets are not difficult to work with. Servicing them involves only a few basic procedures.

To install and adjust the working trim:

Unscrew the spindle. Remove the mounting plates, roses, and thimble from the spindle. Unscrew the sleeve from the inside knob. Remove the spindle from the knob (FIG. 13-35).

Install the rose attaching plates. On wood doors, install flange washers through the mortise, turning the attaching plates into them. It may be necessary to mortise for the washer to permit the mortise lock to clear it. Install the mortise lock.

Align the plates using a No. 028 aligning tool. If not available, assemble the spindles allowing $\frac{1}{16}$ inch gap between halves of the swivel spindles, with mounting plates, sleeve spacer, and adjusting nut in position shown in illustration. Tighten the adjusting nut with fingers. Mark screw holes through both mounting plates. Install the screws.

Reassemble the spindle, tightening the adjusting nut, and back off ¼ turn to line up with the flats of the spacer. The spindle should have a slight end chuck. Try the knob. If the latch binds, back off the adjusting nut another ¼ turn. Disassemble the spindle.

Assemble the roses and thimble. Reassemble the spindle complete with knobs. Tighten the rose thimbles and inside adjusting plate with furnished spanner wrenches. Knobs should turn freely from either side in either direction.

To install and adjust the lever handle trim assembly:

Check to be sure that the handle hole is 1⅛-inch wide. On wood doors, remove the mortise lockset and make additional mortise for retainer on the inside of the case mortise. Replace the locksets and secure with screws.

Place the rose assembly spindle into the lock hub. The label shows the top and latch edge. Spot four screw holes through the assembly plate (FIG. 13-36). Make sure that the latchbolt operates freely. Drill holes very carefully.

Attach the rose assembly to the door. (With a wood door, attach through the door to the retainer.)

Put on the cover plates. Place the lever handle on the assembly and secure with an Allen set screw.

Adjust. If the latchbolt binds, remove the lower plate and slightly change the rose assembly position.

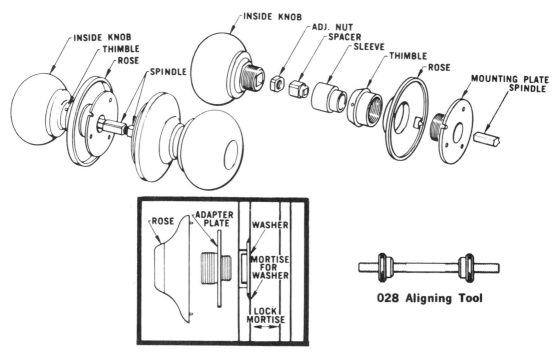

13-35   Removing spindle from knob.

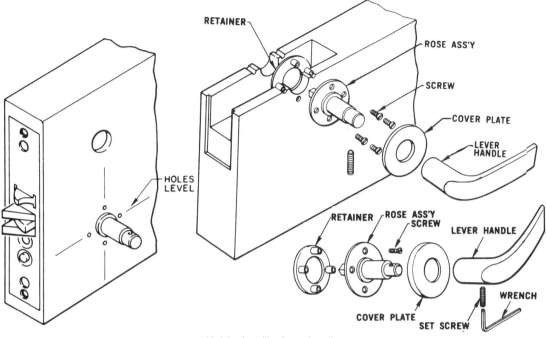

13-36 Installing lever handle.

To reverse the hand:

As you will remember from Chapter 11, the *hand* refers to the position of the hinge on the door and to the direction of swing. Some locksets are universal and fit all four hands. Others must be modified in the field.

Refer to FIG. 13-37. Unfasten the two cap screws, (A) and remove the cap.

Reverse the latchbolt (B). If the lock has an antifriction latch, first remove the L-shaped pin (C) and reinstall it on the opposite side after reversing the bolt. The short leg should be held against the side of the case with the pressure-sensitive decal.

Reverse the auxiliary latch (D) if an auxiliary latch is furnished. To do this, first remove the auxiliary latch lever and spring; then invert the latch so the concave surface is toward the stop in the frame. Then reassemble the lever and spring.

Reverse the front bevel. Loosen three screws (E) in the top and bottom edges of the case, adjust to the proper bevel (or to flat), and retighten the screws securely.

Reverse the knob hubs (F): hub with ⅜-inch square spindle hole toward the outside.

Replace the cap, and securely tighten the screws.

Check for proper operation before installing in the door.

Mounting the Medeco Ultra 7000 deadbolt

Instructions for mounting the Medeco Ultra 700 deadbolt are as follows:

Cut the door using the template as a guide (FIG. 13-38).

Place the wave spring inside the cylinder guard.

Insert the cylinder through the cylinder guard with the wave spring *under* the head of the cylinder.

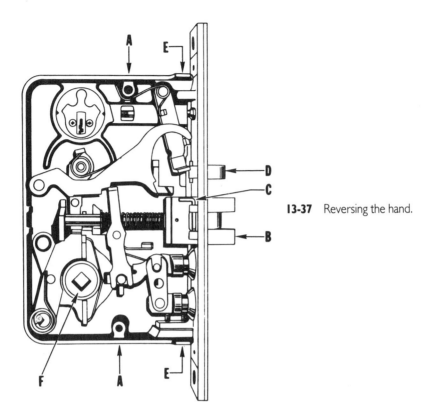

13-37   Reversing the hand.

Thread the cylinder into the lock body securely. The keyway must be horizontal and on the side of the cylinder closest to the bolt. If the keyway does not come to the horizontal, loosen the cylinder as necessary (always less than one turn).

Remove the faceplate and install the four set screws at the top and bottom of the lock. Secure the cylinder with the set screws provided.

If you have to disassemble the lock, insert a wrench into the hole and turn counterclockwise. This will open the lock.

With the lock near or full open, manipulate the lock so the pins slip out of one escutcheon.

13-38   Cut door with template.

13-39 Slide lock into cutout.

Reverse this procedure to reassemble. *Note:* The ³⁄₁₆-inch roll pin must fit into the hole in the opposite escutcheon for alignment.

With the lock assembled and the cover in place, slide the unit into the door cutout (FIG. 13-39). Position it so that the longest escutcheon is flush with the high edge of the door bevel.

Tighten the lock to the door by inserting an Allen wrench in the appropriate holes and turning clockwise. This brings the escutcheons of the lock together so they clamp the door.

Turn the mounting screws finger tight; then give them approximately ½ turn more (FIG. 13-40).

Install the faceplace with the screws provided. (FIG. 13-41).

The Ultra 700 lock can be adapted for a single cylinder on the exterior side and a thumbturn on the interior side of the door. In order for the thumbturn to act as an indicator, the long portion (the actual turning mechanism) must point in the direction of the bolt (extended or retracted position) and the cam must, therefore, be properly positioned on the thumbturn before the entire assembly is installed in the lock.

Check the cam position. If necessary, remove the drive pin and cam, rotate one half turn (180 degrees), and replace the pin. Figure 13-42 shows the thumbturn as viewed from the interior door side. When installing the lock, be sure the cam is inserted between the mechanism arms; then, tighten the cylinder set screws on the top and bottom of the lock.

The instructions are simple, and easy to follow; from a locksmith's viewpoint, the time is well spent in terms of customer relations and business income. In addition, if you are replacing the interior cylinder with the thumbturn, you might well be able to keep the cylinder that was removed, thus adding various parts to your supply stock. (*Note: Never* resell a used lock; the various individual parts of the lock can be used for repair parts, for study, and also for various practical training exercise for new employees.)

Continuing with the MAG Ultra 700 deadbolt, a new series from MAG, the MR series, is an adaption of the original, and one that is well worth investing in. The adaption is for use

13-40 Tightening mounting screws.

13-41   Installing faceplate.

with the Corbin Master Ring cylinder and is now available from your locksmithing distributors nationwide.

The adaptation (FIG. 13-43) enables the same master keying system to be utilized while still providing the high security lock compatible with the present system.

Like the regular Ultra 700, this new series includes no attaching screws on either side of the lock and the key cannot be withdrawn from the lock unless the locking bolt is either fully extended or fully retracted. Also, the cylinder guard is free-turning and recessed into the escutcheon casting for maximum protection.

A new Series "A" deadbolt Ultra 700 is now also available for the locksmith to add to potential sales. The Series "A" is designed for use where one side entrance only is desired (FIG. 13-44). The blank side of the lock is a soldering casting. The key side of the lock is available in stainless steel and uses a 1⅛ inch mortise cylinder.

Let's move on to another MAG lock, the Ultra 800, also a deadbolt. The cylinder used with the lock is a 1 5/32-inch in diameter × 1⅛-inch long mortise cylinder with a standard Yale cam. The lock is set by the factory for 1¾-inch doors, but for other door thicknesses, instructions are included, and are also below.

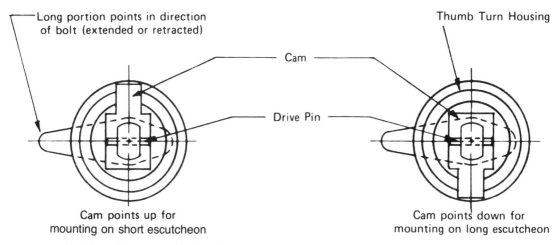

Long portion points in direction of bolt (extended or retracted)

Thumb Turn Housing

Cam

Drive Pin

Cam points up for mounting on short escutcheon

Cam points down for mounting on long escutcheon

13-42   Cam direction positioning when mounting is dependent upon whether a short or long escutcheon is being used. (courtesy M.A.G. Engineering and Mfg., Inc.)

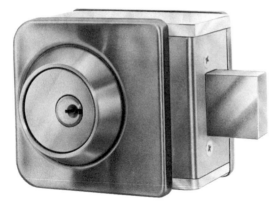

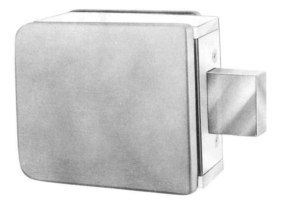

13-43 The Ultra 700 deadbolt adaption for use with the Corbin Master Ring Cylinder. (courtesy M.A.G. Engineering and Mfg., Inc.)

13-44 Ultra 700 deadbolt Series "A" used when one side entrance only is desired. (courtesy M.A.G. Engineering and Mfg., Inc.)

Figure 13-45 illustrates the lock in an exploded view. Notice the bolt; it is a cylindrical deadbolt instead of the standard rectangular bolt mostly associated with deadbolt locks. Also take note of the strike; again while looking like the average mill-run strike, MAG has gone further and ensured that it is ably secured to the frame and supporting stud. Also, from the illustration you can readily see that the interior side of the door will take either a mortise cylinder or thumbturn, providing your customer with an option feature.

13-45 Exploded view of the Ultra 800 deadbolt. (courtesy M. A. G. Engineering and Mfg., Inc.)

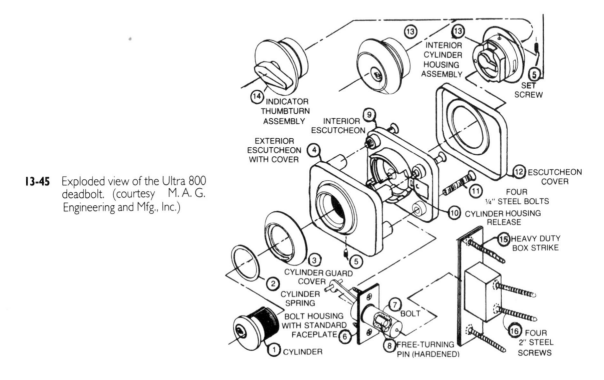

First, you must determine the proper setting for the lock if the door is not 1¾ inch in thickness. Figure 13-46 details adjusting the exterior lock housing piece for use on doors of varying thicknesses. The ring referred to is the circular ring on the reverse side (inside portion) of the exterior lock housing. By looking at the reverse side, at the 8 o'clock position, locate the hole for the pin to be inserted. With this done, proceed to the door preparation, using the template furnished with the lock unit.

Door preparation (wood doors):

Follow the dimensions outlined on the template. Drill ⅛-inch pilot hole through the door face, then drill a 2⅛ inch hole exactly 2¾ inches from the high edge of the door bevel, referring to the template as you do so.

Drill a 1 inch hole through the edge of the door, breaking into the 2⅛ inch hole. Cut out the wood for the bolt housing and faceplate 1⅛ inch wide × 2¼ inches high × ³⁄₃₂ inch deep. This faceplate hole must be centered on the 1 inch hole.

Drill the four ⅛ inch pilot hole for the four corner mounting bolts. With the pilot holes drilled, replace the ⅛ inch drill bit with a ⁹⁄₁₆ inch bit and proceed to put the four ⁹⁄₁₆ inch hole through the door. *Note:* MAG has a drill fixture kit available for fast and accurate installation. If you expect to be performing numerous installations, then the drill fixture kit is a must-have item that becomes a necessity for rapid installation.

Prior to mounting the lock, you should assemble the mortise cylinder to the exterior escutcheon of the lock, as follows. (The numbers refer to the various lock part components in FIG. 13-45.)

Put the cylinder spring (2) over the cylinder (1) then insert the cylinder into the cylinder guard cover (3), and install the escutcheon cover (12).

Now thread the cylinder in until the cylinder is tight (the keyway should be toward the rear of the lock and horizontal). Tighten the set screw. The cylinder guard cover should be tight against the escutcheon cover. The ring should rotate freely one-half turn back and forth. If the cam on the cylinder touches inside the mechanism cover, back off the cylinder one turn.

Put the key in the cylinder and rotate one complete turn, first in one direction and then the opposite direction. This is to check the function of this assembly.

If you haven't adjusted the lock for other than a 1¾ inch door, remove the mechanism cover by first removing the spring assembly. Pry up the cover and remove the ring; then,

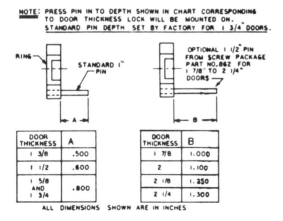

NOTE: PRESS PIN IN TO DEPTH SHOWN IN CHART CORRESPONDING TO DOOR THICKNESS LOCK WILL BE MOUNTED ON. STANDARD PIN DEPTH SET BY FACTORY FOR 1 3/4" DOORS.

DOOR THICKNESS	A
1 3/8	.500
1 1/2	.600
1 5/8 AND 1 3/4	.800

DOOR THICKNESS	B
1 7/8	1.000
2	1.100
2 1/8	1.250
2 1/4	1.300

ALL DIMENSIONS SHOWN ARE IN INCHES

13-46 Pin adjustment settings for doors with thicknesses other than 1 3/4". (courtesy M.A.G. Engineering and Mfg., Inc.)

adjust the pin for the appropriate door thickness, referring to FIG. 13-46 to ensure the details are correct.

Mounting the lock (wood doors):

Insert the bolt assembly into the 1 inch hole.

Insert the exterior escutcheon (with the mechanism ring in the retracted position) into the ⁹⁄₁₆ inch and 2⅛ inch holes. When inserting, make sure the bolt link fits over the mechanism pin and that the *large tab* of the bolt link is towards the hole in the mechanism cover.

Place the interior escutcheon on the door and secure it to the exterior escutcheon with the four ¼ inch in diameter bolts furnished with the lock unit.

Put the cover on the interior escutcheon.

Single cylinder.   Install the thumbturn with the *bolt retracted*. Place the circular wave spring over the thumbturn housing. Be sure the projection on the thumbturn cam and the projection on the housing are together. Place the thumbturn assembly into the interior escutcheon. Make sure the alignment projections line up with the bolt in the escutcheon hole. The thumbturn must be pointing away from the edge of the door. Push the assembly in carefully, making certain the hole in the thumbturn cam lines up with the mechanism pin. Push it in until the catch enters the slot in the thumbturn housing. Check for proper function by rotating the thumbturn and by pulling to be sure the catch has engaged the housing. The long side of the thumbturn should point toward the edge of the door when the bolt is extended, and away from the door edge when retracted.

Double cylinder.   Assemble the mortise cylinder, cylinder spring, and cylinder guard cover to the interior cylinder housing the same as the exterior escutcheon. The keyway must be horizontal and opposite the catch slot. Thread the interior ring onto the cylinder until it stops against the cylinder housing, then, back off one to one and one-half turns prior to assembly into the lock.

Install the interior cylinder housing assembly with the bolt retracted. Insert the assembly into the escutcheon hole, lining up the boss with the slot. Also, line up the ring boss with the escutcheon slot. Press it in fully with the key partially withdrawn, then rotate counterclockwise to stop. Push in on the cylinder until the catch snaps into the slot. Try to turn the cylinder housing clockwise to be sure the catch is engaged to the lock cylinder housing in place. Check for the proper function of the lock, with the key, from both sides.

Insert the bolt housing assembly into the 1-inch bolt hole. *Caution:* Make certain the opening side of the bolt housing is *up* on left-hand doors, and *down* on right-hand doors.

Install the faceplate with the screws provided. Make sure the hole in the faceplate fits over the projection of the bolt housing assembly to center it properly.

Metal doors:

Door preparation and procedures for mounting the lock to hollow metal doors—for doors that have no seam or hole in the center of the door edge—are as follows:

Follow the dimensions outlined on the template. Drill ⅛-inch pilot hole through the face of the door located 2¾-inches away from the high edge of the door bevel. Be sure the pilot hole is the same distance from the high edge of the door on each side. Bore a 2⅛-inch hole through.

Drill four ⁹⁄₁₆ inch holes the same as for wood doors.

Using the template, make three centerpunch marks on the edge of the door; one in the center, one ¹³⁄₁₆-inch above, and one ¹³⁄₁₆ inch below center. Drill ¼-inch holes.

Open the center hole with a ¾ inch hole saw. Remove the burr on the inside edge. Countersink top and bottom holes for 12-24 flathead machine screws.

Install clip nuts to the bolt housing with the long side toward the rear of the housing.

Insert the bolt housing through the 2⅛-inch hole in the door, and place it against inside front edge of the door, making certain the projection of the bolt housing assembly fits into the ¾-inch hole. Put opening on side of housing *up* on left hand door and *down* on right hand *doors*.

Secure the bolt housing to edge of the door with the two, 12/24 screws.

Insert the bolt assembly into the bolt housing.

Follow steps 2 through 5 for wood doors to complete the installation.

For hollow metal doors that have a seam or fold on the edge that would interfere with drilling of holes for mounting the bolt housing, use this procedure:

Cut out door edge as shown (FIG. 13-47) and mount Install-A-Lock. (Depending on whether the door is a 1⅜ inch or 1¾-inch door, you will require a special order; check with your distributor to see what the Install-A-Lock special order number will be for this unit.)

Finish the door preparation for hollow metal doors as previously described.

Mount the lock as per procedures for mounting to wood doors.

Strike installation

Wood jambs:

With the Ultra 800 properly installed, close the door fully and throw the bolt out (by either the key or thumbturn) so that the point on the end of the bolt will make a mark on the jamb. This will be the center mark for the strike mounting.

Open the door. Place the strike plate against the door, ensuring that it lines up with the bolt mark centered on the strike plate hole (FIG. 13-48).

Mark the center hole required to be drilled; also mark around the strike plate edge (FIG. 13-49).

Drill two 1 inch in diameter holes 1⅜ inch deep, one ⅝ inch above and one ⅝ inch below the center mark (FIG. 13-50).

Chisel out the wood in between.

Cut out the wood 1¼ inch wide × 4⅞ inch high × ⅛ inch deep for the strike box and strike plate.

Be sure to drill ⅛ inch pilot hole for all four screws to aid installation.

Insert the strike box. Place the finished strike plate over the strike plate box, and secure with screws provided (FIG. 13-51).

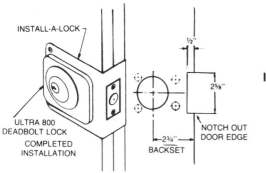

INSTALL-A-LOCK

½"

2⅝"

NOTCH OUT
DOOR EDGE

ULTRA 800
DEADBOLT LOCK
COMPLETED
INSTALLATION

2¾"
BACKSET

13-47 Door edge cutout required for metal door with center seam on the door edge. (courtesy M.A.G. Engineering and Mfg., Inc.)

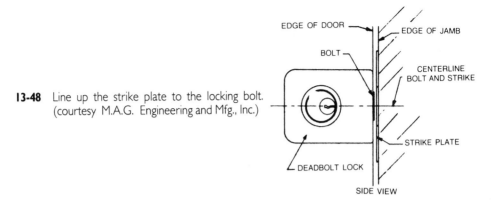

13-48 Line up the strike plate to the locking bolt. (courtesy M.A.G. Engineering and Mfg., Inc.)

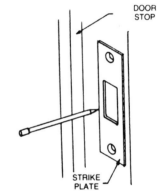

13-49 Mark strike plate. (courtesy M.A.G. Engineering and Mfg., Inc.)

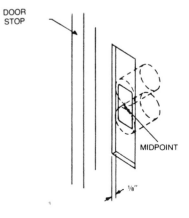

13-50 Ensure the midpoint is centered in the middle of the strike plate hole. (courtesy M.A.G. Engineering and Mfg., Inc.)

13-51   Strike box inserted into the door frame. (courtesy M.A.G. Engineering and Mfg., Inc.)

SIDE VIEW

Within the strike box, drill two ⅛ inch in diameter pilot holes on approximately a 25 degree angle toward the center of the jamb. This will provide for maximum holding power when the two screws are attached to the stud (FIG. 13-52). A side view of the strike box positioning is at FIG. 13-53.

Metal jambs not prepared for a strike:

Mark the jamb with the point on the end of the bolt as described for wood jambs.

Drill a ⅞ inch hole only. No strike plate is required.

If desired by the customer, an Adjust-A-Strike may be installed with the unit. Several models are available. Figure 13-54 refers to the models 450/451—1¼ inch × 2¾ inch lip and deadbolt strike jamb preparation (FIG. 13-55 provides positioning of parts and the strike installation completed for all models described here.)

For 1¾-inch door, start cutout ³⁄₁₆ inch away from the door stop; for 1⅜ inch doors, start the cutout on the edge of the door stop.

Cut out 1¼ inch × 2¾ inch dimension, ⁵⁄₃₂ inch deep.

Cut out 1½ inch dimension, ³⁄₃₂ inch deep.

For 1¾-inch door, ³¹⁄₃₂ inch; for 1⅜ inch door, ²⁵⁄₃₂ inch.

For latches (lip strike), drill ⅝ inch deep; for deadbolts, drill 1 inch deep.

Install the strike and keepers in the center of the cutout. Close the door to check fit. Adjust if necessary.

For the model 452, for a 1¼ inch × 4⅞ inch lip strike, the jamb preparation is as follows:

For 1¾-inch door, start the cutout ³⁄₁₆ inch away from the door stop; for 1⅜-inch door, start cutout on the edge of door stop (FIG. 13-56).

Cut out 1¼ inch × 4⅞ inch dimension ⅛ inch deep.

Cut out 3⅜ inch dimension, ¹⁄₁₆-inch deep.

For 1¾ inch door, ³¹⁄₃₂ inch. For 1⅜ inch door, ²⁵⁄₃₂ inch.

Install the strike and keepers in the center of the cutout. Close the door and check for proper fit, adjusting if necessary.

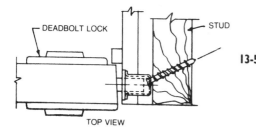

DEADBOLT LOCK

STUD

13-52   Ensure the screws go into the frame stud for maximum holding power. (courtesy M.A.G. Engineering and Mfg., Inc.)

TOP VIEW

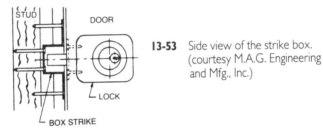

13-53 Side view of the strike box. (courtesy M.A.G. Engineering and Mfg., Inc.)

13-54 Exploded view illustrating parts relationship for the complete installation of the Adjust-A-Strike. (courtesy M.A.G. Engineering and Mfg., Inc.)

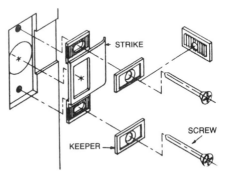

13-55 Jamb preparation for the Adjust-A-Strike models 450 and 451. (courtesy M.A.G. Engineering and Mfg., Inc.)

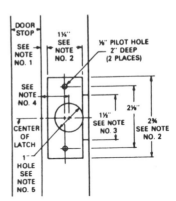

The Schlage "G" series lockset

The mechanics of this lockset have been discussed in an earlier chapter.

Installation:

These instructions, developed from material supplied by Schlage, apply from 1⅜ inch to 2-inch wood doors.

Mark the height line (the centerline of the deadlatch) on the door (FIG. 13-57). It is suggested that the line be 38 inches from the floor.

Mark the centerline of the door thickness on the door edge.

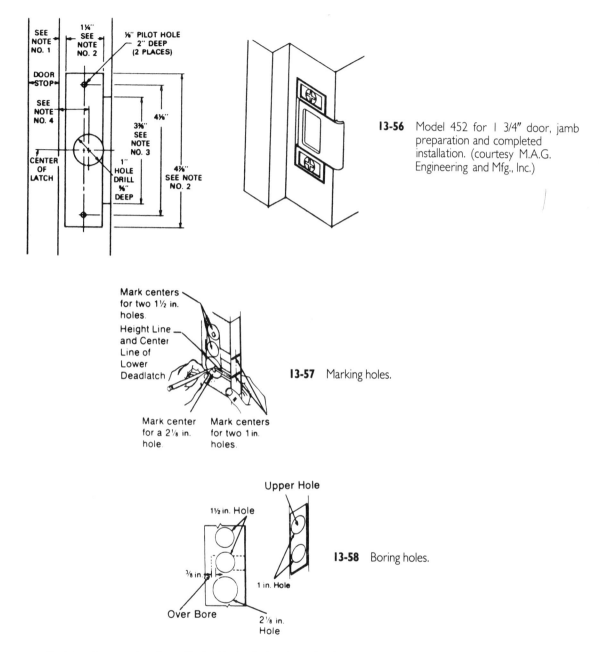

13-56 Model 452 for 1 3/4" door, jamb preparation and completed installation. (courtesy M.A.G. Engineering and Mfg., Inc.)

13-57 Marking holes.

13-58 Boring holes.

Position the template (supplied with the lockset) on the door with the lower deadlatch hole on the height line. Mark center points for one 2⅛-inch hole and two, 1½ inch holes through the template (FIG. 13-57). Mark the centers for two, 1 inch latch holes on the door edge.

Bore one 2⅛ inch hole and two 1½ inch holes in the door panel.

Bore two, 1 inch holes in the edge of the door. The upper hole should be extended ⅜ inch beyond the far side of the middle hole on the door panel (FIG. 13-58). Mortise the edge of the door for the latch front.

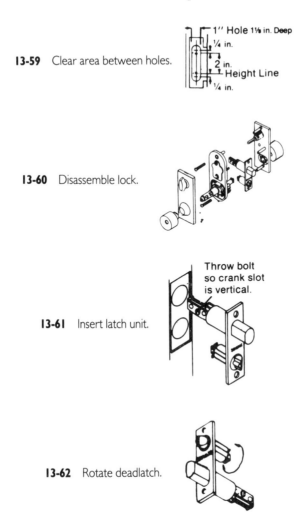

13-59 Clear area between holes.

1″ Hole 1⅛ in. Deep
¼ in.
2 in.
Height Line
¼ in.

13-60 Disassemble lock.

13-61 Insert latch unit.

Throw bolt
so crank slot
is vertical.

13-62 Rotate deadlatch.

Mark vertical and height lines on the jamb exactly opposite the center point of the lower latch hole. Mark a second horizontal centerline 2 inches above the height line for the deadbolt hole. Bore a 1 inch diameter hole 1⅛-inch deep ¼ inch below the height line. Bore a second hole to the same dimensions ¼ inch above the second horizontal line. Clear out area between holes for the strike box (FIG. 13-59).

Install the strike (it can be reversed for either hand). If you have done the work correctly, the strike screws will be on the same vertical centerline as the latch screws.

Disassemble the lock. Remove the inside knob and lift off the outside mechanism (FIG. 13-60).

With the deadbolt thrown and the crank slot in the vertical position, insert the latch unit and secure with the screws provided (FIG. 13-61). The deadlatch can be rotated to match the door hand (FIG. 13-62). The beveled edge of the deadlatch should contact the striker as the door swings shut.

Install the inside mechanism with the knob button released and the crank bar in the vertical position. Insert the end of the crank bar in the slot on the deadbolt (FIG. 13-63).

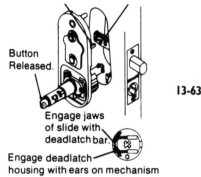

Button
Released.

Engage jaws
of slide with
deadlatch bar.

Engage deadlatch
housing with ears on mechanism

13-63   Install mechanism.

Engage the jaws of the slide with the deadlatch bar, and engage the deadlatch housing with the ears on the inside mechanism. This sounds more confusing than it is; see the insert for FIG. 13-63.

From the outside of the door, turn the cylinder bar slot with a screwdriver to check the action of the deadbolt and deadlatch. Once you are satisfied that the lock works properly, retract the deadbolt and release the knob button. Turn the cylinder bar slot to the horizontal position (FIG. 13-64). The clutch plate should be situated as shown.

Rotate the cylinder bar to the horizontal position with the knob spindle as shown (FIG. 13-65).

Insert the cylinder bar into the bar slot in the top hole. Engage the clutch plate with the outside knob spindle. The spindle must be positioned as shown in FIG. 13-65, and the clutch as shown in FIGS. 13-64 and 13-66.

G85PD locksets require that the cylinder driver bar be flush with the end of the upper bushing (FIG. 13-67).

Install the inside rose with the turn unit in the vertical position and the V-notch on the inside of the turn pointing to the edge of the door (FIG. 13-68). Engage the top of the rose with the mounting plate and, depressing the catch, snap the rose into place. Align the lug on the knob with the slot in the spindle and depress the knob catch. Push the knob home.

Test the lock.

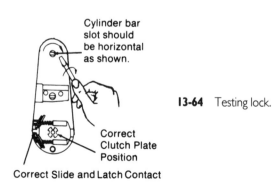

Cylinder bar
slot should
be horizontal
as shown.

13-64   Testing lock.

Correct
Clutch Plate
Position

Correct Slide and Latch Contact

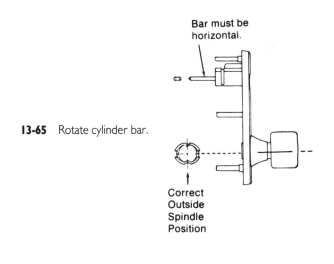

13-65 Rotate cylinder bar.

Bar must be horizontal.

Correct Outside Spindle Position

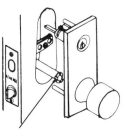

13-66 Positioning clutch.

Correct Clutch Plate Position

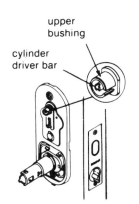

upper bushing

cylinder driver bar

13-67 Driver bar must be vertical.

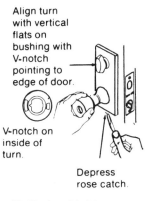

Align turn with vertical flats on bushing with V-notch pointing to edge of door.

V-notch on inside of turn.

Depress rose catch.

13-68 Install inside rose.

Changing the hand:

"G" series locksets are available in right- and left-hand versions. However, the hand can be changed in the field.

Figure 13-69 is an assembled view of a right-hand unit. To convert it to left-hand operation, follow these steps:

Disengage the retaining rings (FIG. 13-70).

Lift off the mounting-plate cover (FIG. 13-71).

Remove the linkage-bar plate, linkage bar, bolt bar, driver bar, knob-driver linkage arm, and lower bushing (FIG. 13-72).

With a factory-supplied wrench or long-nosed pliers, straighten the tabs (FIG. 13-73).

Compress the slide with a screwdriver and pull the assembly out (FIG. 13-74).

Rotate the spindle 180 degrees to relocate the knob catch (FIG. 13-75).

Rotate the slide assembly 180 degrees. With the slide compressed, insert the lugs into slots on the mounting plate (FIG. 13-76).

Bend the tabs to secure the mounting plate (FIG. 13-77).

Install the linkage arm (FIG. 13-78).

Install the lower bushing—the slot is vertical (FIG. 13-79).

Install the retaining ring (FIG. 13-80).

Install the driver bar—the bar is horizontal (FIG. 13-81).

Install the bolt bar (FIG. 13-82).

Install the linkage bar (FIG. 13-83).

Install the knob driver (FIG. 13-84).

Install the linkage-bar plate (FIG. 13-85).

With the upper bushing slot in the horizontal position, install the cylinder driver (FIG. 13-86).

Replace the mounting-plate cover (FIG. 13-87).

Install the inside retaining rings (FIG. 13-88).

Turn the slot in the G85 unit to face the same direction as the slide (FIG. 13-89).

Turn the deadlatch 180 degrees (FIG. 13-90).

Check your work (FIG. 13-91).

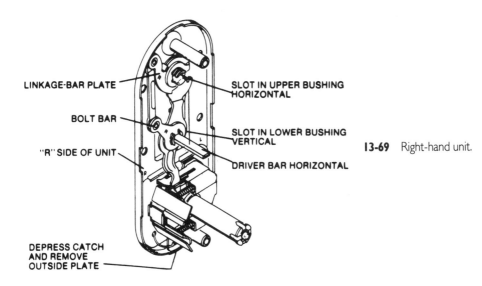

LINKAGE-BAR PLATE

BOLT BAR

"R" SIDE OF UNIT

DEPRESS CATCH AND REMOVE OUTSIDE PLATE

SLOT IN UPPER BUSHING HORIZONTAL

SLOT IN LOWER BUSHING VERTICAL

DRIVER BAR HORIZONTAL

13-69 Right-hand unit.

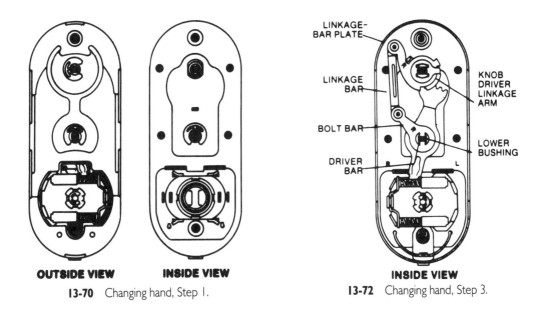

OUTSIDE VIEW   **INSIDE VIEW**

13-70   Changing hand, Step 1.

INSIDE VIEW

13-72   Changing hand, Step 3.

Figure 13-92 is an assembled view of a left-hand unit. To convert it to right-hand operation, follow these steps:

1. Disengage the retaining rings (FIG. 13-93).
2. Lift off the mounting-plate cover (FIG. 13-94).
3. Remove the linkage-bar plate, linkage bar, bolt bar, driver bar, knob-driver linkage arm, and lower bushing (FIG. 13-95).
4. With a factory-supplied wrench or long-nosed pliers, straighten the tabs (FIG. 13-96).
5. Compress the slide with a screwdriver and pull the assembly out (FIG. 13-97).

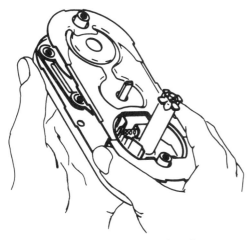

13-71   Changing hand, Step 2.

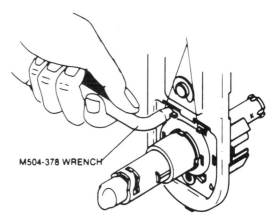

M504-378 WRENCH

13-73   Changing hand, Step 4.

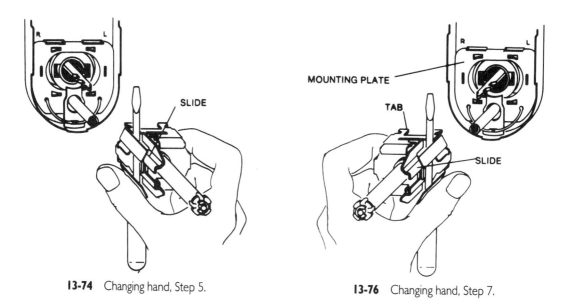

I3-74   Changing hand, Step 5.

I3-76   Changing hand, Step 7.

6. Rotate the spindle 180 degrees to relocate the knob catch (FIG. 13-98).

7. Rotate the slide assembly 180 degrees. With the slide compressed, insert the lugs into slots on the mounting plate (FIG. 13-99).

8. Bend the tabs to secure the mounting plate (FIG. 13-100).

9. Install the linkage arm. (FIG. 13-101).

10. Install the lower bushing (FIG. 13-102).

11. Install the retaining ring (FIG. 13-103).

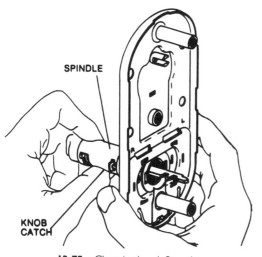

I3-75   Changing hand, Step 6.

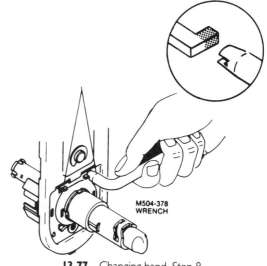

I3-77   Changing hand, Step 8.

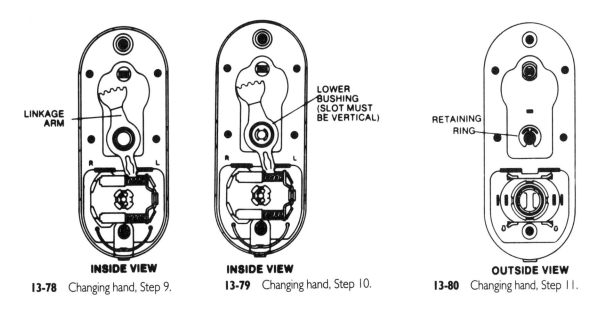

13-78 Changing hand, Step 9.

LINKAGE ARM

13-79 Changing hand, Step 10.

LOWER BUSHING (SLOT MUST BE VERTICAL)

INSIDE VIEW

13-80 Changing hand, Step 11.

RETAINING RING

OUTSIDE VIEW

12. Install the driver bar (FIG. 13-104).

13. Install the bolt bar (FIG. 13-105).

14. Install the linkage bar (FIG. 13-106).

15. Install the knob driver (FIG. 13-107).

16. Install the linkage-bar plate (FIG. 13-108).

17. With the upper bushing slot in the horizontal position, install the cylinder driver (FIG. 13-109).

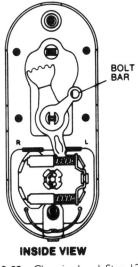

DRIVER BAR (MUST BE HORIZONTAL)

INSIDE VIEW

13-81 Changing hand, Step 12.

BOLT BAR

INSIDE VIEW

13-82 Changing hand, Step 13.

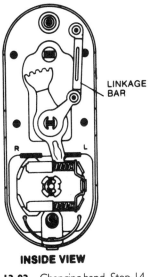

LINKAGE BAR

INSIDE VIEW

13-83 Changing hand, Step 14.

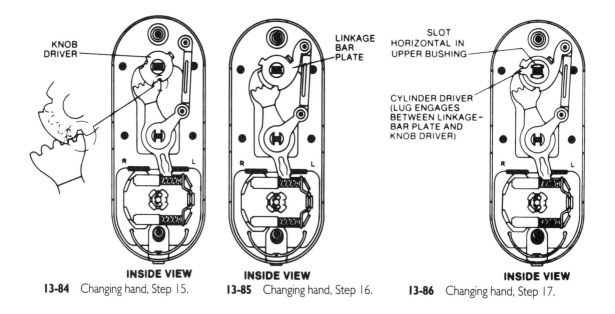

13-84   Changing hand, Step 15.   **13-85**   Changing hand, Step 16.   **13-86**   Changing hand, Step 17.

18. Replace the mounting-plate cover (FIG. 13-110).
19. Install the inside retaining ring (FIG. 13-111).
20. Turn the slot in the G85 unit to face the same direction as the slide (FIG. 13-112).
21. Turn the deadlatch 180 degrees (FIG. 13-113).
22. Check your work (FIG. 13-114).

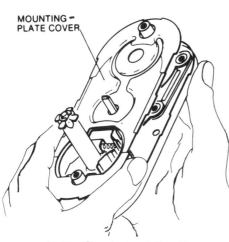

13-87   Changing hand, Step 18.

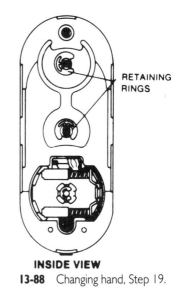

13-88   Changing hand, Step 19.

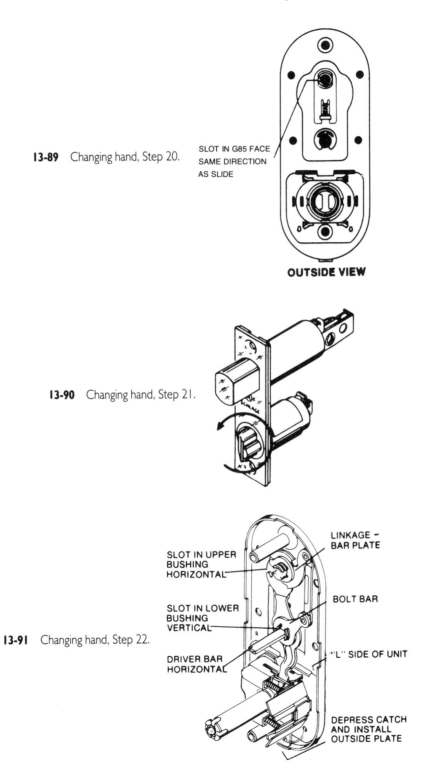

13-89   Changing hand, Step 20.

SLOT IN G85 FACE
SAME DIRECTION
AS SLIDE

OUTSIDE VIEW

13-90   Changing hand, Step 21.

13-91   Changing hand, Step 22.

LINKAGE –
BAR PLATE

SLOT IN UPPER
BUSHING
HORIZONTAL

SLOT IN LOWER
BUSHING
VERTICAL

BOLT BAR

DRIVER BAR
HORIZONTAL

"L" SIDE OF UNIT

DEPRESS CATCH
AND INSTALL
OUTSIDE PLATE

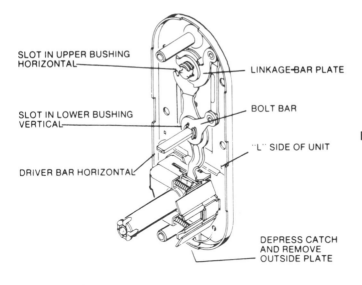

SLOT IN UPPER BUSHING HORIZONTAL

LINKAGE-BAR PLATE

SLOT IN LOWER BUSHING VERTICAL

BOLT BAR

13-92   Assembled left-hand unit.

"L" SIDE OF UNIT

DRIVER BAR HORIZONTAL

DEPRESS CATCH AND REMOVE OUTSIDE PLATE

Attaching and removing knobs:

Figure 13-115 illustrates two of the most frequently encountered methods of securing door knobs. The knob may be threaded over the spindle and held by a set screw, or it may be pinned to the spindle by a screw that passes through the knob and spindle. Another approach is to secure the knob with a retaining lug that extends into a hole in the knob shank. The inside knob can be removed at any time by depressing the retainer; the outside knob can be removed only when the lock is open (FIG. 13-116).

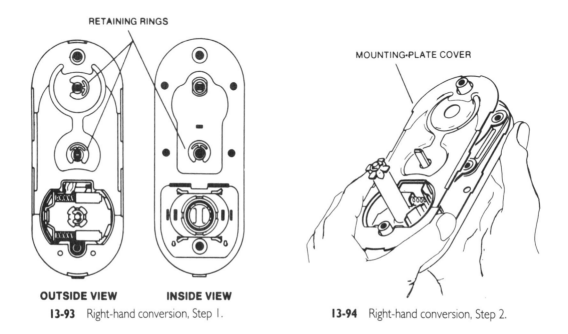

RETAINING RINGS

MOUNTING-PLATE COVER

OUTSIDE VIEW   **INSIDE VIEW**

13-93   Right-hand conversion, Step 1.

13-94   Right-hand conversion, Step 2.

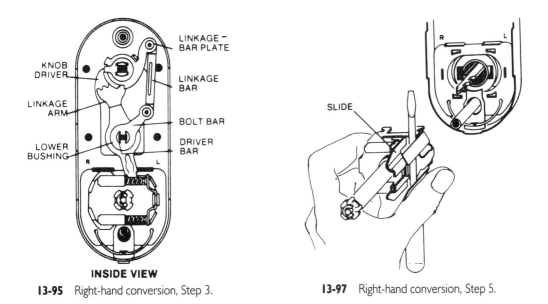

13-95 Right-hand conversion, Step 3.

13-97 Right-hand conversion, Step 5.

Updating a lockset

The following instructions concern the replacement of a worn or outdated lockset with a new one. The replacement is a heavy-duty "G" series Schlage, one of the most secure entranceway locksets made. The work is not difficult.

1. Remove the old lockset (FIG. 13-117).
2. Remove the latch (FIG. 13-118).

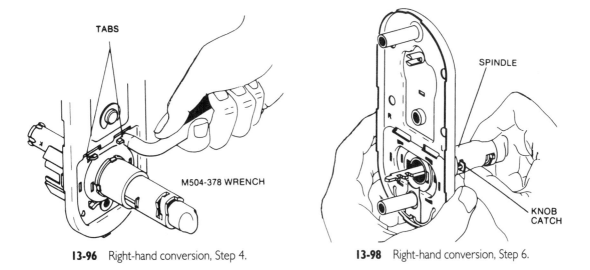

13-96 Right-hand conversion, Step 4.

13-98 Right-hand conversion, Step 6.

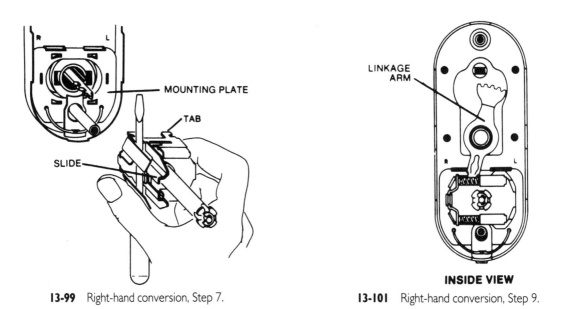

13-99   Right-hand conversion, Step 7.

13-101   Right-hand conversion, Step 9.

3. If a jig is available, use it as a guide for cutting the door; if not, use the template packaged with the lockset (FIG. 13-119 and FIG. 13-120).

4. Mortise the edge of the door to receive the strikeplate (FIG. 13-121).

5. Install the double-locking latch and deadbolt (FIG. 13-122).

6. Install the inside lockface assembly (FIG. 13-123).

7. Install the internal mechanism.

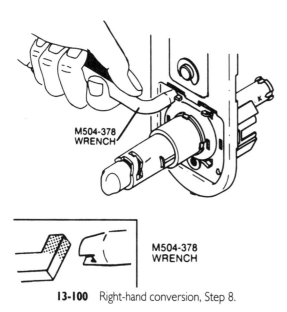

13-100   Right-hand conversion, Step 8.

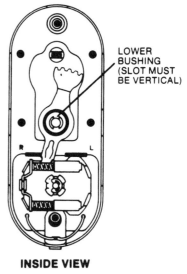

13-102   Right-hand conversion, Step 10.

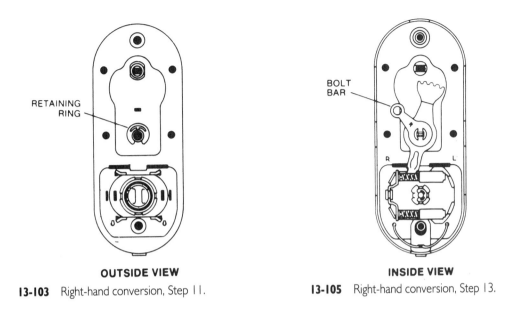

OUTSIDE VIEW

13-103   Right-hand conversion, Step 11.

INSIDE VIEW

13-105   Right-hand conversion, Step 13.

8. Couple the outside lockface assembly with the internal parts (FIG. 13-124).
9. Secure the inside lockface assembly with the screws provided (FIG. 13-125).
10. Snap the inside cover into place (FIG. 13-126). The lockset is now installed (FIG. 13-127).

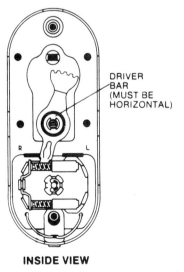

INSIDE VIEW

13-104   Right-hand conversion, Step 12.

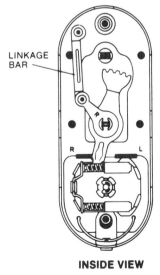

INSIDE VIEW

13-106   Right-hand conversion, Step 14.

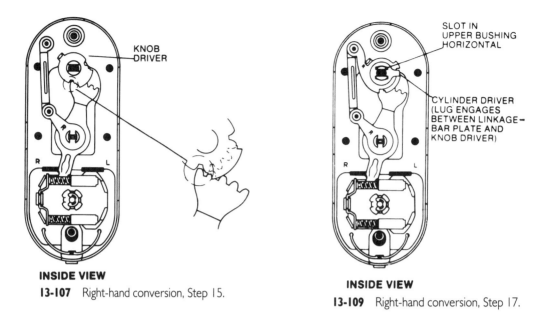

INSIDE VIEW
13-107   Right-hand conversion, Step 15.

INSIDE VIEW
13-109   Right-hand conversion, Step 17.

Strikeplates

Quality locksets are equipped with a deadlocking plunger on the latchbolt. Otherwise, the bolt could be retracted with a knife blade or a strip of celluloid. Nevertheless, it is important to leave very little space between the door edge and the strikeplate. A bolt with a ½ inch throw should have no more than ⅛ inch visible between the door and jamb. If necessary, mount the strikeplate over a steel spacer. This moves the strikeplate closer to the bolt.

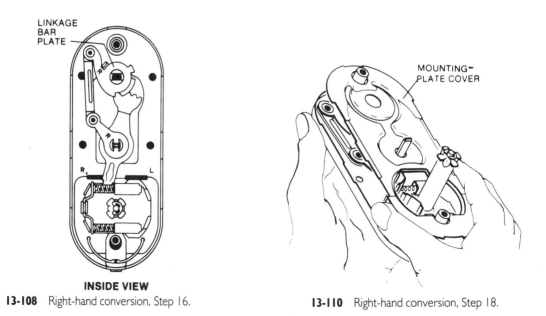

INSIDE VIEW
13-108   Right-hand conversion, Step 16.

13-110   Right-hand conversion, Step 18.

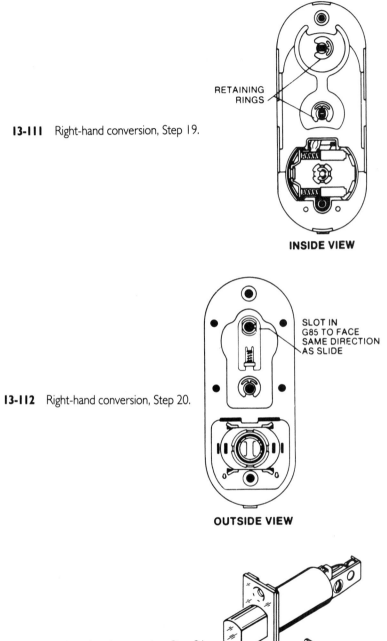

13-111   Right-hand conversion, Step 19.

RETAINING
RINGS

INSIDE VIEW

13-112   Right-hand conversion, Step 20.

SLOT IN
G85 TO FACE
SAME DIRECTION
AS SLIDE

OUTSIDE VIEW

13-113   Right-hand conversion, Step 21.

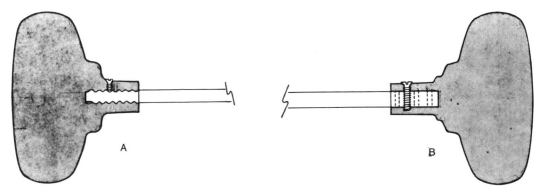

13-114   Right-hand conversion, Step 22.

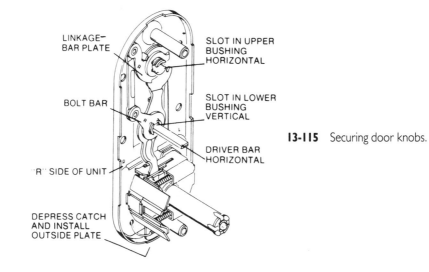

LINKAGE-
BAR PLATE

SLOT IN UPPER
BUSHING
HORIZONTAL

BOLT BAR

SLOT IN LOWER
BUSHING
VERTICAL

DRIVER BAR
HORIZONTAL

"R" SIDE OF UNIT

DEPRESS CATCH
AND INSTALL
OUTSIDE PLATE

13-115   Securing door knobs.

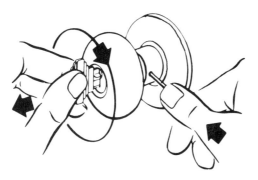

13-116   Removing knob.

13-117 Lockset update, Step 1.

13-118 Lockset update, Step 2.

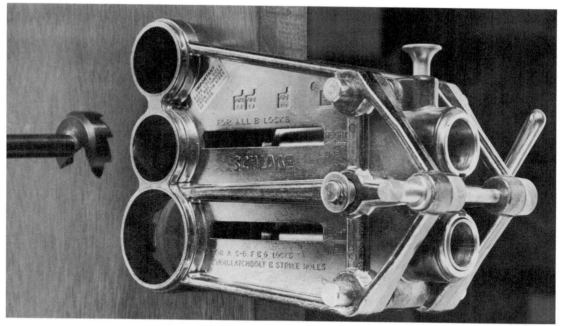

13-119 Lockset update, Step 3.

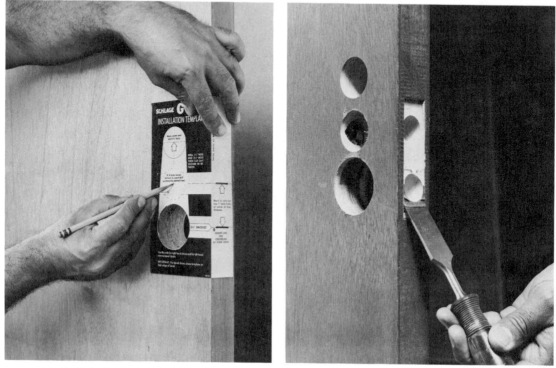

13-120 Lockset update, Step 4.

13-121 Lockset update, Step 5.

13-122   Lockset update, Step 6.

13-124   Lockset update, Step 8.

13-123   Lockset update, Step 7.

13-125   Lockset update, Step 9.

13-126   Lockset update, Step 10.

13-127   Lockset update, Step 10.

The length of the lip is also critical, since a short lip will increase wear on the latchbolt and may frustrate the automatic door-close mechanism, leaving the door unlatched. Mortise strikeplates are mounted on the same vertical centerline as the bolt (or should be). Measure the distance from the centerline of the latchbolt to the edge of the jamb and add ⅛ inch for flat strikeplates and ¼ inch for curved types. Figure 13-128 illustrates a collection of Schlage strikeplates for reference.

Entrance-door security

Security is a function of the door, cylinder, bolt, and strikeplate.

☐ There is *no* security with a hollow wooden door.

☐ Master keying should be kept as simple as possible. Each split pin increases the odds in favor of the lock picker.

☐ Resist demands for extensive cross keying. These demands originate with customers who insist upon single-key performance for executive keys.

☐ Security begins with proper key-control procedures.

☐ It is wise to supply extra change-key blanks during the installation phase of the system. Without readily available blanks, the convenient thing to do would be to cut duplicate change keys from master key blanks. Security could be compromised since the change key could fit other locks.

☐ Double-cylinder locks should be used wherever possible.

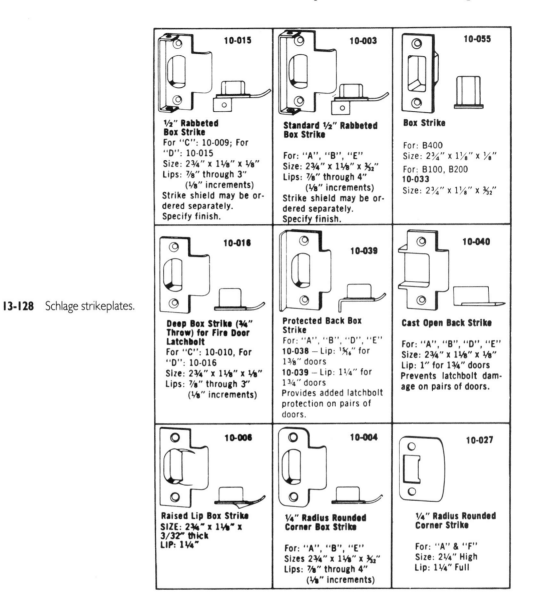

13-128   Schlage strikeplates.

☐ Automatic deadlatches should be specified for all locksets without a deadbolt function. Otherwise the bolt could be retracted by "loiding," that is, by slipping a flat object between the bolt and the strike.

☐ Use the longest bolts available to frustrate attempts to gain entry by spreading the door frame. A strikeplate cover can also have the same function.

☐ Shield the bolts with armored inserts.

☐ Reinforced strikeplate mounts, particularly with metal door frames, add security by increasing the area of contact between the strikeplate and frame.

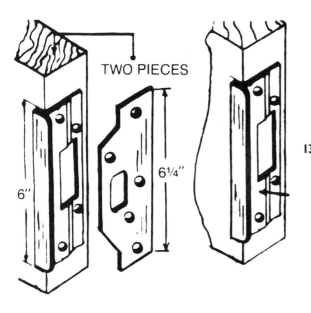

TWO PIECES

6¼"

6"

13-129 The LG-1 and LG-2 units for door units opening outward. (courtesy Sentry Door Lock Guards Co.)

Sentry door lock guards

Integral to improved lock security for home, apartment, or business doors is the choice of the door strike to be used. All locks come with a strike that is especially configured for the lock, but you or your customer may wish to have a better, improved strike that will provide a greater amount of security. The Sentry Door Lock Guards Company, of Dania, Florida, has developed such a line of strike units that will dramatically improve the security posture of individual door units. The designer and manufacturer of these unique lock guards guarantees that they are jimmyproof.

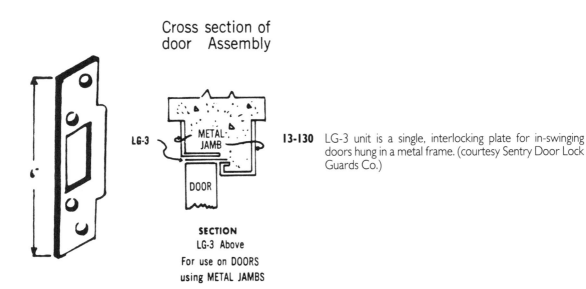

Cross section of
door Assembly

LG-3

METAL
JAMB

DOOR

SECTION
LG-3 Above
For use on DOORS
using METAL JAMBS

13-130 LG-3 unit is a single, interlocking plate for in-swinging doors hung in a metal frame. (courtesy Sentry Door Lock Guards Co.)

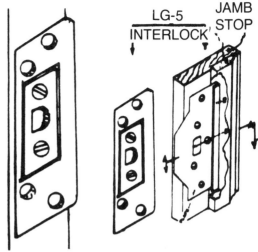

13-131 The LG-4 and LG-5 door guard units.
(courtesy Sentry Door Lock Guards Co.)

Perhaps you've never heard of this company or its president Frank Sushan, but many people and companies have. As the inventor of these unique, impressive, and effective Lock Guard units, he has been written up in several major newspaper articles. In addition, numerous police departments across the nation now demonstrate the Sentry devices as part of local crime prevention lectures and demonstrations. Further, the National Crime Prevention Institute has approved and endorsed the installation of these products.

Figure 13-129 through 13-133 illustrate the various lock guards available.

LG-1 is a maximum security attachment produced for the do-it-yourselfer who may have no metal cutting tools or experience in security applications. You just remove the present

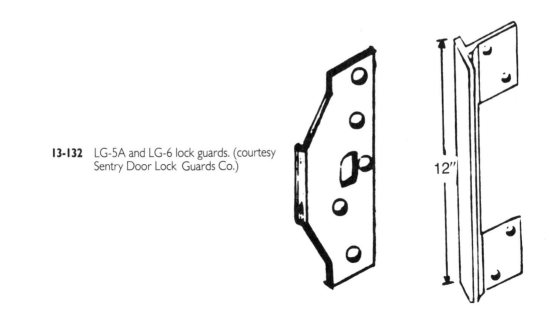

13-132 LG-5A and LG-6 lock guards. (courtesy Sentry Door Lock Guards Co.)

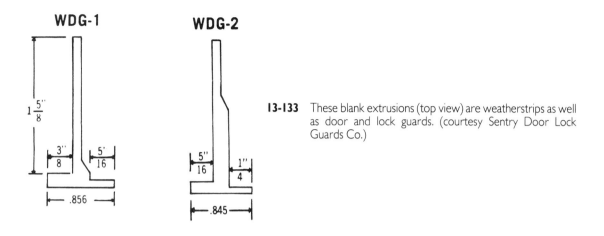

13-133 These blank extrusions (top view) are weatherstrips as well as door and lock guards. (courtesy Sentry Door Lock Guards Co.)

striker and replace it with the fitted unit. The lock guard attachment is installed around the latch faceplate on the door, when such doors open out.

LG-2 is a single plate that is used with an existing striker plate for doors that open out. LG-2A is identical to the LG-2, with the exception of the closing section under the "T." The WDG-1 cross-section provides the extrusion view of the product.

The LG-3 is a single interlocking door plate for use on inswinging doors that are hung on a metal jamb door frame. All that is required is a slot to be cut into the metal door stop to

13-134 Install-A-Lock lock changing kit component parts. (courtesy M.A.G. Engineering and Mfg., Inc.)

receive the tongue of this interlocking door plate. Any door protected by this unit does not require locking with a key from the outside when leaving and the door is closed. The lock cannot be loided, jimmied, or jacked apart. Figure 13-133 shows a cross-section of the installation.

The LG-4 is used similar to the LG-3, and when used alone the difference is that a slot must be made in a wood door jamb to receive the tongue of the unit.

The LG-5 is a two-piece unit for installation on doors that open inward only, and which are of wood. As indicated for the LG-3 and LG-4, such protected locks cannot be jimmied or jacked open. The doors and jambs will remain together and will not separate.

The LG-5A is a replacement safety striker plate for use on a wood jamb door frame in which the striker plate screws will not hold the plates onto the jambs, making the purchase of another new lock unit unnecessary. The LG-5B is a replacement safety striker plate used for the same reason as the LG-5A. This unit provides additional protection where entrance is gained by driving a tool through the molding behind the latchbolt, to force the bolt back to gain unauthorized entrance. The right angle tongue of the strike plate is extended under the mold for added protection.

LG-6 is an enlarged model of the LG-2 unit, for use on any old or new mortise lock application.

The WDG-1 and WDG-2 are seven foot blank extrusions that are weather strips as well as door and lock guards. They require that all holes will be drilled when the unit is installed. The length ensures that it will be long enough to fit any standard door frame, requiring only slight cutting down for perfect fits. Cross sections of these units are provided at FIG. 13-133.

As a point of customer reference, these lock guard units are much stronger and larger than those being used in other applications today. No stronger products are commercially available anywhere.

M.A.G. Install-A-Lock

Lock changing and door reinforcement kits provide an extra measure of security for door locks. For old and new doors alike, these units are providing the ultimate security protection for the lock and also for the area immediately around the locking unit.

M.A.G. Engineering and Mfg. Co. has spent many years perfecting these types of units to provide ultimate security to the users of their products. As a locksmith, you cannot afford to be ill-prepared to meet customer needs when it comes to security applications concerning the locks you install. Because of the ever-increasing need for upgraded security in today's society, these products are widely used and beneficial to all. Figure 13-134 shows just one type of the Install-A-Lock lock changing kits available.

These kits can be applied to cylindrical locks, deadlocks, deadbolt and key-in-knob locksets, and also for mortise locksets (FIG. 13-135, 13-136). The lock changing kit enables you to install a lock on any door within minutes. They are made of heavy gauge stainless steel or brass, and the assembled unit brings the ultimate protection to new or old doors; it also covers damaged portions on burglarized doors. Even worn, old, mistreated doors can take on a new glow when graced with an Install-A-Lock kit. Designed to cover jimmy marks, splits and other defacements around old locks, it also provides new strength in the locking area. With the latch plate secured metal-to-metal—*not* wood—the completed installation is extremely solid and secure.

The one-piece construction provides a great savings in labor time. The doors are then protected, after only a few minutes installation time, right up to the leading edge on both sides of the door.

The security cost is economical and comes in a variety of sizes to meet varying requirements. The cylindrical lock kit will accept locksets with a 2 inch to 2⅛ inch bore;

13-135 Install-A-Lock with a key-in-knob unit, plus a dead-bolt locking unit added for extra security. (courtesy M.A.G. Engineering and Mfg., Inc.)

13-136 Install-A-Lock for a mortise lockset. (courtesy M.A.G. Engineering and Mfg., Inc.)

whereas, the deadbolt and key-in-knob allow for two backsets of 2⅛ inch or 2⅜ inch. Simply put, these are the finest remodeling kits available.

**Installation Instructions—
Deadlock or Night Latch:**

Remove the old lock and all screws from the door.

Bore holes for the new lock, if necessary, using the template furnished with the lockset to be installed in the door.

Chisel out the front edge of the door 1¼-inch wide × 2⅝-inch high × ½-inch deep in the center of the latch hole.

Slide the Install-A-Lock unit and position the large hole in the unit to line up with the hole in the door for the lockset.

After the unit is in position, install the lock in the door, using the 8-32 machine screws that come with the kit. These will secure the latch to the unit.

Install the four screws and washers to secure the unit to the door. Be sure that the face of the unit is flush with the door face.

Install the striker in the door jamb.

For mortise locks, if it is a new installation, follow the manufacturer's installation instructions, then, follow the instructions below for the Install-A-Lock for mortise locks. For preparing a door in which a mortise lock is currently mounted, follow the instructions below only.

Remove the lock from the door.

Cut the door edge ⅝-inch deep and 8½-inch high centered on the lock.

Install two screw fasteners into the recessed area of the Install-A-Lock with the lip facing down.

To determine which side is long, place the unit on a flat surface. The unit will tilt slightly in the direction of the short side. Place the long side on the high edge of the door bevel.

Install the lock in the unit with the cylinder at the top.

(For Sectional Trim)

Select the proper Install-A-Lock insert, and place it between the door and the underside of the Install-A-Lock with the raised area out, fitting it on the large rectangular hole.

Install the cylinder and knobs to hold the Install-A-Lock unit in place.

Check to see that the lock matches the strike correctly.

Be sure the Install-A-Lock unit is tight against the door.

(For Escutcheon Trim)

Place the unit on the door.

Install the cylinder and knob assembly to hold the Install-A-Lock unit in place. *Do not* install the escutcheon retaining screws as yet.

Check to see that the lock matches the strike correctly.

Be sure the unit is tight against the edge of the door.

Mark and drill the top mounting hole closest to the edge of the door, using a ¹³⁄₃₂ inch drill, and install through bolt. Repeat the same procedures on the bottom outside hole closest to the door edge; tighten the bolt securely. Drill and install the remaining three through bolts, installing the center bolt first.

For escutcheon, now drill and install the escutcheon screws as supplied by the lockset manufacturer.

For double-lock units, the following instructions should be followed:

Remove the lock, latch, and all screws (FIG. 13-137).

See FIG. 13-138. Mortise out the front edge of the door: (1) 2⅝-inch high × ½-inch deep in the center of the lock hole (2).

Using the template furnished with the deadbolt lockset, mark and drill holes (3) and (4) 3⅝ inches from the center of the bottom hole.

Slide the unit (5) over the door and position both holes to line up with the holes in the door (FIG. 13-139).

Install both locksets (6) (before installing the mounting screws). Install both striker plates (7) 3⅝-inches between centers. The strike plate installation should be per the specific instructions for the installed lockset (FIG. 13-140).

Close the door and check to be sure that both the latch and deadbolt (8) enter the strikes (7). If they do not enter, move the Install-A-Lock assembly slightly up or down until the latches enter (FIG. 13-141).

Install the mounting screws (9) with finishing washers, being sure that the unit is flush with the door.

EXIT ALARM LOCKS AND PANIC BAR DEADLOCKS

A great concern of businesses that provide for locksmithing sales and servicing is the alarmed and deadbolted exit door lock. These units are used in the main area of controlling, monitoring, or warning of access (either inward or outward) through various doors and also into or out of areas that should be controlled and access limited. One-way (emergency) doors are one example. Controlled access areas, such as high-cost storeroom stockage, back rooms of a small watch or jewelry store, are another example. For a restricted type area (again a watch or jewelry store, or more importantly, any business that has large amounts of money on hand or the drug storage area in a hospital) these lock alarm units are a great convenience and additional security.

The Alarm Lock Corporation, which specializes in these devices, has a long list of satisfied customers, including American industry, the U.S. government, many major chain and department stores, large food markets, major discount centers, factories, hospitals, post offices, many primary and secondary schools and colleges, and finally, small and medium sized businesses in your own city.

The deadbolt exit alarm lock and the double deadlock panic bar lock evolved to meet a major problem among various types of government and private concerns. The problem was that it was (and still is) illegal to lock or bolt an exit or emergency door because it would endanger life. Also, as long as the doors were unlocked, they became an invitation to the unauthorized intrusion from the outside by undesirable elements of society. These intrusions led to the theft of many millions of dollars every year from the concerns that were illegally entered.

The solution to the problem was a locking device that allowed for emergency exit, sounding an alarm when the door was opened from the inside or out, but at the same time

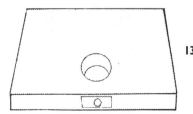

13-137 Door section with the old lock and latch removed. (courtesy M.A.G. Engineering and Mfg., Inc.)

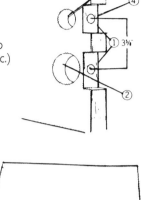

13-138 Two Mortise cuts are required in the door edge in addition to the holes required. (courtesy M.A.G. Engineering and Mfg., Inc.)

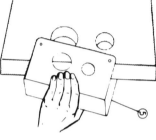

13-139 Line up the Adjust-A-Lock unit when fitting into the door. (courtesy M.A.G. Engineering and Mfg., Inc.)

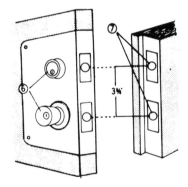

13-140 Install the locksets and strikeplates. (courtesy M.A.G. Engineering and Mfg., Inc.)

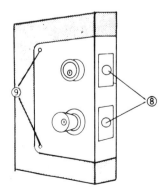

13-141 Install the strikes and mounting screws to complete the installation. (courtesy M.A.G. Engineering and Mfg., Inc.)

ensured that nobody could enter the area anytime they wanted. Such items as the Safetygard 11 Deadbolt Alarmed Security Lock (FIG. 13-142) and the Safegard 70 Panic Alarm Deadlock (FIG. 13-143) were only two of the items that were developed to meet the pressing needs of government and private business.

Safetygard 11 exit alarm lock

The safety alarm lock provides secure full-time locking of emergency exit doors while complying with standard safety, fire, and local building codes. It provides maximum day and night security and a powerful deterrent against pilferage—from within or without—and always remains panic-proof.

The safety alarm lock provides instant automatic exit in case of an emergency. It combines the best features of panic devices, plus a deadbolt, and a warning system. The first person reaching the door releases the lock with a slight pressure on the emergency clapper plate, which automatically releases the lock, opening the door and simultaneously sounding the alarm that the door has been opened.

Alarm lock operation.   The complete lock mechanism retracts from the edge of the door when the clapper arm is depressed, which withdraws the deadbolt and sounds the powerful twin horn alarm. The alarm continues until the lock is reset to the original position with the appropriate key. The alarm can only be bypassed by an individual authorized to carry and use the bypass key. The mounting plate is already predrilled when the unit is received from the factory, so it is easily mounted on all styles of doors. The lock is furnace mounted, and it requires no mortising.

13-142   Safetyguard 111 deadbolt alarm security lock. (courtesy Alarm Lock Corporation)

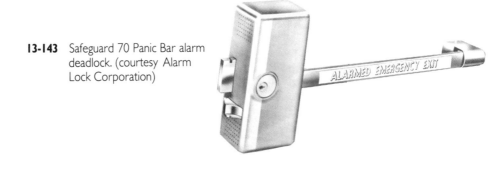

13-143 Safeguard 70 Panic Bar alarm deadlock. (courtesy Alarm Lock Corporation)

Keying. The safety alarm lock unit is received without the cylinders. (This provides a means for you to sell the cylinder along with the unit, possibly keying the cylinder to meet a certain level of master key operation in medium and large size business and manufacturing complexes.) The safety alarm lock will accommodate all standard rim cylinders. This permits the lock to be integrated into an existing or proposed master key system. If special keying or rekeying is not required, the unit should be factory ordered with Alarm Lock's own cylinders; the increased cost is slight, but well worth it.

The lock may be key operated from the outside of the door by mounting an extra rim cylinder through the door. Outside key control cam and cover tamper alarm switch are built into every lock. The outside cylinder, though, must have a flat horizontal tailpiece. The overall length of the cylinder and tailpiece must be a minimum of 3½ inches in length to suit a 1¾-inch door.

Safegard 70 panic alarm deadlock

This is the world's only available double deadlock panic lock, and it is the best. It provides for anti-pilferage alarm security from within and deadlock security against intrusion from the outside; also, it meets all requirements as an emergency exit device.

With deadbolt locking, the Safegard 70 is designed to deter the unauthorized use of the emergency and fire exit doors. Instant emergency panic exiting is provided, though, as authorized by law. An added advantage is that a remote electrical control for this unit is available.

The unit can be operated in four different modes:

Maximum Security Locking—Once the door is closed, the bolt is projected by use of the key, firmly securing the door by deadbolting it. The crossbar must be depressed to release the deadbolt and deadlatch to open the door. In doing this, the alarm is activated and remains so, until the door is closed and relocked by the authorized relocking key.

Outside Key and Pull Access—By adding an outside cylinder and door pull, you have another mode of access. The deadbolt is retracted by the key and the deadlatch is released by the pull to open the door. Entry remains unrestricted from both sides until the deadbolt is relocked by the key from either the inside or the outside.

Outside Key Only Access—By adding the outside cylinder, in this mode the key must retract the deadbolt and release the deadlatch momentarily to open the door. The door relatches with each closing; the key is required for each re-entry. To relock the door and rearm the alarm, the deadbolt must be projected by use of the key.

Bar Dogging Function—This provides for free passage from both sides of the door.

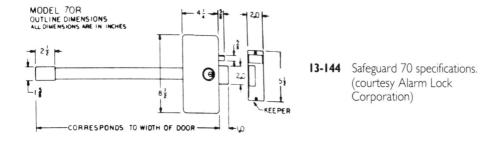

13-144 Safeguard 70 specifications. (courtesy Alarm Lock Corporation)

The Safegard 70 dimensions are shown at FIG. 13-144. Modes of operation for single doors, pairs of doors, and pairs of doors with center mullions are shown at FIG. 13-145.

Narrow stile glass doors

Narrow stile glass doors have evolved the need for a more secure type of locking device than is used in other applications. The architectural trend for such types of doors will continue for quite a number of years to come. As an individual concerned with security, you must be extremely aware of what is available to meet a variety of locksmithing situations, both in the home and the general business environment.

Initially, realize that because of this architectural trend, the aluminum used in narrow stile glass doors is shrunk in width and material distance from that of the wooden door that would have a large piece of glass mounted in it. Because the aluminum is narrower and of less thickness, it naturally becomes slightly more flexible. At the same time, of course, the space to house the lock also becomes smaller. To retract a long bolt in the conventional matter is impossible since there is no place for it to go. The conventional lockbolt in an aluminum frame of a glass door has a ½ inch to ⅝-inch throw with an effective penetration of the strike, in some instances, of as little as ¼ inch. This door can be sprung open by just about anyone with a couple of screwdrivers or, if the frame is weak or worn, with a well-directed frontal attack by a moving body (the intruder). This highly attractive situation means that the lock installed must be able to stop the hammer and bar and general forced entry.

To an extent, the maximum security considerations for glass doors that *slide* must also be considered. These doors have spawned some misconceptions, some fostered by locksmiths who really don't know what they are talking about because they haven't studied the doors, their locks, and the other applications.

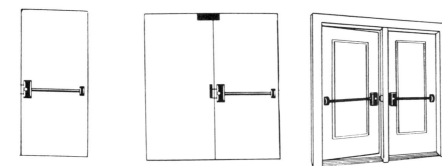

13-145   Modes of door operation with the Safeguard 70 installed. (courtesy Alarm Lock Corporation)

This confusion has resulted in some of these doors being installed still lacking the most basic elements of protection that any door should have. Chief among these is a *real* lock—a secure lock, a lock that can be operated by a key if desired and also one in that the key is compatible with the system of other locks in the same building.

Convenience *and* security are both necessities to the owner/user of the lock, and this is where we now consider the specialized locks that offer the maximum security and protection available for these types of doors. Only one company specializes in locks and locking devices for narrow stile doors, and that is the Adams Rite Manufacturing Company, of California. The next several pages will cover their products.

Deadlocks

The MS1850A deadlock (FIG. 13-146) is the first in a series of specialized locks developed especially with the narrow stile door in mind. This lock uses a high bolt of laminated steel, nearly three inches long and actuated by an uncomplicated pivot mechanism. This lock has made the Maximum Security Deadlock the standard for the entire narrow stile door industry. The length of the bolt provides the maximum security for any single leaf door, even a tall and flexible one. It is also excellent for installation applications where the gap between the door and the jamb is greater than it should be.

The lock case measures 1 inch × 5 3/16 inches × the desired depth. The case depth varies with the backset (FIG. 13-147). The cylinder backset can be 7/8 inch, 3 1/32 inch, 1 1/8 inches or 1 1/2 inches. Remember that in measuring the backset, it is from the centerline of the faceplate to the centerline of the cylinder.

The bolt is 5/8 inch × 1 3/8 inches × 2 7/8 inches with a 1 3/8 inch throw. It is constructed of a fly ply laminated steel, the center ply having an Alumina-Ceramic core to defeat any hacksaw attack, including rod-type "super" hacksaws.

The faceplate for the MS1850A comes in five variations, as illustrated in FIG. 13-148. Figure 13-149 shows the dimensions (in inches and millimeters), illustrates the stile preparation for the lock, lock and cylinder installation, and information about the cylinder cam.

13-146 The MS1850A deadlock for narrow stile doors. (courtesy Adams Rite Mfg. Co.)

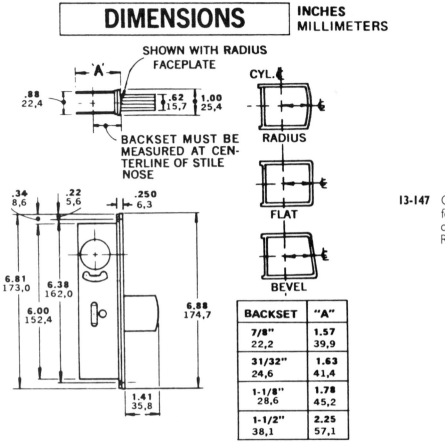

DIMENSIONS

INCHES
MILLIMETERS

SHOWN WITH RADIUS FACEPLATE

CYL.

BACKSET MUST BE MEASURED AT CENTERLINE OF STILE NOSE

RADIUS

FLAT

BEVEL

BACKSET	"A"
7/8" 22,2	1.57 39,9
31/32" 24,6	1.63 41,4
1-1/8" 28,6	1.78 45,2
1-1/2" 38,1	2.25 57,1

13-147 Critical lock dimensions for the SM1850A deadlock. (courtesy Adams Rite Mfg. Co.)

Latch locks

The MS+1890 latch/lock (FIG. 13-150) is a deadlocked Maximum Security unit offering the highest type of protection after-hours for the businessman or homeowner, in addition to traffic control convenience for management (in businesses) during the business day.

A typical installation in a bank or a store could require any of three modes of door control: (1) both the lock and latchbolts retracted for unrestricted entry and exit during

FLAT: MS1850A	
RADIUS: MS1851A	
RADIUS WITH WEATHERSTRIP: MS1851AW	
L.H. BEVEL: MS1852A(LH)	
R.H. BEVEL: MS1852A(RH)	

13-148 MS1850 A faceplate variations. (courtesy Adams Rite Mfg. Co.)

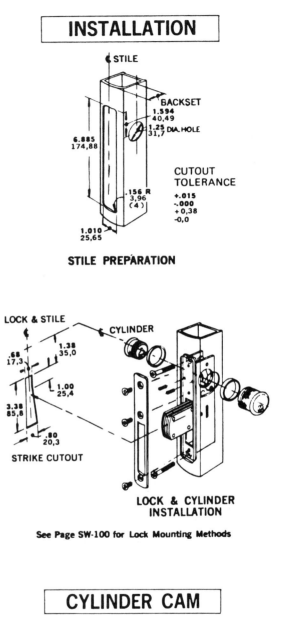

INSTALLATION

STILE

BACKSET
1.594
40,49

1.25 DIA. HOLE
31,7

6.885
174,88

.156 R
3,96
(4)

CUTOUT
TOLERANCE

+.015
-.000
+0,38
-0,0

1.010
25,65

STILE PREPARATION

LOCK & STILE

CYLINDER

.68
17,3

1.38
35,0

1.00
25,4

3.38
85,8

.80
20,3

STRIKE CUTOUT

**LOCK & CYLINDER
INSTALLATION**

See Page SW-100 for Lock Mounting Methods

13-149 Lock and cylinder preparation, stile preparation and critical cam dimensions for the MS1850A. (courtesy Adams Rite Mfg. Co.)

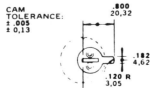

CYLINDER CAM

CAM
TOLERANCE:
± .005
± 0,13

.800
20,32

.182
4,62

.120 R
3,05

13-150 MS1890 Maximum Security Latch/Lock unit, with an armored strike and a lever handle. (courtesy Adams Rite Mfg. Co.)

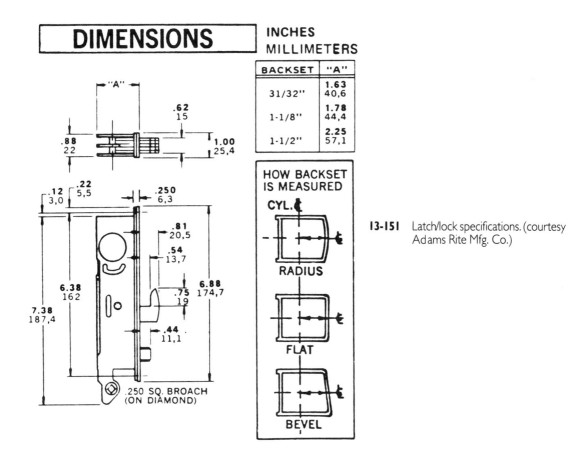

DIMENSIONS

INCHES
MILLIMETERS

BACKSET	"A"
31/32"	**1.63** 40,6
1-1/8"	**1.78** 44,4
1-1/2"	**2.25** 57,1

HOW BACKSET IS MEASURED

CYL. ℄

RADIUS

FLAT

BEVEL

13-151 Latch/lock specifications. (courtesy Adams Rite Mfg. Co.)

INSTALLATION

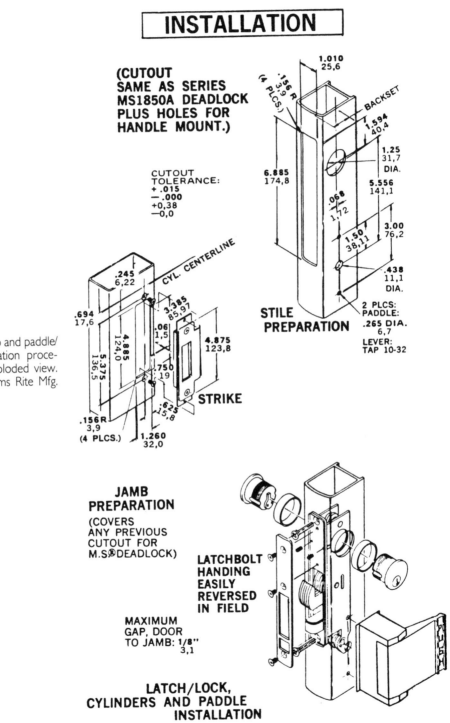

(CUTOUT SAME AS SERIES MS1850A DEADLOCK PLUS HOLES FOR HANDLE MOUNT.)

CUTOUT TOLERANCE:
+ .015
— .000
+0,38
—0,0

STILE PREPARATION

STRIKE

13-152 Metal stile jamb and paddle/cylinder installation procedure in an exploded view. (courtesy Adams Rite Mfg. Co.)

JAMB PREPARATION
(COVERS ANY PREVIOUS CUTOUT FOR M.S.® DEADLOCK)

LATCHBOLT HANDING EASILY REVERSED IN FIELD

MAXIMUM GAP, DOOR TO JAMB: 1/8"
3,1

LATCH/LOCK, CYLINDERS AND PADDLE INSTALLATION

business hours; (2) handle-operated latch for exit-only traffic; and (3) maximum security hookbolt for overnight lockup.

This has the operational advantage not necessarily provided by others which may seem similar. It requires a 360 degree turn of the key to throw or retract the hook-shaped deadbolt. When unlocked, a 120 degree further turn of the key will also retract the latchbolt as well. When using the door handle that can also be obtained for the lock, the handle retracts the spring-loaded latchbolt only.

Figure 13-151 shows the overall dimensions of the lockset. Figure 13-152 has the stile preparation, strike location and dimensions necessary, and the latch/lock, cylinders and paddle installation. Latchbolt holdback information is at FIG. 13-153.

Threshold bolts

Along with a lock installation, you might also be required to install a corresponding threshold bolt mechanism in a narrow stile door (FIG. 13-154). The use of the threshold bolt allows maximum security by the turn of a single operating key by the user for pairs of doors. With the simultaneous dropping of a hardened steel hexbolt into the threshold and pivoting of the doors' stile, the two-point lock secures the entire double door entranceway. The threshold bolt is harnessed to the rear of the pivoted bolt. The standard threshold bolt rod is sufficient for a cylinder height up to 53 $\frac{7}{17}$ inch maximum. It is fully threaded, and it can be cut off to allow for low cylinder heights.

A full turn of the key (or a thumbturn if one is installed on the interior side of the door) throws the counterbalanced bolt into the opposite door and the drop bolt into the threshold. The key can only be removed when the bolts are in a positive locked or unlocked position—not halfway! Figure 13-155 shows the dimensions, FIG. 13-156 the installation of the threshold bolt, and FIG. 13-157 illustrates the header bolt installation.

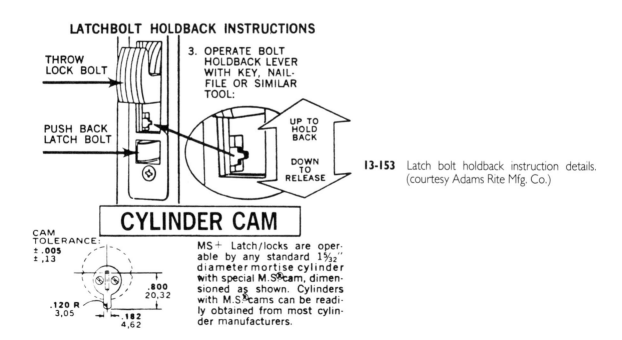

LATCHBOLT HOLDBACK INSTRUCTIONS

THROW LOCK BOLT

3. OPERATE BOLT HOLDBACK LEVER WITH KEY, NAIL-FILE OR SIMILAR TOOL:

PUSH BACK LATCH BOLT

UP TO HOLD BACK

DOWN TO RELEASE

CYLINDER CAM

CAM TOLERANCE:
± .005
± ,13

.800
20,32

.120 R
3,05

.182
4,62

MS+ Latch/locks are operable by any standard 1⁵⁄₃₂″ diameter mortise cylinder with special M.S. cam, dimensioned as shown. Cylinders with M.S. cams can be readily obtained from most cylinder manufacturers.

13-153 Latch bolt holdback instruction details. (courtesy Adams Rite Mfg. Co.)

13-154 Adams Rite threshold bolt (left) and header bolt (right); the unit has the MS1850A maximum security lock installed. (courtesy Adams Rite Mfg. Co.)

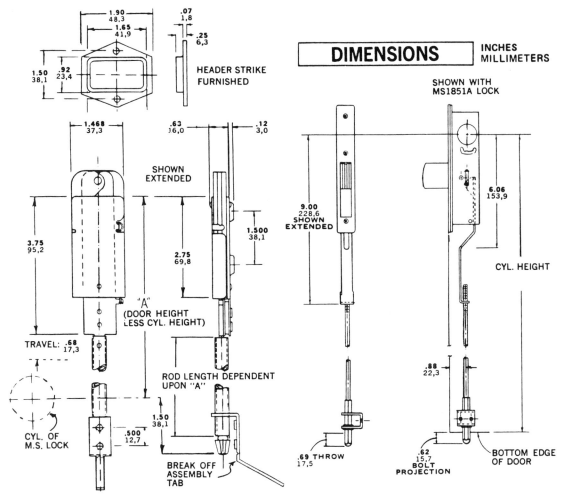

13-155   Threshold bolt and header critical dimensions. (courtesy Adams Rite Mfg. Co.)

Narrow stile door deadlock strikes

Three basic strikes are available. The majority of Adams Rite deadlocks are installed in metal construction where the strike cutout can be simply a slot. For aesthetic reasons, or, in the case of certain locks, for added security, specific strikes should be used.

The trim strike (FIG. 13-158) is a simple strike plate that can be surface-mounted on a wood or metal jamb or mortised flush. It uses #10 flathead machine screws and/or two, #10 wood screws. It is made of aluminum.

The box strike has a dust box added (FIG. 13-159). It is customarily used only for wood construction where the dust box prevents chips, sawdust, and other debris from entering the strike.

The armored strike (FIG. 13-160) has the basic trim plate, but this is backed up by a massive steel doubler designed to prevent the method of forced entry known as "jamb

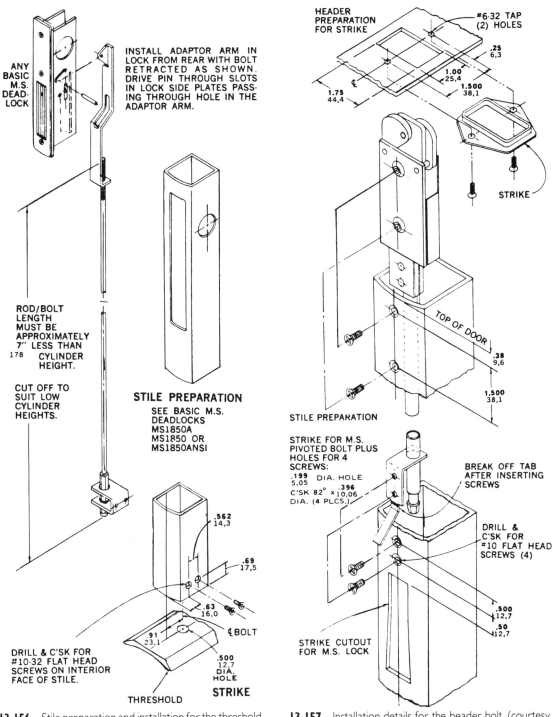

ANY BASIC M.S. DEAD-LOCK

INSTALL ADAPTOR ARM IN LOCK FROM REAR WITH BOLT RETRACTED AS SHOWN. DRIVE PIN THROUGH SLOTS IN LOCK SIDE PLATES PASS-ING THROUGH HOLE IN THE ADAPTOR ARM.

ROD/BOLT LENGTH MUST BE APPROXIMATELY 7″ LESS THAN CYLINDER HEIGHT.
178

CUT OFF TO SUIT LOW CYLINDER HEIGHTS.

STILE PREPARATION
SEE BASIC M.S. DEADLOCKS MS1850A MS1850 OR MS1850ANSI

.562
14,3

.69
17,5

.63
16,0

¢ BOLT

DRILL & C'SK FOR #10-32 FLAT HEAD SCREWS ON INTERIOR FACE OF STILE.

.91
23,1

.500
12,7
DIA.
HOLE

THRESHOLD

STRIKE

13-156   Stile preparation and installation for the threshold bolt. (courtesy Adams Rite Mfg. Co.)

HEADER PREPARATION FOR STRIKE

#6-32 TAP (2) HOLES

.25
6,3

1.00
25,4

1.500
38,1

1.75
44,4

STRIKE

STILE PREPARATION

TOP OF DOOR

.38
9,6

1.500
38,1

STRIKE FOR M.S. PIVOTED BOLT PLUS HOLES FOR 4 SCREWS:
.199 DIA. HOLE
5,05
C'SK 82° x .396
10,06
DIA. (4 PLCS.)

BREAK OFF TAB AFTER INSERTING SCREWS

DRILL & C'SK FOR #10 FLAT HEAD SCREWS (4)

.500
12,7

.50
12,7

STRIKE CUTOUT FOR M.S. LOCK

13-157   Installation details for the header bolt. (courtesy Adams Rite Mfg. Co.)

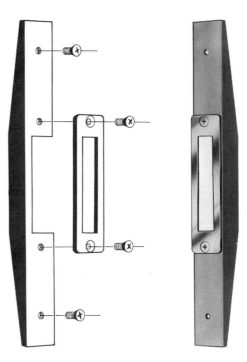

13-158 Simple trim strike plate for deadlocks in a metal stile door unit. (courtesy Adams Rite Mfg. Co.)

13-160 Armored strike unit, essentially used to prevent jamb peeling types of forced entry. (courtesy Adams Rite Mfg. Co.)

13-159 Box strike, customarily for wood frames where the dust box permits debris from entering the strike. It can also be used on metal door frames. (courtesy Adams Rite Mfg. Co.)

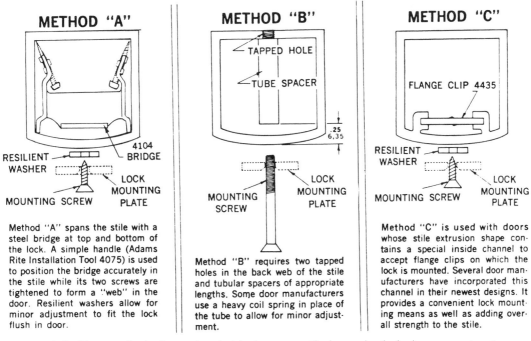

METHOD "A"

RESILIENT WASHER

4104 BRIDGE

LOCK MOUNTING PLATE

MOUNTING SCREW

Method "A" spans the stile with a steel bridge at top and bottom of the lock. A simple handle (Adams Rite Installation Tool 4075) is used to position the bridge accurately in the stile while its two screws are tightened to form a "web" in the door. Resilient washers allow for minor adjustment to fit the lock flush in door.

METHOD "B"

TAPPED HOLE

TUBE SPACER

.25
6,35

LOCK MOUNTING PLATE

MOUNTING SCREW

Method "B" requires two tapped holes in the back web of the stile and tubular spacers of appropriate lengths. Some door manufacturers use a heavy coil spring in place of the tube to allow for minor adjustment.

METHOD "C"

FLANGE CLIP 4435

RESILIENT WASHER

LOCK MOUNTING PLATE

MOUNTING SCREW

Method "C" is used with doors whose stile extrusion shape contains a special inside channel to accept flange clips on which the lock is mounted. Several door manufacturers have incorporated this channel in their newest designs. It provides a convenient lock mounting means as well as adding overall strength to the stile.

13-161   Three methods of mounting a lock in the narrow stile doors using the basic component parts of the Adams Rite installation kit. (courtesy Adams Rite Mfg. Co.)

13-162   Cylinder pull set for both sides of a sliding door. (courtesy Adams Rite Mfg. Co.)

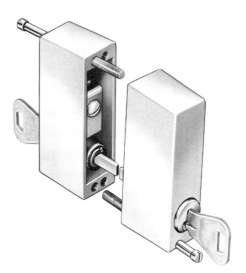

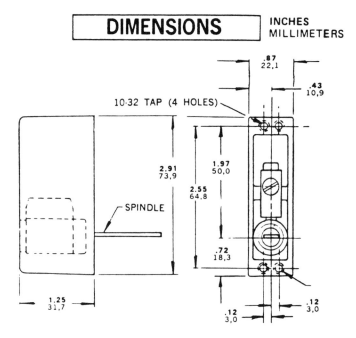

DIMENSIONS

INCHES
MILLIMETERS

10·32 TAP (4 HOLES)

.87
22,1

.43
10,9

2.91
73,9

1.97
50,0

2.55
64,8

SPINDLE

.72
18,3

1.25
31,7

.12
3,0

.12
3,0

STANDARD PULLS		SPECIAL PULLS	
PULL NO.	CYLINDER INCLUDED:	PULL	FURNISH CYLINDER:
4025	ADAMS RITE	4025-3	LOCKWOOD R-71, H-71, S-71
4025-1	NONE (DUMMY)	4025-4	SARGENT SIXLINE
4025-5	SCHLAGE 22-001	4025-8	CORBIN 471AR
4025-6	WEISER KEYWAY	4025-9	RUSSWIN AR848
		4025-10	DEXTER 9094
		4029	BEST 5A5A2
		4025-15	MEDECO 20 800

13-163 Cylinder pull dimensions breakout of cylinders that can be used with the unit. (courtesy Adams Rite Mfg. Co.)

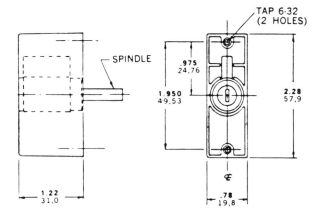

TAP 6·32
(2 HOLES)

SPINDLE

.975
24,76

1.950
49,53

2.28
57,9

1.22
31,0

.78
19,8

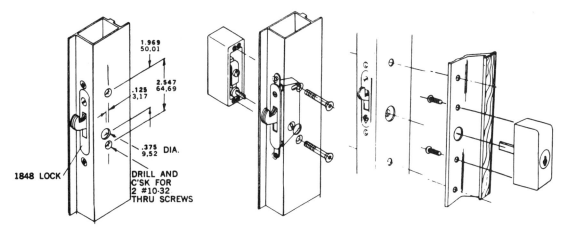

13-164 Installation of the cylinder pull; two different pull variations shown. (courtesy Adams Rite Mfg. Co.)

peeling." It fits within aluminum or other hollow jamb sections; with a trim face flush, the steel is completely hidden.

Figure 13-161 shows mounting methods for narrow stile aluminum doors. The three most prevalent methods of mounting locks and latches within the hollow tube stile of a door are basically easy.

Method A uses the steel bridge spanning the stile. A simple handle, the installation tool, is used to position the bridge accurately in the stile while its two screws are tightened to form

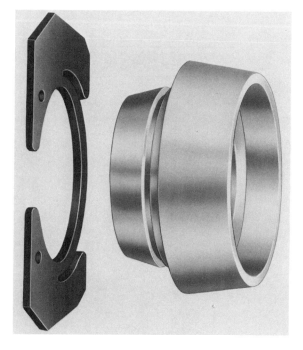

13-165 Hardened steel free-swiveling security ring and retainer plate to provide protection against direct attack on the locking unit. (courtesy Adams Rite Mfg. Co.)

a web in the door. Resilient washers allow for minor adjustment to fit the lock flush with the door.

In method B, two tapped holes in the back web of the stile and tubular spacers of the appropriate lengths are required. Some door manufacturers use a heavy coil spring in place of the tube to allow for minor adjustments.

Method C is for doors whose stile extrusion shape contains a special inside channel to accept flange clips on which the lock is mounted. Several door manufacturers have included this channel in their latest designs, so it is something you must be aware of and look for prior to installing the lock. It provides a convenient lock mounting means, as well as adding overall strength to the stile. *Note:* Refer to the business chapter for the Adams Rite Locksmithing Installation Kit which discusses some of the specific tools necessary for the three methods of mounting locks.

Cylinder pulls

The cylinder pulls (FIG. 13-162) are designed for certain deadlocks and latches. These surface-mounted pulls allow a sliding glass door to be keyed in to other types of doors which have a key-in-knob or mortise cylinder lock installed. Dimensions for the cylinder pull are at FIG. 13-163 and installation instructions shown at FIG. 13-164.

Lock cylinder guards

The standard mortise cylinder is made of brass and is literally the "soft spot" in the narrow stile door's security. Using special pliers, or other leverage devices, a burglar can tear the cylinder out of the door, leaving an opening through which he can operate the deadlock by hand. Addressing this problem is the cylinder guard by Adams Rite (FIG. 13-165). It offers a three-way defense. First, the outer shield ring offers a poor purchase for either prying or twisting. Second, the ring is of hardened steel so that the combination of shape and hardness

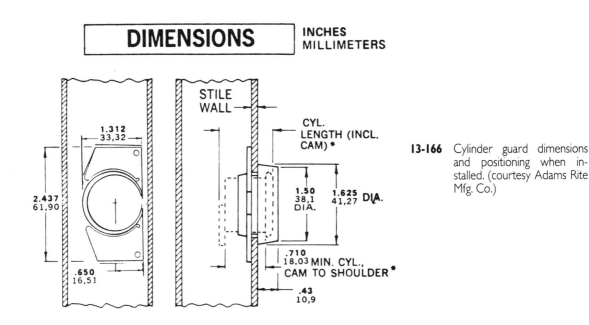

13-166 Cylinder guard dimensions and positioning when installed. (courtesy Adams Rite Mfg. Co.)

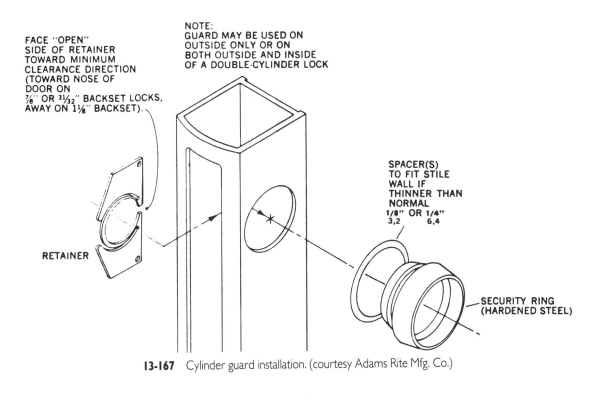

FACE "OPEN"
SIDE OF RETAINER
TOWARD MINIMUM
CLEARANCE DIRECTION
(TOWARD NOSE OF
DOOR ON
7/8" OR 31/32" BACKSET LOCKS,
AWAY ON 1 1/8" BACKSET).

NOTE:
GUARD MAY BE USED ON
OUTSIDE ONLY OR ON
BOTH OUTSIDE AND INSIDE
OF A DOUBLE-CYLINDER LOCK

RETAINER

SPACER(S)
TO FIT STILE
WALL IF
THINNER THAN
NORMAL
1/8" OR 1/4"
3,2   6,4

SECURITY RING
(HARDENED STEEL)

13-167   Cylinder guard installation. (courtesy Adams Rite Mfg. Co.)

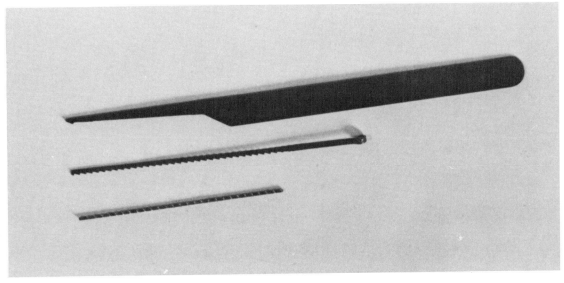

13-168   Three homemade key extractors.

makes it virtually impossible to grip, even with sharpened tools. Third, in the event that a prying tool, such as a cold chisel, can be (if ever) driven into the stile behind the shield ring, the would-be burglar is obliged to pull a heavy steel plate through the round hole in the metal door stile. This degree of leverage is far beyond that available with most hand tools.

The standard package includes the security ring which is of hardened steel (and free-swiveling when properly installed) and the steel retainer plate to permit the security ring to swivel. The latter is hardened to act as an anti-drill shield. The package also contains spacers for flush fitting in thin stile walls.

Figure 13-166 shows the cylinder guard dimensions. Figure 13-167 provides installation information.

Extracting broken keys

Figure 13-168 illustrates three homemade extractor tools. The first type, made from a length of spring stock, is inserted with the hook down. The round tip of the tool nudges the pins up, and the hook fits into the most forward key cut. The second two are made from coping or fret saw blades. They are inserted, so the teeth snag the edges or top of the key blade.

14

Auxiliary door locks

*T*he term *auxiliary* applied to door locks has two distinct meanings. It can mean a second lock acting in conjunction with a first and, hopefully, increasing the security from intruders. These locks are usually surface-mounted, in which case they are known as *rim locks or locking bodies*. Locks used on interior doors are also known as auxiliary locks. Most of these are mortise locks, i.e., they mount inside the door.

LOCKING BODIES

Locking bodies come in a variety of styles and shapes. A few of the options are shown in FIG. 14-1. The bolt may lock automatically when the door is closed. The two locking bodies at the top of FIG. 14-1 employ these self-locking latchbolts. Other locking bodies use deadbolts that must be manipulated to lock. Either type may be worked only from the inside or have a keyway on the outside of the door.

Keyless locking bodies with latchbolts have one big disadvantage: they invite lockouts.

Construction

The basic mechanism is a case (or cover), a bolt, and a bolt-actuating mechanism. Latchbolts are spring-loaded and are checked by a safety catch. Engaging the safety from the inside puts the bolt and spring in check. The door can be shut without latching.

Locking bodies are adaptable to many different doors. Since they mount on the inside surface of the door, the bevel of the door edge is unimportant. But there are times when you will have to choose a particular locking body because of the configuration of the door frame. If at all possible, use the strike intended for this lock since strikes and locks work best as a matched team. If this is not possible, be certain that the strike you use works properly. FIGURE 14-2 illustrates two of the many strikes available.

Installation

Every locking body worthy of the name comes with installation instructions, but you will sometimes find yourself working "blind." In the absence of manufacturer's instructions, follow this procedure:

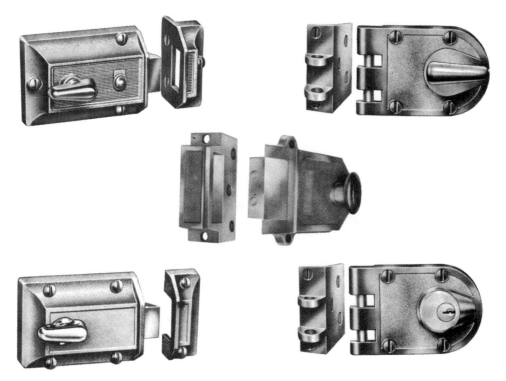

14-1   A collection of locking bodies.

Position the locking body on the door and scribe its corners and locking body screw holes.

Remove the locking body and align the strike on the door frame.

Tentatively mark the strike position.

Assuming that an outside cylinder will be fitted, drill a hole through the door at the estimated point of contact of the tailpiece (the extension on the back of the cylinder) and the bolt-activating mechanism.

Measure the diameter of the cylinder and chuck up the appropriate bit or hole saw. Most cylinders are 1⅜ inches in diameter.

Drill the hole from the outside using the first hole as a guide.

Break off the splinters and smooth the edges of the hole.

14-2   Two strikes by Corbin.

Mount the cylinder over the collar and attach it to the retaining plate on the inside of the door.

Hold the latch in position and check that the cylinder works properly. You may have to trim the tailpiece.

Check the screw holes against the locking body.

Start the screw holes with a small nail and drill 1/16 inch pilot holes 1/3 inch deep.

Mount the locking body and tighten the screws.

It may be necessary to chisel out part of the frame behind the strike. Some are surface-mounted.

The customer may want a set of instructions so he can do the work himself. In this case, the following instructions would be helpful:

14-3   The Corbin 7000 series mortised lockset.

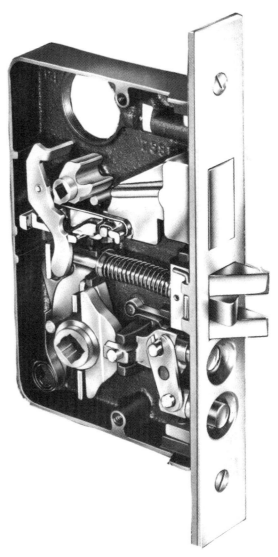

14-4   A massive heavy-duty bolt deadlock. (courtesy Taylor Lock Company)

Drill a cylinder hole in the door 2⅜ inches from the edge for doors opening inward, and 2½ inches for doors opening out.

Install the cylinder. Make sure the pins are on top of the keyway.

Position the locking body even with the edge of the door and check that the tailpiece operates the latch when the key is used.

Attach the locking body.

Install the strike.

INSIDE DOOR LOCKSETS

The Corbin 7000 series is typical of the better locksets (FIG. 14-3). Designed for exterior or interior use, it has these features:

- ☐ Case—cast iron and lacquered.
- ☐ Deadbolt—forged brass with a ⅝ inch throw. Hardened steel inserts are optional.
- ☐ Latchbolt—extruded brass with a ⅝ or ¾ inch throw, depending upon application. Antifriction latchbolts are available.
- ☐ Latch holdback (safety catch)—optional.
- ☐ Hub—cold-forged bronze.
- ☐ Strike—wrought brass or bronze with the edge of the lip ⅛ inch from the centerline as standard.

Installation

Installation of Corbin and other locksets has been discussed in the previous chapter.

The Taylor 850 (FIG. 14-4) is a massive bolt rim deadbolt specifically designed for the extra measure of protection required on doors in a home or business. The bolt measures

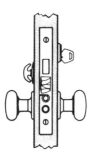

14-5   Apartment lock. (All drawings in this series courtesy of the Emhart Corporation.)

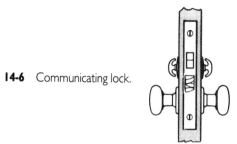

14-6 Communicating lock.

1¹¹⁄₁₆-inch high by ⅝-inch thick, which is twice the size for a conventional bolt in a locking unit of this type. In addition to a full one inch extension beyond the case, the bolt also possesses two hardened steel inserts which are intended to block sawing the bolt in half.

The strike, like the lock itself, is also massive in design and size and unlike more conventional strikes, uses five screws set on two face angles to provide for an optimum penetration and holding strength once applied.

From the inside, the deadlock is operated by a thumbturn, and from the exterior by a key. Normally, the unit is randomly keyed at the factory, but when several are required, they can be keyed alike (KA) or master keyed (MK), as circumstances warrant.

Like other units of this general type, lock installation is easy and takes an extremely short time to mount and put into operation.

VARIANTS

Indoor locksets have evolved according to function:

Apartment Lock.—The latchbolt is released by either knob, unless the outside knob is locked by the stop button or deadbolt. The stop button can be released manually at the button or by turning the inside knob. Turning the key retracts the deadbolt; turning the inside knob retracts the deadbolt and the latchbolt (FIG. 14-5).

Communicating Lock.—The latchbolt is released by either knob. A split deadbolt allows operation by either turnpiece (FIG. 14-6).

Deadlock.—There are three variations to the familiar deadlock: deadbolt operation by key from one side only (FIG. 14-7A); deadbolt operation by key from both sides (FIG. 14-7B); deadbolt operation by key on one side and turnpiece on the other (FIG. 14-7C).

14-7 Dead locks. The lock may be controlled by a single key (A), two keys (B), or by a key and a turnpiece (C).

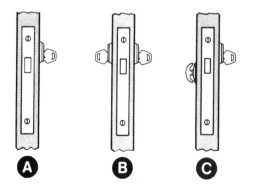

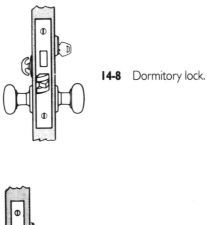

14-8   Dormitory lock.

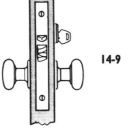

14-9   Entrance or classroom lock.

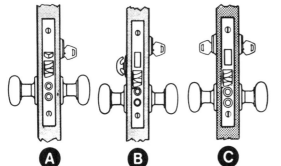

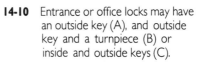

14-10   Entrance or office locks may have an outside key (A), and outside key and a turnpiece (B) or inside and outside keys (C).

Ⓐ   **Ⓑ**   **Ⓒ**

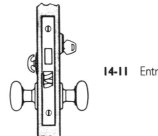

14-11   Entrance or storeroom lock.

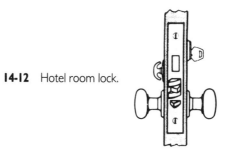

14-12   Hotel room lock.

14-13   Passage or closet latch.

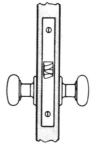

14-14   A privacy lock.

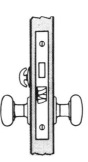

14-15   Storeroom or closet latch.

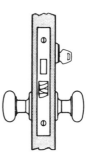

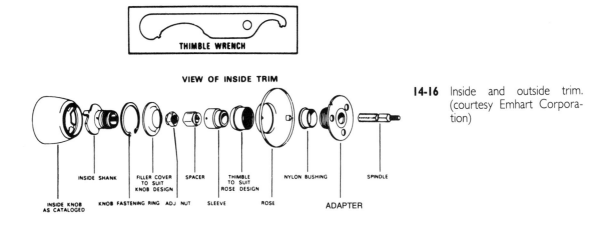

THIMBLE WRENCH

VIEW OF INSIDE TRIM

INSIDE SHANK   FILLER COVER   SPACER   THIMBLE   NYLON BUSHING   SPINDLE
                TO SUIT   TO SUIT
                KNOB DESIGN   ROSE DESIGN

INSIDE KNOB   KNOB FASTENING RING ADJ NUT   SLEEVE   ROSE   ADAPTER
AS CATALOGED

14-16 Inside and outside trim. (courtesy Emhart Corporation)

Dormitory Lock.—The latchbolt is released by either knob, unless the outside knob is locked by the deadbolt. The deadbolt deadlocks the latchbolt. The deadbolt is activated by the inside knob (which also retracts the latchbolt and unlocks the outside knob) by turning the turnpiece or the key (FIG. 14-8).

Entrance or Classroom Lock.—The latchbolt is released by the knob on either side, unless the outside key locks the outside knob. In this mode the inside knob still opens the door (FIG. 14-9).

Entrance or Office Lock.—There are three variations. The first does not have a turnpiece (FIG. 14-10A). The latchbolt is released by the key from the outside and either knob, unless the outside knob is locked by the key. Alternately, this same type of lockset may be fitted with a turnpiece on the inside (FIG. 14-10B). In this case, the latchbolt is worked by the key on the outside and the knob on either side, unless the outside knob is locked by the key. The deadbolt is released by the inside turnpiece and the outside key. The third variation has a key on both sides (FIG. 14-10C) and works like the lock just described, except that the key, rather than the inside turnpiece, works the deadbolt.

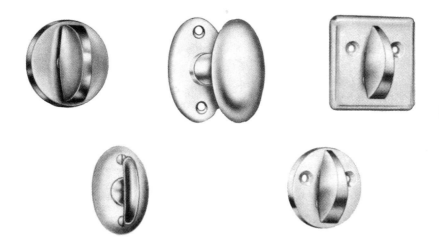

14-17 Turnpieces by Corbin.

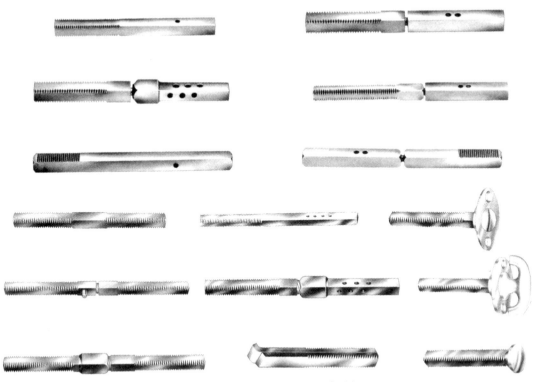

14-18 Knob and lever spindles by Corbin.

Entrance or Storeroom Lock.—The latchbolt is activated by the knob on either side. The deadbolt operates by the outside key or the inside turnpiece (FIG. 14-11).

Hotel Room Lock.—The deadbolt is activated by the turnpiece on the inside. The latchbolt is retracted by the inside knob and by the guest, master, and grand master keys. The inside knob retracts the deadbolt and latchbolt simultaneously. Only the emergency key can release the deadbolt from the outside (FIG. 14-12).

Passage or Closet Latch.—Turning either knob releases the latchbolt (FIG. 14-13).

Privacy Lock.—The latchbolt is released by either knob. The deadbolt is activated by the inside turnpiece or by the inside knob (FIG. 14-14).

Storeroom Lock or Closet Lock.—The latchbolt is released by either knob. The deadbolt is activated by a key on the outside (FIG. 14-15).

HARDWARE

Figure 14-16 illustrates a sample of inside and outside door trim offered by one manufacturer. Figure 14-17 shows five turnpiece (thumb knob or turn knob) options from the hundreds that are available. Many options are purely cosmetic; others are mechanical parts that must be replaced with the exact equivalent. Prime examples are knob and lever spindles (FIG. 14-18). You would do well to stock a variety of these parts (FIG. 14-19).

Every locksmith, especially the beginning one, must have a variety of spindles and knob screws on hand, for repair, direct sales, or for specific job requirements. With so many

FOR WOOD DOORS

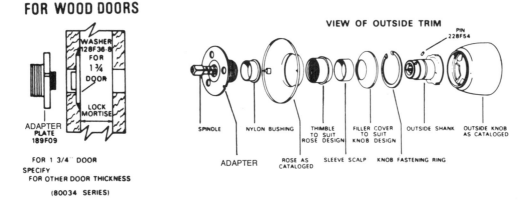

14-19   Knob mechanism, exploded drawing.

SPINDLE STYLES AND KNOB SCREWS

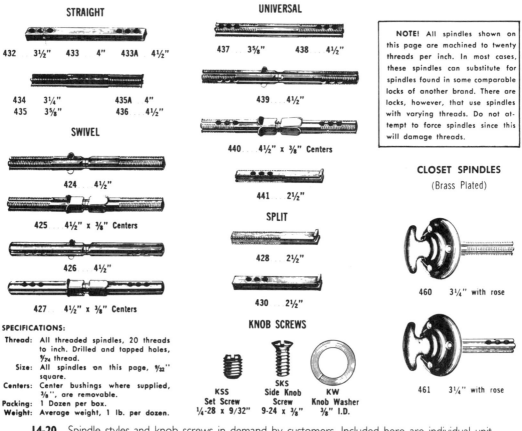

14-20   Spindle styles and knob screws in demand by customers. Included here are individual unit dimensions allowing you to select a correct spindle to fit a specific lockset. (courtesy Tylor Lock Company)

different types available, you cannot survive without the variety immediately required. Figure 14-20 shows the technical specifications of the various types of spindles and knob screws.

These spindles are threaded at 20 threads to the inch. They have drilled and tapped holes at $\%_{24}$ thread and are $\%_{32}$-inch square. In most cases, these spindles can substitute for spindles found in other comparable locks of various brands. There are locks, however, that use spindles with varying threads. Do not attempt to force spindles, because this will damage the threads.

The set screws, knob washers, and closet spindles are quite often used for specific lock installations, and these sizes should be kept on hand, usually in the same storage box with or next to the spindles, because they are closely related and used together.

15

Office locks

*O*ffice locks represent an important and lucrative market and one that is easy to overlook. An important part of the security of every business depends upon office locks. A profusion of manufacturers are in this field, making thousands of different locks, each with particular features. But almost all of these locks fall into three groups: rotary-cam locks, sliding-bolt locks, and plunger locks.

ROTARY-CAM LOCKS

Simplest of office equipment locks, most rotary-cam locks employ disc tumblers, although you will occasionally run across one with pin tumblers. The cam is oval and rotates with the cylinder. Those with a 90 degree throw must be locked for key removal; 180 degree throw locks can be unkeyed in the locked or unlocked position (FIG. 15-1).

The lock body may be threaded (as shown in the illustration), or it may be secured by a retaining clip. The cam may be integral with the lock, or it can be secured by a screw and washer. A removable cam is an advantage, since it allows the same lock body to be used in a variety of applications. Locks that turn 90 degrees have a single tumbler chamber; locks that turn 180 degrees employ a double chamber and may have a stopping pin, notch, or blocking member at the rear that limits rotation. In some cases, it is possible to change the rotation in the field. The lock may then be used in other applications.

SLIDING-BOLT LOCKS

Most of these locks are disc tumblers; although, pin tumblers are not unknown. There is no cam; the bolt is grooved to accept a projection on the back of the plug (FIG. 15-2). The projection engages the groove and converts the rotary motion of the plug into reciprocating (to and fro) motion. The bolt is generally heavier and more durable than those of other office locks. This arrangement allows the key to rotate 180 degrees; the key can be withdrawn in either the locked or unlocked position.

The most serious sliding-bolt lock is known as the T-bolt because of its shape (FIG. 15-3). The locks almost always employ a disc tumbler cylinder with the bolt-actuating pin cast as part of the plug. The plug can generally be released with a probe wire.

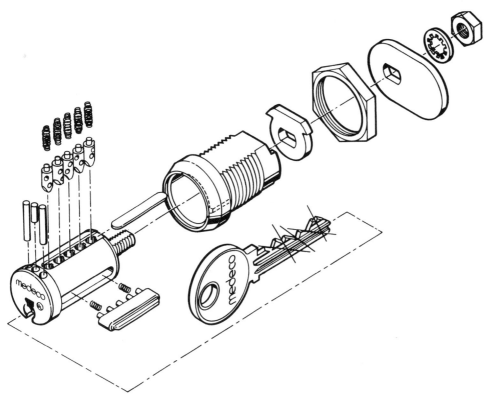

15-1   A Medeco high-security cam lock.

If the key is not available and the bolt is thrown, you have three options. The best is to pick the lock. Once the drawer is open, removing the retainer is no problem. Another approach is to drill through the drawer. Figure 15-4 shows the approximate drilling point. Drill no deeper than ⅞ inch to avoid damage to the lock itself. Finally, you can drill out the plug, destroying the lock in the process.

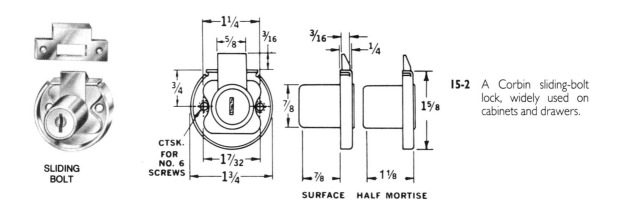

SLIDING
BOLT

CTSK.
FOR
NO. 6
SCREWS

SURFACE   HALF MORTISE

15-2   A Corbin sliding-bolt lock, widely used on cabinets and drawers.

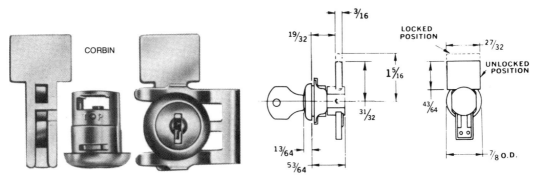

15-3 A Corbin T-bolt lock, generally stronger than other sliding-bolt locks.

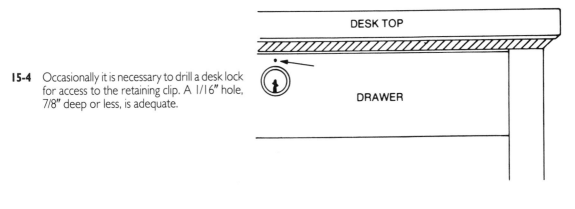

15-4 Occasionally it is necessary to drill a desk lock for access to the retaining clip. A 1/16" hole, 7/8" deep or less, is adequate.

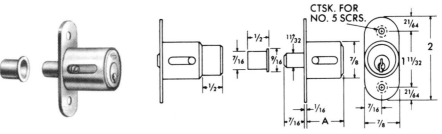

15-5 A Corbin plunger lock, intended to secure sliding doors.

PLUNGER LOCKS

Plunger locks are usually found on sliding cabinet doors. These locks are mounted on the outside door and the bolt, extending from the rear of the plug, engages a hole in the other door (FIG. 15-5). The doors are locked against each other, so neither can be moved. A spring returns the plug to the open position when the key is turned. Although disc tumbler types once were popular, pin-tumblers are almost universal today. Locksmiths new to the trade might never have seen a disc-plunger lock.

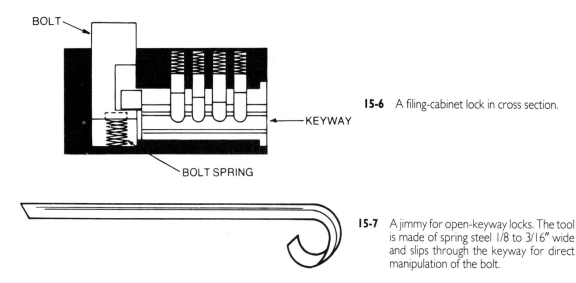

15-6   A filing-cabinet lock in cross section.

15-7   A jimmy for open-keyway locks. The tool is made of spring steel 1/8 to 3/16″ wide and slips through the keyway for direct manipulation of the bolt.

The plug is usually secured by a screw located at the side of the cylinder and set in a notch in the lock (FIG. 15-5). The screws serves a double purpose; it determines the throw of the bolt, and aligns the plug with the cylinder.

The bolt is a projection from the rear of the plug unit. Without the key you can pick the lock or force the doors far enough apart to disengage the bolt. The latter method is the faster of the two.

FILING-CABINET LOCKS

Filing-cabinet locks are fairly well standardized and use a bolt extending from the top of the lock body (FIG. 15-6). The bolt manipulates the control bars that lock each drawer of the cabinet.

The bolt is spring-driven; unless the plug is turned, the bolt extends out of the locking body. In other words, the drawers are automatically locked.

The bolt may be round or rectangular in cross section. The lower face of the bolt—the part buried in the cylinder—is recessed for the spring. The plug cam works in a notch on the side of the bolt.

Open keyways are generally used, as opposed to blind, or masked, keyways. The keyway runs from the face through the plug body. You can open the lock with relative ease. Use the tool shown in FIG. 15-7 to work the bolt directly.

During the past decade, new standards have been set for filing-cabinet locks; many manufacturers have revised their dies to do away with the open keyway and the breach in security it represents. This means that today many filing-cabinet locks have bolts that are isolated from the keyway. A few manufacturers use their original dies but add a fifth pin. This pin is longer than the others and has no upper chamber. It cannot be lifted, and so it blocks access to the bolt. Another option has been to block the keyway end with a horizontal pin. The jimmy no longer works, and we must turn to the more traditional methods of opening a keyless lock. There are three choices: pick the lock, drill the plug, or force the bolt.

Picking is the most desirable of these alternatives, but requires patience and skill. Drilling is fast and means the loss of the lock, but it is preferable to forcing the bolt. If the customer accepts the risk, pry one drawer open far enough to manipulate the bolt with a letter opener.

16

Automotive locks

While the automobile has been with us since the start of the century, the lock was slow to be adopted. However, by the late 1920s nearly every auto had an ignition lock, and closed cars had door locks as well. Current models may be secured with half a dozen locks.

SIDEBAR LOCKS

While automobiles may be fitted with pin- or disc tumbler cylinders, the sidebar lock is most typical and the only one that merits special attention. The other two have been discussed earlier.

FIGURE 16-1 illustrates this lock in simplified form. Notice that the wafers have V-shaped notches in their sides. Inserting the correct keys aligns the notches, and the spring-loaded sidebar moves out of the cylinder and into the plug. Once the sidebar passes the shear line, the plug is free to rotate.

KEY FITTING

Key fitting for disc and pin-tumbler locks has been described in earlier chapters. The techniques used for automotive variations of these locks are no different.

Key fitting sidebar locks requires that the lock be disassembled. Remove the lock from the vehicle and, with a small screwdriver or other sharp tool, release the staking that holds the sidebar and associated spring in place.

Remove the tumblers and their springs. Insert the appropriate blank into the keyway and replace the tumbler at the far end of the plug. Working from the sidebar chamber, scribe a mark on the blank where the blank touches the tumbler. This is the guide mark for the cut. File the mark lightly and reinsert the blank; the file mark should be at or under the tumbler. File slowly until the V-notch on the side of the tumbler appears to align with the sidebar. Hold the sidebar and spring temporarily in place with your fingers and try the key. If all is right, the sidebar will extend into the notch, freeing the plug. The key should turn.

Repeat this operation for each of the remaining tumblers.

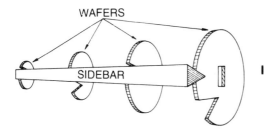

WAFERS

SIDEBAR

16-1   A simplified drawing of a sidebar lock.

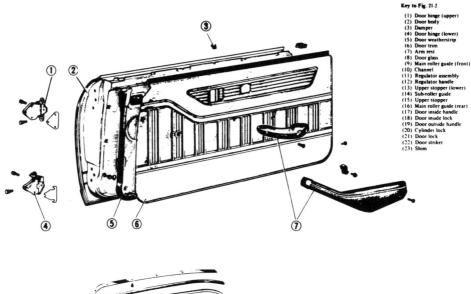

Key to Fig. 21-2

(1) Door hinge (upper)
(2) Door body
(3) Damper
(4) Door hinge (lower)
(5) Door weatherstrip
(6) Door trim
(7) Arm rest
(8) Door glass
(9) Main roller guide (front)
(10) Channel
(11) Regulator assembly
(12) Regulator handle
(13) Upper stopper (lower)
(14) Sub-roller guide
(15) Upper stopper
(16) Main roller guide (rear)
(17) Door inside handle
(18) Door inside lock
(19) Door outside handle
(20) Cylinder lock
(21) Door lock
(22) Door striker
(23) Shim

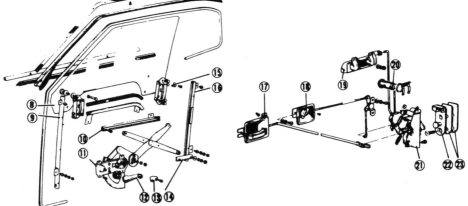

16-2   A typical front-door assembly. (courtesy Chrysler Corp.)

DOOR LOCKS

Modern door locks are disassembled from inside the door. Figure 16-2 shows a typical example. Remove the arm rest (shown as No. 7 in the drawing) and inside handles. Most handles are secured by a spring clip that is accessible with the tools shown in FIG. 16-3. Note the positions of the handles on their spindles for assembly.

Remove the Phillips screws holding the door trim (6) to the door body (2). Gently pry the trim away from the body and set it aside. The door lock mechanism is visible through access holes in the inner door-body panel.

Most problems involve the lock mechanism and not the cylinder. Check for broken springs, missing retainer clips, and rust on working parts. The cylinder (20) is secured by a U-shaped retainer clip. Cylinders can be disassembled, although the modern practice is to replace the cylinder as a unit, rather than to attempt repairs.

The striker can be moved to align the door with the body panels and to compensate for door sag. Loosen the mounting screws a quarter turn or less and, using a soft mallet, tap the striker into position (FIG. 16-4). Shims are available from the dealer to move the striker laterally, out from the door post.

While this is out of the province of locksmithing, as such, it is useful to know that the door can be adjusted at the hinge on all automobiles except Vega (FIG. 16-5). Loosen the pillar-side bolts slightly and manipulate the door.

IGNITION LOCKS

Early production ignition locks are secured to the dash by bezel nuts or retainer clips and feature a "poke hole" on the lock face or in the back of the cylinder that gives access to the retaining ring. In a few cases, the lock is held by screws from the underside. Late-model locks, (the kind that lock the steering wheel as well as the ignition) are not something that

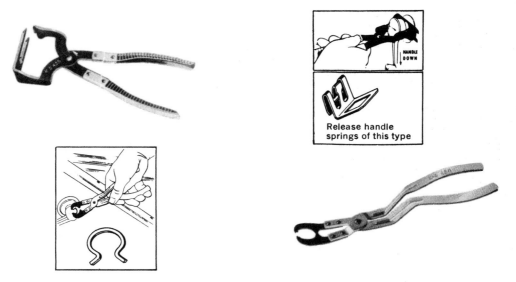

16-3 These K-D door handle tools are available from auto parts stores. View A illustrates the tool for Chrysler retainers; view B, the tool for Ford and General Motors.

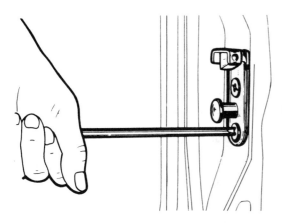

16-4　Door strikes are adjustable. (courtesy Chrysler Corp.)

the average locksmith should become involved with. These locks can be serviced, but special tools and special knowledge are required. The upper portion of the steering column must be dismantled and, in some cases, the steering column must be dropped from the dash hanger. A mistake in assembly—even so much as using the wrong screw—can defeat the collapsible steering column. If you want to get into this aspect of the trade, by all means spend a few weeks at dealer schools and invest in the proper tools.

GLOVE COMPARTMENT LOCKS

Modern glove-compartment locks are typically secured by a retainer accessible from the front of the lock. Figure 16-6 shows how this is done. Make a ¼″ hook in the end of a piece of Bowden (hood-latch) cable. Insert the hook into the keyway and work the retainer free. Most are installed with the open ends toward the passenger door; nudge the retainer toward the center of the car to disengage it.

Extract the cylinder with the key and release the bolt by hand. The cylinder is stamped with a numerical keying code—the same code that is used for the trunk or, in the case of station wagons, the tailgate.

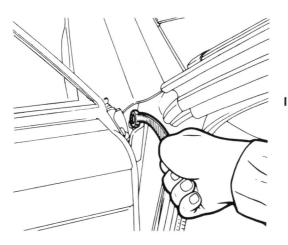

16-5　The door adjustment is at the pillar side of the hinges for most vehicles. (courtesy Chrysler Corp.)

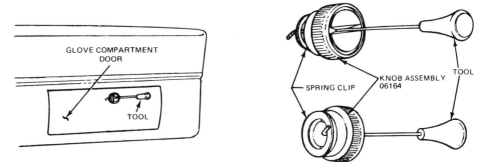

16-6 The glove-box cylinder can be removed from the outside with the tool shown. (courtesy Ford Motor Co.)

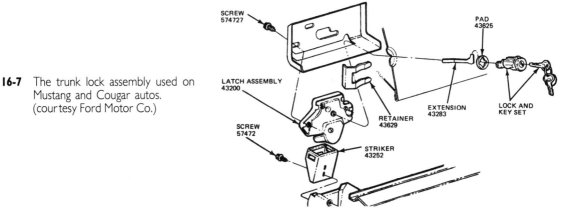

16-7 The trunk lock assembly used on Mustang and Cougar autos. (courtesy Ford Motor Co.)

TRUNK LOCKS

Figure 16-7 shows a luggage-compartment latch, typical of those found on middle-priced automobiles. More expensive cars use a more complex lock mechanism that may be solenoid-tripped from inside the vehicle. The basic parts and their relationships are the same. The cylinder is secured to the trunk lid by means of a retaining clip and works the lock assembly through a tailpiece or extension. (A few cars mount the cylinder on the lower body panel and the striker on the lid.) The striker is located on the aft body panel.

Most lock assemblies are adjusted only at the striker plate. The plate is mounted on elongated bolt holes and can be moved up, down, left, and right. Other types have an additional adjustment at the latch. The idea is to position the parts so the lid closes without interference from the striker and tightly enough to make a waterproof seal between its lower edge and weatherstripping.

17

Decoding, picking and emergency entry

Another skill required by the locksmith is the ability to cut different key types when only the lock manufacturer and key number are known. The locksmith might never see the lock or key; at best he might have only the lock. Certain information may be required of the customer, depending on city or county ordinances. In most cases you will need to find out who owns the lock. Has the owner given authorization to have a key made for the lock? Requests involving high-security locks must be forwarded to the factory together with a written release from the owner.

The code varies with each manufacturer, but includes this information:

□ Key blank type.

□ Cut spacing.

□ Cut depths.

READING CODES

Direct-digit and indirect codes are most popular. The direct-digit code consists of one or two letters followed by a series of numbers. The letters identify the lock series, and the numbers refer to the depths of the cuts. The series may start at the last tumbler (the one closest to the key tip), or at the first (next to the key shoulder). Indirect codes are safer; the numbers must be decoded before the cuts can be identified. Professional locksmiths have the necessary code books.

The smallest number—either 0 or 1—in both kinds of codes represents the shallowest cut; 9, the last of the series, represents the deepest cut. Once you have the number, you can usually determine the type of lock. For example, key code LL76 is limited to desks. This fact can be useful; in that, it narrows the range of possible blanks.

The code book says that LL76 is a Yale desk lock and that you can use ILCO 01122A, National Y12, or Yale 9278A blanks. The depth and spacing charts show five depths, each 0.020-inch apart, and five spacings. Counting from the tip to the shoulder, the first four cuts are spaced 0.094-inch apart; the last cut is 0.140-inch from the shoulder. Code number LL76 translates as 24253, and it is read from the tip to the shoulder. To cut a key with this code, place a Yale No. 2 depth and spacing key in the vise, together with the appropriate blank. Cut the No. 2 bits. Repeat the operation with Nos. 3, 4, and 5 depth and spacing keys.

Depth and spacing keys

Depth and spacing keys simplify the work. These keys are milled to various depths and spacings. Because manufacturers vary these dimensions, you need a complete key set, or at least one that covers the major American locks.

Schlage wafer-tumbler keys

Schlage keys are easy to cut by code since the cuts all have the same depths. Three Schlage keys and one depth key are required.

LOCKPICKING

Lockpicking is the ultimate test of a locksmith; it tries his knowledge as well as his skill. To many customers, lockpicking is the one ability that distinguishes a true locksmith from someone who merely tinkers with locks.

LOCKPICKING TOOLS

Basic lockpicks are illustrated in FIG. 17-1 and include the following:

- Half-round feeler
- Round feeler
- Rake
- Half-rake
- Diamond
- Double-diamond
- Circular
- Reader
- Extractor
- Mailbox
- Flat lever

You can make your own lockpicks from flat cold-rolled steel, 0.020 to 0.025-inch thick. The strip should be about 6-inches long and at least ⅓-inch wide. One end may be fitted with a handle, or both ends may carry working surfaces.

Picks for warded locks come in various designs also. A few of them are shown in FIG. 17-2. It's usually a simple matter to make one of these picks. Sometimes a standard precut key shaved to pass the keyhole wards is all that is needed.

Tension tools for use with various picks are made of hardened spring steel. Normally they are from three to five inches in length, excluding the parts that actually fit into the keyway opening.

A letter box tension tool is made from a thin strip of spring steel about four inches long and has a ⅛ inch to ¼ inch probe at its working end. It is best to have several with different probe lengths to ensure that you always have the correct one on hand.

Picks and tension tools for double-sided disc tumbler locks are unique in their design (FIG. 17-3). They should be purchased from your locksmith supply house; it's cheaper to buy them than to make them.

Lever lockpicks come in two varieties—those for true lever locks and those that double as key picks. The true lever lockpick is made from 0.095 inch, or thinner spring steel, and it

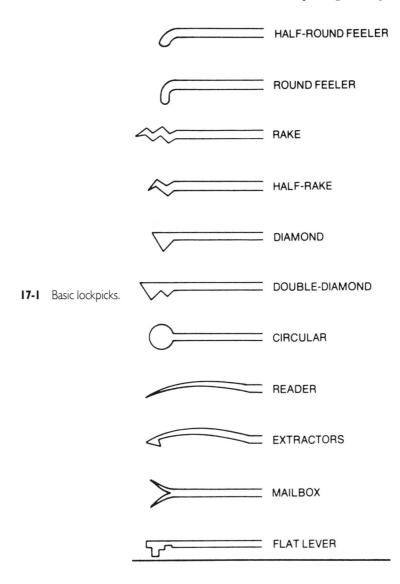

17-1 Basic lockpicks.

HALF-ROUND FEELER

ROUND FEELER

RAKE

HALF-RAKE

DIAMOND

DOUBLE-DIAMOND

CIRCULAR

READER

EXTRACTORS

MAILBOX

FLAT LEVER

has a tension tool that is the same thickness (FIG. 17-4). The second variety is cut from spring steel, or it is made from standard key blanks.

Note: It is illegal to possess lockpicks without a locksmith's license in many jurisdictions. The penalties are severe, even for first offenders.

LOCKPICKING TECHNIQUES

The prerequisite for successfully picking a lock is a sound knowledge of how the lock mechanism works. Without such an understanding, picking a lock becomes a matter of sheer luck.

17-2   Picks for warded locks.

Picking the pin-tumbler lock

To pick a pin-tumbler lock you must be able to determine when the cylinder pins are at the shear line. Apply just the right amount of pressure on the cylinder core with a tension tool. Figure 17-5 shows a feeler pick raising a pin to the shear line. To raise the pins in this way requires constant practice. Too much or too little pressure on the pick will cause the pin to stick at the wrong place (FIG. 17-6).

Factory precision in drilling the cylinder pin holes is never perfect. An enlarged view of the top of a plug would show that the holes are not really in alignment (FIG. 17-7). This is of help in opening the lock. Since the holes are not perfectly aligned, you can hold one pin at the shear line while working on others. Another built-in advantage is the fact that locks

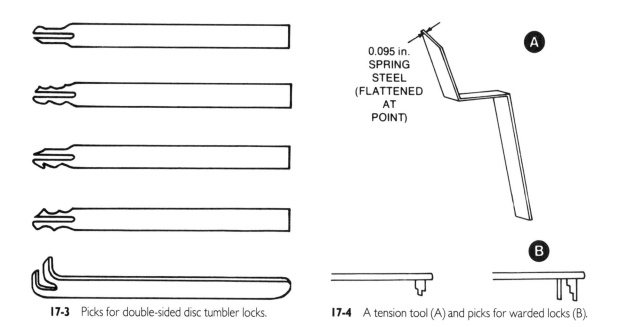

17-3   Picks for double-sided disc tumbler locks.

17-4   A tension tool (A) and picks for warded locks (B).

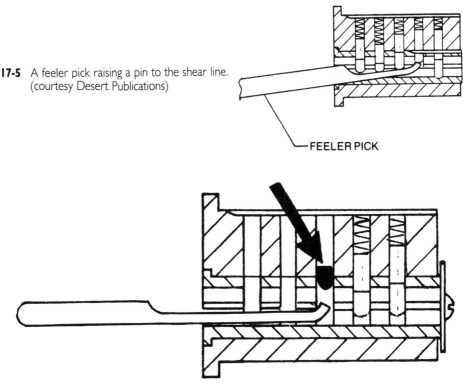

17-5 A feeler pick raising a pin to the shear line. (courtesy Desert Publications)

FEELER PICK

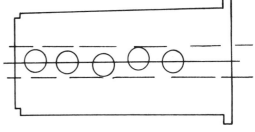

17-6 Too much pressure with a pick will raise a pin above the shear line.

17-7 Cylinder pin holes are never in perfect alignment.

always have at least some clearance between moving parts—and a little working space is all a good locksmith needs to pick a lock.

In raising the various pins, it is preferable to start with those of the greatest length. By using this method, you progress from the smallest amount of pick movement up to the greatest. As each pin reaches the shear line, the core moves ever so slightly; you will be able to feel this movement when it is transferred to your tension tool.

Using a feeler pick, move one pin at a time. Since the purpose of the pick is to move the individual pins up *to* the shear line, not *above* them, you must take your time. If you feel that all but one of the pins have reached the shear line, and the last one just won't go, you have probably pushed the last pin above the shear line. Release the tension and start again, with slightly less tension than before.

Remember that you must vary the amount of pressure you apply to the pick, depending upon the amount of resistance you receive from each pin. Also, different locks require different amounts of pressure. What works on one lock will not necessarily work on the next. Even two locks of the same make and style may not require the same amount of pressure.

It is sometimes possible to use picks (especially the rakes) to "bounce" the pins to the shear line. To do this, insert the pick fully then withdraw it quickly while holding light tension on the plug. Simply forcing the pick rapidly in and out of the cylinder won't do the job: Such a technique merely bounces the pins *above* the shear line.

Picking warded locks

Warded locks are very easy to pick. Sometimes you can even use a pair of wires: one for throwing the bolt, the other for adjusting the lock mechanism to the proper height for the bolt to be moved. Here, as with any other kind of lockpicking, it is a matter of practice on a variety of locks.

Warded padlocks have only one or two obstacles, thus simplifying your choice of pick. Some warded locks (such as the Master) which have keyways with corrugated cross sections can be opened in no time at all by having several precut blanks available. The regular picks are usually just a fraction of an inch too wide to fit into the keyway.

Smaller locks with correspondingly smaller keyways require that you either get a second set of tools and cut them down to size, or use several blanks to obtain the proper picks for these locks.

Picking lever-tumbler locks

Figure 17-8 shows a lock with the frontplate removed, indicating the position of the tension tool in relation to the bolt. Holding the levers with the tension tool (as shown in the side view) enables you to manipulate the levers with the pick.

In learning to pick the lever lock, it is best to start with a lock with only one lever in place. At first, work on the lock with the faceplate removed so you can get an idea of how much pressure to apply to the bolt, and how much movement is required of the pick to move the lever into position for the bolt to move through the lever gating.

As you progress to locks with several levers, keep the faceplate on. When you encounter a problem, remove the faceplate so you can see what you're doing wrong. Always remember to insert the tension tool first. Push it to the lowest point within the keyway so the pick will have maximum working space.

As with the pin-tumbler lock, one tumbler tends to take up most of the tension; work on this one first. As you move this tumbler slowly upward to the proper position for the bolt to

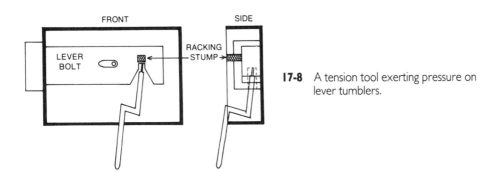

17-8 A tension tool exerting pressure on lever tumblers.

pass through the gate, you will feel a slight slackening in tension from the bolt as it attempts to force its way into the gating. This tension will be transmitted to you through the tension tool. Stop at this point and do the same to the next lever with the greatest amount of tension. Continue until all the levers reach this point. Then by shifting the tension tool against the bolt, the bolt will pass through the gate, opening the lock.

There must be at least some pressure on the tension tool at all times. A lack of pressure will cause any levers in position to drop back into their original locations.

Picking disc tumbler locks

Standard disc tumbler locks (those using a single-bitted key) take the same picks as pin-tumbler locks. Also like the pin-tumbler lock, a disc tumbler lock can be picked by bouncing the tumblers to the shear line. Usually a rake is the best tool for this job, but other picks can be used too.

In order to develop your proficiency, try opening the disc tumbler lock with the feeler pick, working on each tumbler individually. By doing this you learn which picks you are most comfortable with, and which ones you prefer for different types of locks.

When working on double-bitted disc tumbler locks, the bounce method is best. Insert the tension tool into the keyway and apply a slight pressure to the core as the pick is pulled out of the lock. These locks can also be opened with a standard pick set and tension tools, but it can take from ten minutes to half an hour.

LOCKPICKING CONSIDERATIONS

The following are points to remember when picking any lock:

□ Use the narrowest pick available in order to give yourself maximum working space.

□ Hold the pick as you would a pen or pencil. Don't use wrist action; your fingers work much better in manipulating the pick in the lock.

□ Steady your hand with your little finger against the door. When working on a key-in-knob cylinder, steady your hand against the edge of the knob.

□ Your pick should enter the keyway above the tension tool without moving any of the tumblers. If it doesn't, the tension tool is either too high in the lock, or the keyway grooves are such that the tension tool must go at the very top of the keyway.

□ Worn cylinders and loose plugs frequently open easier and quicker with the bounce method—normally within three to five tries.

EMERGENCY ENTRY PROCEDURES

Sometimes locksmiths gain entry into locked premises by picking open the lock. But a good locksmith can also gain entry through special techniques called Emergency Entry Procedures (EEP). In some instances, a lock resists picking. It is then that the EEP are used.

Emergency Entry Procedures are also called *special techniques, emergency measures,* and *quick entry*. Much of the information and illustrations in this chapter come from Desert Publications' book *Lock Out*.

As a professional locksmith, you should have the proper equipment to make emergency entries as quickly and as neatly as possible. Some entry tools are manufactured; you should purchase a set of these for your use. Some tools you can fashion yourself. Experience will show you which tools and techniques work best for you.

Using Emergency Entry Procedures can account for a substantial percentage of a locksmith's income; thus, it is wise to be knowledgeable in a variety of entry techniques. It might not seem like good locksmithing practice to force a lock open, but economics, time, and other requirements may make it necessary.

Several preliminaries should be undertaken when preparing to make an entry using EEP. Try to obtain as much information as possible. With this information, you can decide what methods will work. Ask questions about the type of lock the key numbers, nearby open windows, a possible extra key held by someone else, the condition of the lock, alarms and other devices attached to the door, etc. All this information will aid you in preparing for the entry. Fortunately, most lockouts require only that you pick the lock or make a new key.

DRILLING PIN-TUMBLER LOCKS

Drill a lock only as a last resort. There are two drilling methods used with pin-tumbler locks. The first method is the drilling of the cylinder plug. This has the advantage of saving the cylinder itself. The inner core can be replaced. This method destroys the lower set of pins just below the shear line, allowing the plug to be turned. Follow these procedures for this method:

Drill the plug below the shear line.

Insert a key blank and a wire through the drill hole to keep the upper pins above the shear line and the destroyed pins below it.

Turn the cylinder core and open the lock.

Once the door is open, dismantle the lock; remove and replace the core and fit new pins to the lock. If you used the plug follower when removing the core, you will only have to fit new lower pins to the lock to match a key.

The second method involves drilling just above the shear line into the upper pins.

Insert a key blank into the lock first to force all the pins to the upper parts of their chambers.

Drill about ⅛ inch above the shear line or shoulder of the plug and directly above the top of the keyway (FIG. 17-9). Use a drilling jig to ensure drilling at the proper point on the lockface. Use a ⅛ inch or ³⁄₃₂-inch drill.

Poke a thin wire or needle into the hole and withdraw the key to just inside the keyway. Then, use the tip of the key to turn the core. Removing the key partway allows the bottom

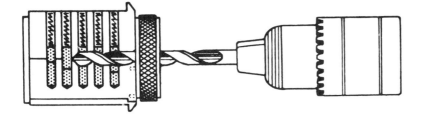

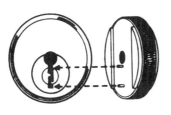

17-9 A cylinder removal clamp. (courtesy Desert Publications)

17-10 A cylinder removal clamp. (courtesy Desert Publications)

pins to drop below the shear line, while the wire or needle keeps the upper pins above the shear line.

Once the door is open, dismantle the lock, remove the cylinder, and replace it with a new one.

CYLINDER REMOVAL

Sometimes the cylinder must be removed in order to open a lock. This means shearing off the screws that hold the cylinder in place. Several different rim cylinders removal tools are available. The one in FIG. 17-10 can be made in your own shop. It's called a cylinder removal clamp.

To make one, follow these steps:

Use a piece of steel tubing about ⅛-inch larger in diameter than the cylinder. The tubing should be no more than 2½-inches long.

Cut into the tubing, as shown in FIG. 17-10 about 1³⁄₁₆ inches.

Cut a center hole at the end of both cuts ¼ inch to ⅓ inch in diameter, and insert a steel rod about 10-inches long. The rod becomes your handle for turning the cylinder.

Drill two holes in the tubing, so you can insert a bolt perpendicular to the handle. Ensure that the bolt will be ½ inch in front of the handle. The bolt threads must extend on both sides of the tubing.

To remove the cylinder, follow this procedure:

Set the removal clamp over the edge of the cylinder after first removing the cylinder rim collar. Tighten down the tension bolt; this provides the gripping pressure to the cylinder.

Twist the cylinder clamp by the handles, and force the cylinder to rotate, shearing off the retaining screw ends.

Remove the cylinder from the lock. When removed, reach in and open the door by reversing the bolt.

You can also use the standard Stillson wrench to remove the cylinder.

Cylinders can also be pulled out of the lock unit, but this method ruins the cylinder threads, requiring you to replace the entire unit. The tool used is called a nutcracker. The sharp pincer points are pushed in behind the front of the cylinder face. Clamp down and pull the cylinder out. Many times an unskilled locksmith can ruin the entire lock and a portion of the door by not knowing how to use this tool properly.

Some cylinders have to be drilled. Follow this procedure:

Drill two ³⁄₁₆ inch in diameter holes about ⅞ inch to 1⅛ inch apart on the face of the cylinder.

Drive two heavy bolts into these holes so that at least 1½″ is sticking out.

Place a pry bar or heavy screwdriver between the bolts and use it as you would a wrench to force the cylinder to turn, shearing off the long screws.

WINDOW ENTRANCES

Window entrances are relatively easy. The old butter knife trick is usually successful. Since most window latches are located between the upper and lower windows, sliding a knife up between the windows allows you to open the latch.

Should the area be too narrow for a knife, shim, or other device, drill a ¹⁄₁₆-inch hole at an angle through the wood molding to the base of the catch. Insert a stiff wire, and push back the latch.

OFFICE LOCKS

Most office equipment can be opened using a few basic methods and a handful of tools.

Filing cabinets

Though most filing cabinets have the lock in essentially the same position, the locking bar arrangement for the various drawers will vary. These variations have to be considered in working on the locks.

One method is to work directly on the lock itself:

Slide a thin strip of spring steel ⅛-inch thick into the keyway, and pull the bolt downward. This will unlock the lock (FIG. 17-11).

If the lock has a piece of metal or pin blocking access to the locking bolt, use a piece of stiff wire with one end turned 90 degrees. Insert the wire between the drawer and the cabinet face, and force the bolt down with the wire. This will allow you to open the drawers.

If there is not enough room for you to work with the wire, use a piece of thin steel or a small screwdriver to pry back the drawer from the cabinet face to allow you to see and work with the wire.

You can also open the drawers individually if necessary:

Use a thin piece of spring steel or a wedge to spread the drawer slightly away from the edge.

As your opening tool, use another strip of steel about 18-inches long, ½ inch to 1-inch wide, and 0.020-inch thick.

Insert the opening tool between the drawer catch and the bolt mechanism (FIG. 17-12).

With a healthy yank, pull the drawer open. The opening tool creates a bridge for the drawer catch to ride upon and pass the bolt.

Other filing cabinets can be inverted to release gravity-type vertical engaging bolts. When the lock mechanism itself is fouled up, the best way to proceed would be to drill out the cylinder and replace the entire assembly.

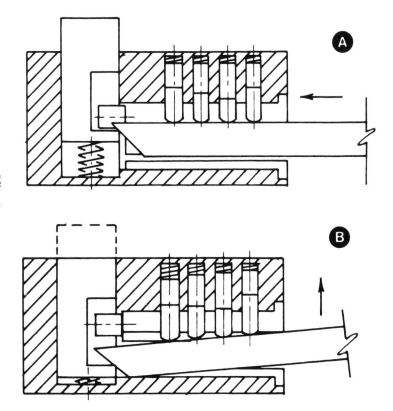

17-11 Sometimes a thin strip of spring steel inserted into the keyway of a filing cabinets lock (A) can be used to pull the bolt down (B). (courtesy Desert Publications)

Desks

Desks with locking drawers controlled from the center drawer can be opened in a couple of ways besides picking and drilling.

Look at the desk from underneath to see what the locking mechanisms for the various drawers looks like. Notice that the locking bar engages the desk by an upward or downward pressure, depending upon the bolt style. The bolt is pushed into the locked position by the motion made when closing the desk drawer all the way. Herein lies the weakness in the desk's security. The bolt usually needs to be pushed up from under the desk by hand to open most of the drawers. The center one has its own lock.

In other desks, you might have to use a little force and pull outward on the center drawer to push the bolting mechanism downward slightly to open the various drawers.

To open the center lock, follow this procedure:

Use two screwdrivers and some tape or cardboard. Put the cardboard or tape between the drawer and the underside of the desk top so you don't mar the desk.

Insert a screwdriver and pry the drawer away from the desk top.

With the other screwdriver pull the drawer outward to open it. With practice, this can be done with only one screwdriver.

If you drill a small hole in the drawer near the lock, you can insert a stiff piece of wire, such as a paper clip, to push down the plug retainer ring. In doing so, you pull the plug free of the lock, causing the bolt to drop down into the open position.

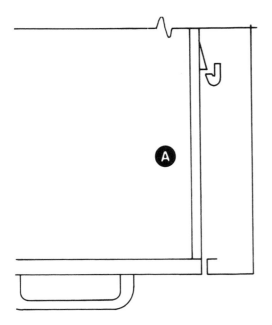

17-12 Sometimes the drawer of a filing cabinet (A) can be opened by wedging the drawer away from the cabinet frame (b) and inserting a strip of steel between the drawer catch and the bolt mechanism (C). (courtesy Desert Publications)

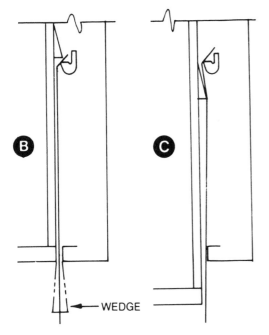

WEDGE

DOORS

Possibly the simplest way to open most doors is with a pry bar and a linoleum knife:

Insert the linoleum knife between the door and the jamb with the point tipped upward.

Insert the pry bar as shown in FIG. 17-13. Exerting a downward motion on the pry bar spreads the door slightly and allows the locking safety latch to be disengaged.

When this is done, bring the linoleum knife forward, pushing the latchbolt into the locking assembly, and opening the door. If there is no safety catch, the knife alone can be used to move the bolt inward. It's also possible to use a standard shove knife or even a kitchen knife.

Sometimes the deadlatch plunger is in the lock but there isn't room to insert a pry bar. What do you do? Use wooden wedges. Insert one of them on each side of the bolt, about 4 to 6 inches from the bolt assembly. Spread the door away from the jamb. Then use a linoleum knife to work the bolt back.

Some doors and frames have such close clearances that you cannot insert a wooden wedge or pry bar. In an instance such as this, use a stainless steel door shim. Merely force it into the very narrow crevice between the door and the frame and work back the bolt.

Often a door lock can be opened with a Z-wire. This tool is made from a wire at least 0.062-inch thick and 10 to 12-inches long (FIG. 17-14). The Z-wire is inserted between the door and the jamb. When the short end is all the way in, it is rotated toward you at the top. As you do this, the opposite end will rotate between the door and the jamb. It will contact the bolt and retract it. If the bolt should bind, exert pressure on the knob to force the door in the direction required.

Sometimes, you may be required to open locked chain latches. You can, of course, force the door and break the chain, but there are better ways. The rubber band technique works most of the time:

Reach inside and stick a tack in the door behind the chain assembly (FIG. 17-15).

Attach one end of a rubber band to the tack; attach the other end to the end of the chain.

Close the door. The rubber band will pull the chain back. If it doesn't pull the chain off the slide, shake the door a little.

If the door's surface will not receive a tack, use a bent coathanger to stretch the rubber band (FIG. 17-16). Make sure the coathanger is long enough and bent properly, so you can close the door as far as possible.

 **17-13** Opening a door using a pry bar and linoleum knife. (courtesy Desert Publications)

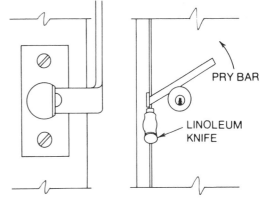

PRY BAR

LINOLEUM KNIFE

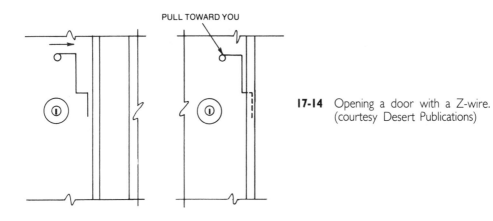

17-14   Opening a door with a Z-wire. (courtesy Desert Publications)

If the door and jamb are even and there is enough space, a thin wire can be inserted to move the chain back.

Sometimes you can open a door very easily if it has a transom. Use two long pieces of string and a strip of rubber inner tubing. The tubing should be 8 to 10 inches long. Attach a string to each end of the tubing, so you can manipulate the tubing from the open transom. Lower the tubing and wrap it around the knob. Pull up firmly on both strings to maintain tension and turn the knob. This method can be used with either a regular door knob or an auxiliary latch unit.

AUTOMOBILES

Automobile EEP can be used on front and rear ventilation windows, door windows, doors, and trunks. Automobile locks vary considerably in their design and reliability. Despite this,

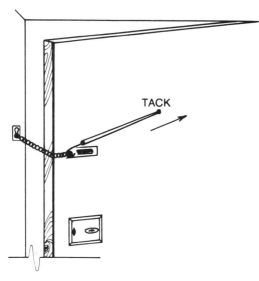

17-15   The rubber band technique is usually the easiest way to unlock most chain locks.

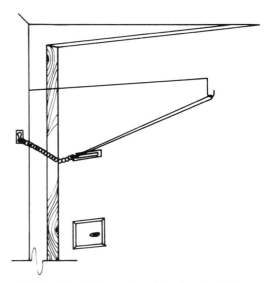

17-16   A variation on the rubber band technique.

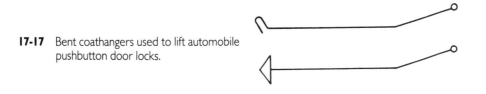

17-17 Bent coathangers used to lift automobile pushbutton door locks.

you can use several methods to gain access. Most of them are faster and easier than picking the lock. In some automobile locks, picking must be ruled out entirely because of the sidebar cylinder which makes picking a very time-consuming, tedious job.

Keyless entry into an automobile is most commonly done at the windows, the most vulnerable part of a car's security. The door release pushbutton lever is often the easiest area to attack, especially on the newer car models that have a single pane of glass. The procedure is fairly simple:

Use a straightened coathanger with a loop or triangle in its end as shown in FIG. 17-17.

If the window is rolled up tight, force a paint scraper between the edge of the window and the weatherstripping (FIG. 17-18).

Insert the coathanger in the crevice made by the paint scraper, and pull the pushbutton lever up with the loop or triangle. Of course, a coathanger isn't the only tool that could work here. Any instrument that could slip through the crevice and lift the lever could do the job.

Two .22 caliber gun cleaning sets can help you lift either the pushbutton lever or the door handle (FIG. 17-19). The cleaning tools come in sections so you can make a variety of rod lengths.

Use the two sets to make a rod long enough to reach across the inside of the car.

Use the slotted cleaning attachment and some fishing line to make a loop in the end of the rod (FIG. 17-20).

Work on the window or door opposite the side you are on. Raise the lever or the handle with the loop by pulling on the line and lifting the rod up.

17-18 Sometimes a paint scraper can be used to make room for a coathanger that can pull open a car pushbutton lock. (courtesy Desert Publications)

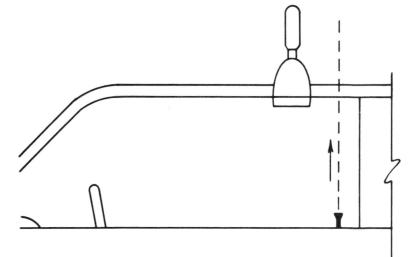

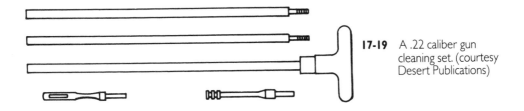

17-19 A .22 caliber gun cleaning set. (courtesy Desert Publications)

The rear ventilation, or wing, window comes in two varieties: with and without a locking button on the swivel. Without the locking button, gaining entry is only a matter of forcing the swivel lock up into the unlocked position.

To open rear ventilation windows without locking buttons on the swivel, follow this procedure:

Insert a putty knife or paint scraper between the window and the frame, bending the knife slightly to allow the unlocking tool to enter. In this case, the unlocking tool could be a thin piece of wire.

Loop the wire around the swivel latch and pull upward. The wire should be bent slightly on the end to ensure a better upward motion without it slipping off the lever.

Most cars now have locking pushbuttons in addition to the swivel levers. To unlock these mechanisms you need standard automobile opening tools. These tools are inserted on different sides of the lock (FIG. 17-21). One is used to depress the button by pulling the tool towards you. The other is twisted slightly to push the lever into the unlocked position.

The front ventilation window gives you access to the window roller handle, and the door handle. Two different tools can be used, depending upon the amount of space you have to work with (FIG. 17-22). Both tools work something like the one shown in FIG. 17-23. The window must be pried slightly to insert the tool.

A Lenco car opener tool can also be used from the front ventilation window. You can get the tool through your local locksmithing supply house. The tool can pay for itself with the first lockout that you are called upon to open.

How you gain entry into the trunk of a car depends upon where the lock is located. Drilling is an acceptable practice here. Remember that door locks and the ignition usually take the same key; the glove compartment and the trunk usually take the same key also. Sometimes you can open the trunk lock by inserting the door key and pulling it out as you turn it.

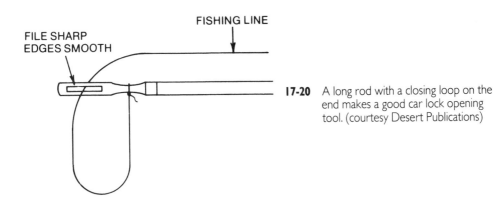

FISHING LINE

FILE SHARP
EDGES SMOOTH

17-20 A long rod with a closing loop on the end makes a good car lock opening tool. (courtesy Desert Publications)

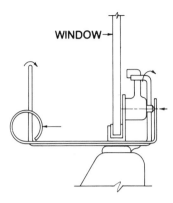

WINDOW→

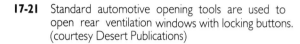

17-21 Standard automotive opening tools are used to open rear ventilation windows with locking buttons. (courtesy Desert Publications)

Picking the trunk lock is simple, but it can be time consuming if you have not worked with very many automobile locks. You can knock the lock out by forcing it inward with a hammer and a heavy screwdriver, but this ruins the lock and, sometimes, associated parts on the inside.

Drilling can also have its drawbacks. You can drill out the plug by shearing off the pins. This means replacement of the lock unit. Lately, manufacturers have improved the security of trunk locks by installing a steel plate in front of the catch to prevent drilling. However, it is possible to drill about 2 inches to the side of the catch bolt. This enables you to use a bent piece of wire to force the catch backwards to open the trunk.

In some trunk cylinders there is a retainer. An L-shaped wire inserted into the keyway will catch the retainer, and you can pull it down and pull the plug out of the lock (FIG. 17-24). After that a small screwdriver or an awl can be used to push back the catch mechanism.

The easiest of all automobile locks to open is the glove compartment lock. Insert an L-shaped wire into the keyway and pull down the retainer. This wire actually works on two distinct types of locks. In one, you remove the plug; in the other, the wire engages the glove compartment catch mechanism and opens it with a downward movement.

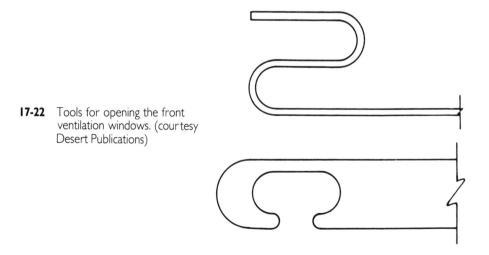

17-22 Tools for opening the front ventilation windows. (courtesy Desert Publications)

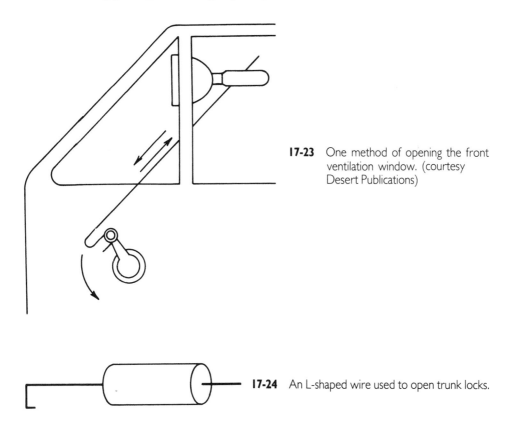

17-23 One method of opening the front ventilation window. (courtesy Desert Publications)

17-24 An L-shaped wire used to open trunk locks.

FORCED ENTRY

Forced entry in place of proper professional techniques is never recommended, except in emergencies or when authorization is given by the owner. There are some simple rules about forced entry you should consider:

☐ Attempt a forced entry only as a last resort. Try other techniques first.

☐ If you must jimmy the door, do it carefully. Today antique doors and frames can be only part of a valued collection that may be in a home. Don't risk a costly lesson by being too hasty.

☐ If you must break a window to gain access to a lock, break a small one. Replacing broken windows can be expensive.

☐ Don't saw the locking bolt. This is very unprofessional.

Remember that lockpicks and many other entry tools are considered burglar tools in some jurisdictions. In many cases you must have a license to carry them. Check with your local police, keep your locksmith's license current, and when a tool is worn out, destroy it.

Be aware of the trust which your city and customers put in you by allowing you to be a locksmith; you possess certain tools and knowledge that others do not have. Don't abuse this privilege.

18

Combination locks

Like other locks, the combination lock is available in a variety of types and styles. In this chapter we will consider a number of representative types. Some of these locks are more secure than others, but each has its role. For example, it would be inappropriate to purchase a high-quality combination padlock for a woodshed; by the same token, it would be foolish to protect family heirlooms with a cheap lock. Even the best lock is only as good as the total security of the system. A thief always looks for the easiest entry point.

PARTS

Though combination locks have some internal differences, all operate on the same principle—rotation of the combination dial rotates the internal wheel pack. The pack consists of three (sometimes four) wheels. Each wheel is "programmed" to align its gate with the bolt-release mechanism after so many degrees of rotation. Programming may be determined at the factory, or it may be subject to change in the field. As the wheels are rotated in order (normally three turns, then reverse direction for two turns, and then reverse direction again for one turn), the gates are aligned by stops, one for each wheel and one on the wheel-pack mounting plate. When all the gates are aligned, the bolt is free to release. Padlocks are built so the shackle disengages automatically, or manually by pulling down on the lock body. The captive side of the shackle, the part known as the *heel*, has a stop so the shackle will not come free of the lock body.

One weakness of low-priced locks is their ease of manipulation. With practice you can discriminate between the clicks as the wheels rotate. Once you learn to do this, manipulation of the lock is child's play. Fortunately, manufacturers of better locks make false gates in the wheels, and these gates make manipulation much more difficult. True, a real expert will not be thwarted, but experts in this arcane art are few and far between. It takes training and intensive practice to distinguish between three or more false gates on each wheel and the true gate.

MANIPULATION

Manipulation of combination locks is a high skill—almost an art. A deep understanding of combination-lock mechanisms and many, many hours of practice are needed to master this skill. But the rewards—both in terms of the business this skill will bring to your shop and the personal satisfaction gained—are great.

No book can teach you to manipulate a lock any more than a book can teach you to swim, box, or do anything else that is essentially an exercise in manual dexterity. But a book can teach the rudiments—the skeleton, as it were, that you can flesh out by practice and personal instruction from an expert.

At the risk of redundancy, allow us to repeat that manipulation is a matter of touch and hearing. It also involves an understanding of how the lock works. Electronic amplification devices are available to assist the locksmith. These devices are useful, especially as a training aid.

Padlocks usually do not have false gates, and so are ideal for beginners (FIG. 18-1). Pull out the shackle as you rotate the wheels. This tends to give better definition to the clicks. Work only for one number at a time. Stop when the bolt hesitates and touches the edge of the gate. The bolt has touched the far side of the gate, so move back a number on the dial and note it. Turn the dial in the reverse direction, going past the first number at least twice. Pull on the shackle and slowly continue to turn the dial. When you sense that the bolt has touched the gate, note the number that came up just before the bolt responded. This is the second number of the combination. The third number comes easier than either the first or second one. The dial may stop at either the first or second number. Since many of the earlier combination locks did not have the accuracy of modern locks, the bolt catches at any one of the three numbers at any time.

You may have the three numbers, but not in their correct order. Vary the sequence of the numbers until you hit the right one.

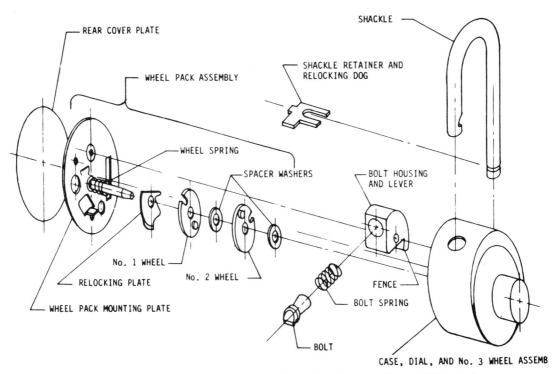

18-1   A typical combination padlock in exploded view. (courtesy Desert Publications)

Because of the imperfections in older and inexpensive modern locks, the bolt may stop at points other than the gates. Only through practice can you learn to distinguish between true and phantom gates.

For practice, obtain two or three locks of the same model. Disassemble one to observe the wheel, gate, and bolt relationship and response. These insights will help you manipulate the other two locks.

DRILLING

All combination locks can be opened by drilling as a last resort. To drill these locks follow this procedure:

Drill two ⅛-inch holes in the back of the lock.

Turn the dial and determine that the gate of one wheel is aligned with one of the holes. You might be able to see the gate. If that fails, locate the gate with a piece of piano wire inserted through one of the holes.

When the gate and the hole are aligned, note the number on the dial face. Determine the distance, as expressed in divisions on the dial, between the hole and bolt.

Subtract this distance from the reading when the gate is aligned with the hole. The result is the combination number for that wheel.

Reverse dial rotation and find the number for the second wheel. Do the same for the third.

CHANGING COMBINATIONS

Many locks are designed for combination changes in the field. We will discuss three of them.

Sargent and Greenleaf

S & G padlock combinations are changed by key.

Turn the numbers of the original combination to the change-key mark located 10 digits to the left of the zero mark on the dial face.

Raise the knob on the back of the lock to reveal the keyway. Insert the key and turn 90 degrees.

Repeat step 1, but use the new combination and the change-key mark as zero.

Remove the change key and test the new combination.

Simplex

The Simplex is a unique combination lock, employing a vertical row of pushbuttons rather than the more usual dial (FIG. 18-2).

1. Turn the control knob left to activate the buttons.
2. Release the knob and push the existing combination (FIG. 18-3).
3. Push down the combination change slide on the back of the lock (FIG. 18-4).
4. Turn the control knob left to clear the existing combination (FIG. 18-5).
5. Push the button for the new combination—firmly and in sequence (FIG. 18-6).
6. Turn the control arm right to set the new sequence (FIG. 18-7).

18-2 The Simplex pushbutton combination lock. (courtesy Simplex Access Controls Corp.)

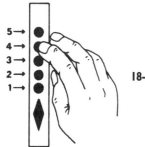

18-3 Changing Simplex combination, Step 2.

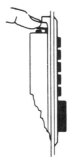

18-4 Changing Simplex combination, Step 3.

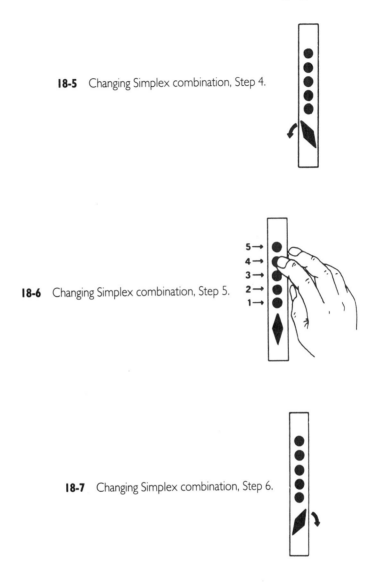

18-5   Changing Simplex combination, Step 4.

18-6   Changing Simplex combination, Step 5.

18-7   Changing Simplex combination, Step 6.

Dialoc

Partial disassembly is required to change the combination. Refer the parts numbers in the follow steps to FIG. 18-8.

To disassemble:

Remove the Dialoc from the door by withdrawing the two mounting screws 487-1, holding the inside plate 821 against the door.

Place the outside plate 820 face down on a smooth work surface so that housing 414 is facing up.

Remove the three nuts 484-5 holding the housing cover 745.

After removing the nuts, gently lift the housing cover up and away from the housing.

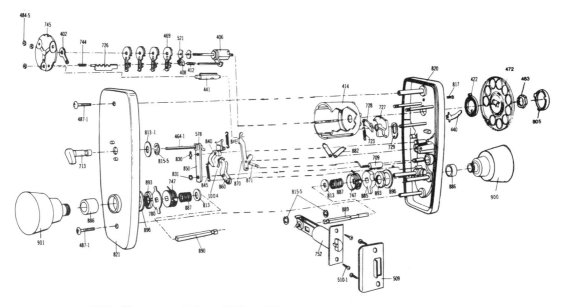

18-8 The very sophisticated Dialoc 1400. (courtesy Dialoc Corporation of America)

Take care that control-bar spring, 744 is not dislodged and lost in the process. Lay it where it won't get lost.

Now, remove the secondary arm 402.

Grasp the end of the control bar 726 and lift out the four ratchet assemblies. Observe how they are arranged before dismantling.

Now you can rearrange these four ratchet assemblies to give you a new combination. (Should you want to use numbers that are not in the present combination, your distributor or the Dialoc Corporation can readily supply these.) Remove the ratchet assemblies 469 from the control bar and change your sequence as you desire. Put the ratchet assemblies back on the control bar. When replacing the ratchet assemblies on the control bar, make sure that they go on with the ratchet part of each assembly down, and that the first digit of the combination is put on first, then the remainder of the combination in order from the first ratchet assembly outward.

To assemble:

Place the four ratchet assemblies 469 on the control bar 726 and, grasping the assembled control bar and the ratchet assemblies, slip them over the center spindle of the primary arm 406. The ratchet assemblies can be guided over the lower mounting pin protruding from the outside plate 820.

Care should be taken that the control bar 726 has passed through the elongated hole in the bottom of the housing 414.

Replace the secondary arm 402 on the primary arm 406. Make sure that the finger 408 and the finger follower spring 412 are in position.

If the cam pawl trip 441 has been dislodged during the combination change, it can now be replaced by inserting it into the housing with the small protruding end down. The blade on this part must be against the ratchet pawls of the ratchet assemblies.

Make sure that the secondary arm 402 is flush with the top of the square bar. This allows a spacing of approximately 0.020 inch between the secondary arm and the top ratchet assembly.

After the secondary arm is properly placed, return control spring 744 to the end of the control bar 726.

Replace the housing cover 745. Make sure that the cam pawl trip 441 and the control bar 726 are in their respective holes in the housing cover. A straight pin is helpful in guiding these parts into holes. Be sure reset arm 870 is inside the housing 414.

Replace the cover-retaining nuts 484-5. These nuts are to be turned down snugly with the fingers, plus approximately ½ turn with a wrench.

The lock should now be operated before replacing on the door to ensure its proper operation on the new code settings.

Master 1500 padlock

The Master 1500 combination padlock is covered with heavy-gauge sheet steel, rolled and pressed at the edges. The net result of this somewhat unorthodox construction is a tight-fitting lock and one that is not pried open easily. Disassembly is recommended only as a training exercise. It is not practical to open this lock for repair.

Three wheels are employed, each with a factory-determined number. If you receive the lock with the shackle open you can determine the combination by looking through the shackle hole. Note the dial reading as you align each wheel. If you read the numbers properly and if the wheel gates are in alignment, you should only have to add 11 to the dial readings to get the true combination. If you are slightly off, compensate by adding 10 or 12 to the original readings.

The Master 1500 series may be master keyed (FIG. 18-9). This lock is popular in schools. While the combination can be obtained by manipulation, picking the keyway is faster. Once the lock is open, determine the combination by the method described above.

PUSHBUTTON KEYLESS LOCKS

Combination locks are those devices that do not require a key to open. Although many people think of such things as safes and vaults or perhaps a bicycle padlock requiring a three-digit combination as the only type of locks that fall under the classification "combination locks," this is not true. Any device that does not require a key, but instead relies on a single or series of digits or letters in a certain order to open the mechanism, is classed as a combination locking device.

This concluding section of the chapter will discuss, illustrate, and provide specific details concerning several of the most popular and best keyless locking devices available today. Remember that the products discussed here are not the only types, nor are they representative of the only manufacturers of such products; they are the products that have proven themselves to a wide variety of customers in varied and numerous types of installations. As such, they are recommended for your consideration as some of the highest quality, professional, and most secure keyless type of combination locking devices on the market today.

KEYLESS LOCKING DEVICES

Keyless entry locks provide exceptional security in a variety of modes. Motels, hotels, various private and government institutions are using these locks with increasing success (FIG. 18-10). The main manufacturer and proponent of keyless locking devices is Simplex Security

18-9   The Master 1525 master keyed combination lock. (courtesy Master Lock Company)

systems. Their locks are commonly used in a variety of security-related situations where more than one entrance control device is required. As an example, consider the case of a company research and development facility. For added security, a keyless lock is added to each door. At night, then, two locks secure the door. During the day, the main lock is in the unlocked position, but the pushbutton lock is readily available for the use of the R&D employees since they are the only individuals who know the particular combination to gain access.

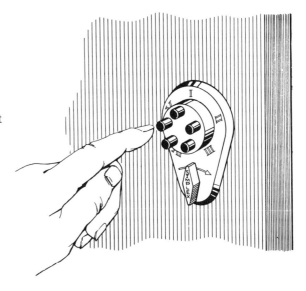

18-10 Simplex 5-pushbutton deadbolt lock. (courtesy Simplex Access Controls Corp.)

In motels and hotels, the use of keyless locks has increased security, cut costs, and saved time and manpower. How? By using the Simplex lock, in one example, one motel stopped losing more than 50 keys a month and saved the costs of continually having duplicate keys made and locks rekeyed. Best of all, thefts were almost entirely eliminated.

Time and manpower were also saved with the installation of the Simplex lock. The desk clerk no longer had to hunt a duplicate key or go to a room and unlock it because the key was locked inside. All that is required is to give out the specific room combination, just as was done when the individual checked into the motel.

Once the guest checks out, it requires only three minutes to change the combination; this keeps the room security integrity very high.

The Simplex lock is extremely easy to operate and maintain, despite its very high security. The lock has five pushbuttons arranged in a circle and numbered from one to five. The combinations can be set at random; to operate the lock, the various buttons are pushed in sequence, in unison, or a combination of both to affect entry.

Installation of the lock is also relatively easy. The only tools required are a screwdriver, electric drill and a hole saw set.

Simplex

Before we get into the various installation procedures, let's look at the specific advantage of a Simplex lock so you will understand, as a locksmith, why many businesses and homeowners are installing these locks.

- ☐ No keys are required. For businesses, this means none are issued or controlled, and there are none to recover when the employee leaves.
- ☐ They are burglar-proof since there is no keyhole to pick.
- ☐ There are thousands of possible combination options.
- ☐ You can change the combination in seconds whenever there is a change in personnel, tenants, or for other reasons of security.
- ☐ Money is saved since there are no keys or cylinders to be changed or replaced.

□ The locking units are extremely durable, being wear-tested for the equivalent of 30 years of intensive use.

□ It is an extremely easy-to-install lock, more so than any other existing lock or knob type lockset. It is completely mechanical and requires no electrical wiring or specialized accompanying equipment.

The pushbutton combination lock (FIG. 18-11) comes in several models: the deadlock with a full 1-inch deadbolt, the automatic spring latch, the key bypass, and the special security model (for use with Department of Defense or industrial complexes requiring the security necessary for the protection of closed or restricted areas).

The deadlock, model DL, requires only the combination pressed in the right order and the turn knob moved in order to open the lock from the outside. The lock remains in the open position, with the bolt retracted, until either the inside lever or outside knob is turned to throw the bolt into the lock position.

The automatic spring latch, model NL, locks automatically. It opens with a finger lever from the inside. A thumb slide on the inside holds the retracted latch bolt in the unlocked position, if desired (FIG. 18-12). Lacking the latch holdback is a variation, the model NL-A. A deadlocking latch is standard for these units; this will prevent prying the latch back on the closed door. The latchbolt is reversible without affecting the deadlatching function.

The key bypass models (FIG. 18-13), DL-M and NL-M, are the DL and NL models with an optional key cylinder. They provide the flexibility of a master key system and the security advantages of a combination lock. Employees, tenants, and others use the combination. The key ensures entry, regardless of the combination setting used, and it is given only to selected and designated management personnel. This eliminates the need to maintain a central listing of combinations for reference use by top management personnel. It is completely safe as there is no widespread proliferation of keys, and key distribution can be severely restricted.

The key bypass model is extremely popular in facilities working for and with the Department of Defense and other government agencies, due to its excellent security advantages.

The special security model, NL-A-200-S, is designed for complete compliance with the Industrial Security Manual of the Department of Defense. The model is an approved

18-11   Deadbolt with a key-bypass. (courtesy Simplex Access Controls Corp.)

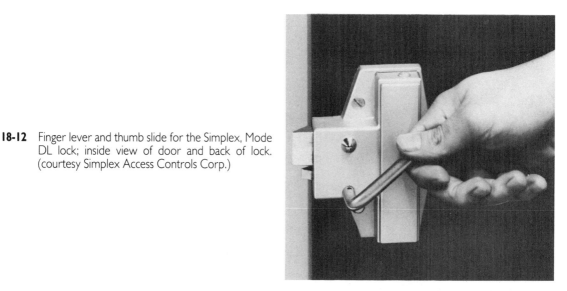

18-12 Finger lever and thumb slide for the Simplex, Mode DL lock; inside view of door and back of lock. (courtesy Simplex Access Controls Corp.)

substitute for the expensive guard forces in the protection of closed and restricted areas in contractor facilities. It has all the features of the model NL automatic spring latch, plus:

- □ A face plate shield to prevent observations of pushbutton operation and maintain the secrecy of the combination (FIG. 18-14).
- □ The combination change access can be padlocked to ensure authorized combination changes only.
- □ The latch holdback feature is eliminated so the lock can never be kept open. It is in the locked condition at all times.

18-13 The key bypass is also available with the standard latch bolt model Simplex lock. (courtesy Simplex Access Controls Corp.)

18-14 Face-plate shields will prevent the observation of the combination code by unauthorized individuals. (courtesy Simplex Access Control Corp.)

All of the various Simplex models have the same basic three-piece self-aligning assembly and installs easily and quickly. *Note:* All Simplex door locks require a minimum flat surface of 3½″ on the inside of the door to mount.

Simplified, the procedures for installation are as follows:

Wood and door frame installation: Just drill two holes, ¾ inch and 1⅝ inches, a distance of 2⅝ inches from the door edge. Key bypass models will require an additional ⅞-inch key cylinder hole. All parts are provided for a complete wood door and frame installation. Included are two styles of strikes (standard) to suit any wood door frame ("A" in FIG. 18-15).

Metal door and frame installation: An adapter kit containing a special surface-mounted strike ("B" in FIG. 18-15) is available. The special strike is fastened to the inside surface of the metal frame and eliminates the need to cut or mortise the frame for the standard strike. The adjustable riser plate raises the lock housing to ensure the proper alignment of the latch or deadbolt with the special strike.

Installation instructions—models DL and NL:

Using the template provided with the lock (FIG. 18-16), cut the template from the sheet and fold on the printed line as shown by the arrow. With the door closed, tape the template to the inside of the door with the fold line aligned with the edge of the door (or door stop if one is present). The same template applies to both inswing and outswing doors (FIG. 18-17). Using a center punch, mark the centers of the two large and three small holes.

Using a ⅛ inch in diameter drill, drill the centers of the ¾-inch in diameter and 1⅝-inch in diameter holes all the way through to the front surface of the door. Drill the three mounting screw holes approximately ½″ deep.

With a 1⅝ inch in diameter hole saw, open up the larger hole by drilling from both the outside and the inside surfaces, meeting in the middle. Follow the same procedure with a ¾-inch in diameter drill for the smaller hole. *Note:* With the model NL, the latch is reversible by removing the two large and two small screws on the undersurface and disassembling the lock housing.

Slide the lock housing into the holes in the door as shown in FIG. 18-18. If the holes are drilled at an angle, and the undersurface of the lock is not in full contact with the door, file out the holes to correct. Screw the lock housing to the door.

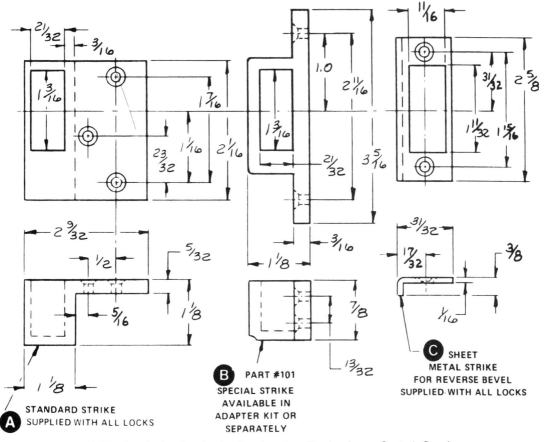

18-15 Standard and optional strikes. (courtesy Simplex Access Controls Corp.)

Place the holding bracket on the front surface of the door with the slotted legs engaging the aligning pin on the barrel. With the large radius of the holding bracket mating with the radius of the barrel, fasten with the screws provided.

Remove the lock housing from the door.

Slide the holding edges of the face plate behind the formed up edges of the holding bracket. Make sure the face plate is securely held on both sides.

Replace the lock housing while holding the control knob, marked "SIMPLEX," in the vertical position. Turn the knob to ensure proper engagement and fasten the lock housing to the door. *Note:* It is very important to try the combination of the unit several times before locking. It is also a necessity to have your customer try and open the lock several times before leaving. If the lock does not fit exactly to the door, there is an adjustable riser plate available from the factory. With minimum maintenance, this unit will now provide a very long and extremely useful service to your customer.

Key bypass installation instructions:

Follow the above instructions for installing the standard DL and NL locks, with these exceptions:

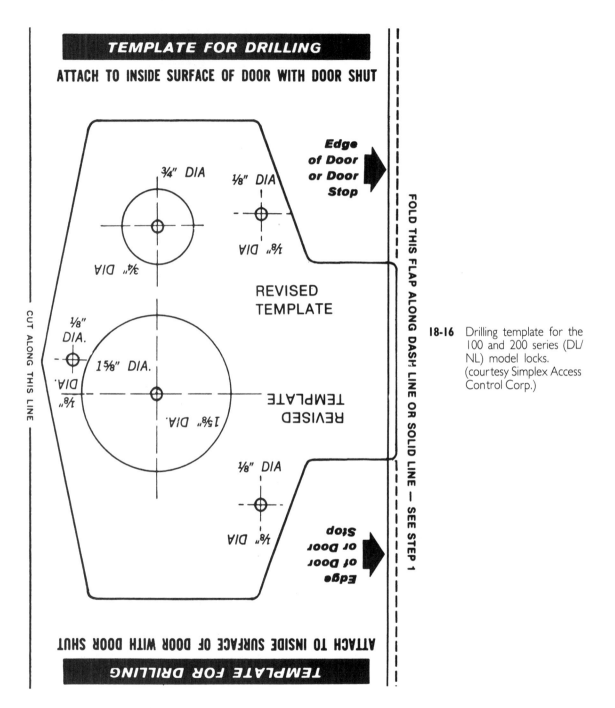

18-16 Drilling template for the 100 and 200 series (DL/NL) model locks. (courtesy Simplex Access Control Corp.)

18-17 This visual provides for the proper determination of which template line will be used for positioning the lock properly to the door. (courtesy Simplex Access Controls Corp.)

Fold template on **Dash Line** for additional clearance required on outswing doors or when using Part #100 Riser Plate on inswing doors.

Fold template on **Solid Line** on inswing doors when **not** using Riser Plate.

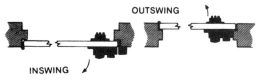

Use the template (FIG. 18-19) for positioning ⅞ inch in diameter hole for the key lock cylinder.

With two pairs of pliers, break the tail piece of the key cylinder at the proper location as indicted (FIG. 18-20). (1) for the Simplex 100 or 200 series locks when the riser plate is not used; (2) for 100 series when either ¼-inch or ⅜-inch side of the riser plate is used, or with the 200 series locks when ¼-inch side of the riser plate is used; (3) for 200 series locks with ⅜ inch side of riser plate, use full length of tail piece.

Assemble the key lock cylinder in the ⅞ inch in diameter hole. *Note:* The hole in the flange of the cup must be positioned over the ⅛ inch in diameter mounting screw hole used to secure the lock housing to the door (FIG. 18-21).

Proceed with the last step of the basic instructions.

You may wish to substitute a standard rim cylinder for the ⅞-inch in diameter cylinder supplied with the Simplex key bypass models DL-M and NL-M. Because of space limitations, the Simplex face plate with Roman numerals must cover a portion of the standard key cylinder (FIG. 18-22).

When a standard key cylinder will be used, you as the locksmith must let the factory know this; in your order you must specify the Simplex model to be used (DL-M or NL-M, 200

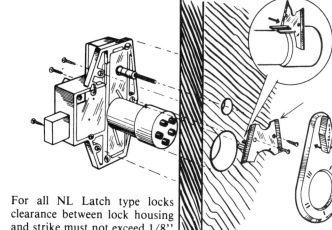

18-18 Lock housing and holding bracket positioning for proper lock installation. (courtesy Simplex Access Controls Corp.)

For all NL Latch type locks clearance between lock housing and strike must not exceed 1/8'' to ensure deadlatching.

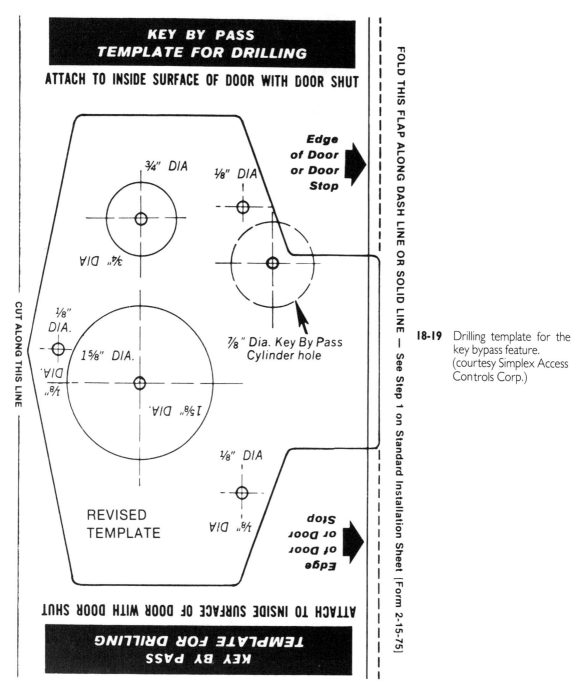

18-19 Drilling template for the key bypass feature. (courtesy Simplex Access Controls Corp.)

18-20   Tailpiece break points. (courtesy Simplex Access Control Corp.)

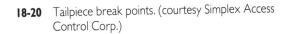
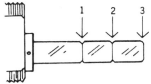

18-21   Bypass lock assembly procedure. (courtesy Simplex Access Control Corp.)

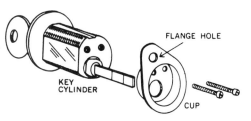

18-22   Simplex lock using a standard sized rim cylinder. Note how the Simplex overlaps a portion of the cylinder. (courtesy Simplex Access Controls Corp.)

18-23   Standard lock cylinder positions with the door front surface. (courtesy Simplex Access Controls Corp.)

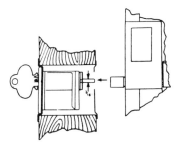

series, and whatever finish is desired) plus the very important notation ". . . for use with standard rim cylinder. . ." placed prominently on your order blank.

Make sure you get the *rim* cylinder, *not the mortise*. In the proper keying of the unit, your local or regional distributor or other supply source will probably be contacted also, unless you have a wide variety of various lock parts in stock. You must furnish the cylinder. An escutcheon (or bezel) must be used that allows the nose of the cylinder to be mounted flush with the front surface of the door (FIG. 18-23). Cut the length of the tailpiece to suit. Cut or grind the width of the tailpiece to ³⁄₁₆ inch to fit the tubular coupler on the Simplex lock.

Use the Simplex key bypass template (Simplex Form 2-15-65A) for drilling. Use the center of the ⅞ inch in diameter hole for drilling the appropriate size hole for the standard rim cylinder and escutcheon. Assemble the cylinder, escutcheon and back plate to the door *before* mounting the Simplex lock.

Now install the Simplex by following the standard installation instruction procedures. The Simplex face plate (with Roman numerals) and holding bracket will cover part of the cylinder. It might be necessary to notch the edge of the escutcheon to clear one of the two screws used to fasten the holding bracket to the front surface of the door. One of the two lock housing screws nearest the door edge must be fastened to the back plate of the rim cylinder (FIG. 18-24). Using the hole in the lock housing as a guide, drill a .161 inch in diameter hole (#20 size drill) in the back plate for the #10 sheet metal screw.

After the Simplex lock is installed, you might have a service and maintenance contract for the numerous Simplex locks in the building. During the course of your locksmithing career, you will be called upon many times to carry out your portions of the service contract with the building manager. Among the several procedures most common for the Simplex lock, you may be changing the combination for one or many of the locks or be required to disassemble the lock for one reason or another. You will also be expected to be able to find an unknown combination. The procedures for these various parts of the locksmith's job are detailed below.

Combination changing:

With the door in the open position and the lock in the locked position, push the existing combination. The hexagon screw at the top of the lock housing (FIG. 18-25) is then removed with an Allen wrench. After removing the screws, the wrench is inserted into the hole and a red button inside is depressed. The wrench is then removed.

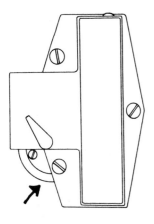

18-24   Arrow indicates lock housing screw that must be attached to the cylinder back plate. (courtesy Simplex Access Controls Corp.)

Keyless locking devices 309

18-25 Use the Allen wrench to remove the access screw in order to change the combination of the lock. (courtesy Simplex Access Controls Corp.)

The front control knob (below the pushbuttons and marked "Simplex") should then be turned to the left. This removes the existing combination.

The pushbuttons on the front are now depressed in the sequence desired for the new combination. (This can be up to five single numbers or two at a time simultaneously.)

It is advisable to write down the combination sequence immediately and present it to your customer for safekeeping. The front knob marked Simplex should now be turned to the right. The new combination is now set in the lock. The combination should be tried several times with the door in the open position before attempting it with the door in the closed position and the door, for all intents and purposes, locked. (At this point, stand inside the door and let the customer operate the combination at least once from the outside.)

The Simplex Unican 100 series lock (FIG. 18-26) is a heavy-duty lockset ideal for securing high traffic areas in commercial, institutional, and industrial buildings, apartment lobbies, hotels, motels, luxury homes, and restricted areas in commerce and industry. Like other simplex high quality products, the Unican 1000 has the following features:

☐ Built-in security with thousands of possible combinations.

☐ No keyway to pick.

☐ Weatherproofing which allows for all types of installations.

18-26 Simplex Unican 1000 Series Heavy Duty Lockset. (courtesy Simplex Access Control Corp.)

□ A ¾-inch deadlocking latch with a 2¾-inch backset.

□ Completely mechanical.

□ No electrical wiring.

□ One-hand operation.

The unit is 7¾-inches high, 3-inches wide, and 3¹³⁄₁₆-inches deep. Latch front is 1⅛ inches × 2¼ inches, with a backset of 2¾ inches. The Unican lock will fit doors with thicknesses ranging from 1⅜ inches to 2¼ inches.

A key bypass option is available, mainly for master keying purposes, using the Best or Falcon removable core cylinders in the outside knob (FIG. 18-27). The removable core cylinders are not available with the lock when ordered, but must be obtained separately. On a special order basis, the Russwin/Corbin removable core capability can be obtained.

The second optional feature is the passage set function. When activated by an inside thumb turn or key cylinder, the outside knob can be set to open the lock without the need for using the pushbutton combination (FIG. 18-28). This is designed for use in offices, etc., when the door is to be left unlocked for an extended period of time.

Both 9-inch and 14-inch high "mag" type filler plates are available to cover up previous lock installation holes on 1¾-inch doors. These three-sided steel plates wrap around the door extending 5½ inches on both the outside and inside surfaces.

18-27   Unican 1000 with key bypass fester. (courtesy Simplex Access Controls Corp.)

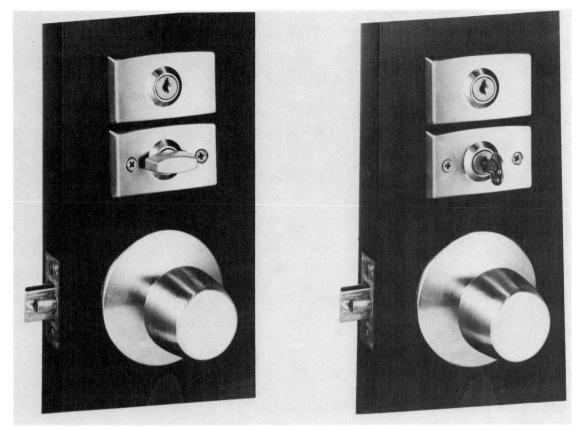

18-28 Key bypass or thumb turn can be employed with the unit. (courtesy Simplex Access Controls Corp.)

Unican 1000 series installation.

The Unican 1000 uses an ASA 161 cutout (2⅛ inches in diameter through hole with a 2¾-inch backset) as found with most other heavy-duty cylindrical locksets (FIG. 18-29). In addition, two ¼ inch in diameter holes are required to bolt the top of the lock case to the back side of the door, and a 1-inch in diameter hole for the key operated combination change mechanism to operate from the back side as well. The Unican can be installed in wood or metal doors from 1⅜ inches to 2¼-inches thick.

In addition to the special tool provided with each lock, you, as the installer, will also need the following tools:

☐ Drill (½ inch in diameter electrical drill preferred).

☐ 2⅛ inch in diameter hole saw (for metal doors) or spur bit (for metal doors).

☐ 1 inch in diameter hole saw or spur bit.

☐ ¼ inch in diameter drill.

☐ Two wood chisels (wood doors) ¼ inch and 1-inch wide.

☐ Phillips and standard screwdrivers.

18-29   Exploded view of Unican 1000 positioning for a wood door. (courtesy Simplex Access Controls Corp.)

□ Center punch and hammer.

□ Standard 12-inch ruler.

Before starting the actual installation procedures, let's look at a portion of the Unican three-part 1000 template that will be used. (Figure 18-30 shows the template and proper dimensions with FIG. 18-31 the door edge portion of the template.) Figure 18-32 provides the information and specifications for the latch cutout required. *Caution:* When using this type template, always apply the template and drill from the outside, being sure to compensate for the door bevel, if any (FIG. 18-33).

Initially, make all necessary corrections to ensure the door is properly hung in the frame. Apply the template to the high edge of the door bevel and tape it in place. Should a metal door frame preposition the strike location, be sure to position the template to center the 1 inch in diameter latchbolt hole in the strike cutout in the frame. With a center punch and hammer, mark the centers of all required holes.

With the ¼-inch drill, drill the centers of the 2⅛ inches and 1-inch holes all the way through the door. Should you wish to center mark the hole on the other side of the door (the low side of the bevel) using the template, be sure to move the "fold" line to shorten the backset to compensate for the bevel. Laying a carpenter's square on the high edge of the bevel will show how much compensation to make.

It is a good practice to bore the 2⅛ inch and 1-inch in diameter holes halfway in from both the front and back surfaces of the door.

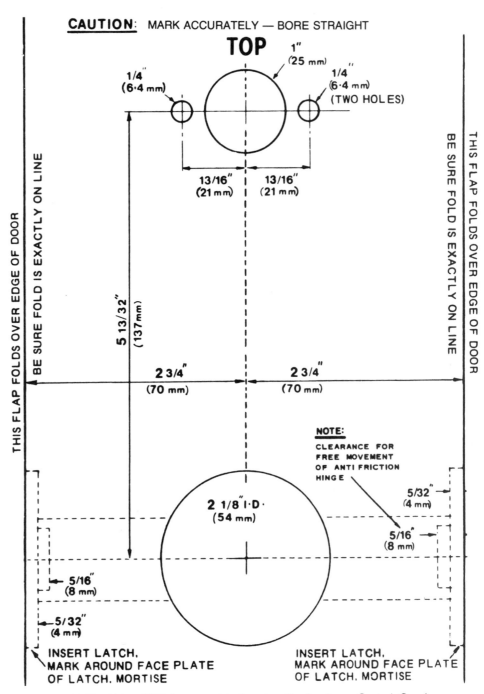

CAUTION: MARK ACCURATELY — BORE STRAIGHT

TOP

1" (25 mm)

1/4" (6·4 mm)

1/4" (6·4 mm) (TWO HOLES)

13/16" (21 mm)   13/16" (21 mm)

THIS FLAP FOLDS OVER EDGE OF DOOR — BE SURE FOLD IS EXACTLY ON LINE

THIS FLAP FOLDS OVER EDGE OF DOOR — BE SURE FOLD IS EXACTLY ON LINE

5 13/32" (137 mm)

2 3/4" (70 mm)   2 3/4" (70 mm)

NOTE: CLEARANCE FOR FREE MOVEMENT OF ANTI FRICTION HINGE

5/32" (4 mm)

2 1/8" I·D· (54 mm)

5/16" (8 mm)

5/16" (8 mm)

5/32" (4 mm)

INSERT LATCH, MARK AROUND FACE PLATE OF LATCH, MORTISE

INSERT LATCH, MARK AROUND FACE PLATE OF LATCH, MORTISE

18-30   Unican 1000 door template. (courtesy Simplex Access Controls Corp.)

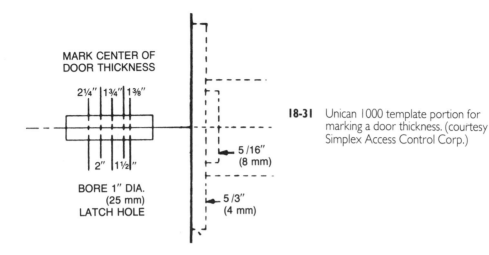

MARK CENTER OF
DOOR THICKNESS

2¼" 1¾" 1⅜"

2" 1½"

BORE 1" DIA.
(25 mm)
LATCH HOLE

5/16"
(8 mm)

5/3"
(4 mm)

18-31 Unican 1000 template portion for marking a door thickness. (courtesy Simplex Access Control Corp.)

If the door was originally prepared for a 2⅜-inch backset, you can file out the 2⅛-inch hole by an additional ⅜ inch as the lock case and inside rose are wide enough to cover.

Be sure to drill the 1-inch latchbolt hole on the door edge straight and level to prevent binding of the latch assembly. Carefully drill the two ¼-inch mounting holes through the door. It cannot be overemphasized that a drill jig is of great assistance in keeping your mounting hole straight and true.

Care must also be taken to allow clearance for the hinge that operates the antifriction device on the Unican latch. The latchbolt hole must be relieved adjacent to the antifriction trigger to permit the hinge to rotate freely and prevent the binding of the latch.

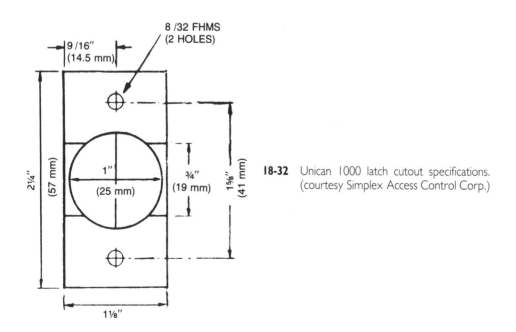

8 /32 FHMS
(2 HOLES)

9/16"
(14.5 mm)

2¼"
(57 mm)

1"
(25 mm)

¾"
(19 mm)

1⅝"
(41 mm)

1⅛"

18-32 Unican 1000 latch cutout specifications. (courtesy Simplex Access Control Corp.)

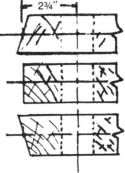

18-33 Details of the bevel compensation for the Unican 1000. (courtesy Simplex Access Controls Corp.)

CAUTION:

APPLY TEMPLATE AND DRILL FROM OUTSIDE BUT COMPENSATE FOR DOOR BEVEL IF ANY.

It is imperative for a proper Unican 1000 installation to have a freely operating latch properly positioned to the strike. Excessive binding or dragging of the latch can cause the front knob clutch to "freewheel," preventing the latch withdrawal. Also, the deadlatching plunger must be held depressed by the strike plate to be operable. The gap between the door and the frame must never be more than a ¼ inch to ensure proper deadlatching. If door silencers are present, allowances must be made in positioning the strike.

To properly position the strike, install the latch assembly into the edge of the door. On wooden doors, mark around the faceplate of the latch and mortise in with a chisel. Remove ⁵⁄₁₆ inch minimum for antifriction trigger hinge and ⁶⁄₃₂ inch of material for the faceplate assembly. When inserted, the latch assembly should be flush with the nose of the door. Close the door to the point where the latch touches the door frame, and mark the frame at the top and bottom lines previously scribed. Mark the screw holes. When the door is closed, the screw holes of the strike should be even with the screw holes of the latch plate. Cut or mortise the frame to a minimum depth of ¾ inch to ensure a full throw of the latch when the door is closed. Combinations cannot be cleared or changed if the full travel of the latch is impaired. The Unican 1000 comes factory packed with a 2¾ inch strike (and strike box for metal frames). An ASA 4⅞ inch strike is also available as a accessory from the manufacturer. Do not use a strike with a lip radius different than that supplied by the manufacturer. A short radius on the lip can damage the latch assembly.

Unless otherwise specified in your order, all Unican 1000 series locks are factory assembled for left-hand operation. To change it to right-hand operation, remove the back plate containing the cylindrical unit from the lock case by removing the Phillips head screws. Remove the four Phillips head screws holding the cylindrical unit at the back plate (FIG. 18-34) and rotate the cylindrical unit so that the cutout for the latch faces in the opposite direction. Reattach the cylindrical unit using the four screws (with two lock washers for each screw), making sure the unit is centrally positioned to the hole in the back plate and rotates freely when the spindle is turned. To prevent internal jamming, two lockwashers must be used with each screw. Remount the back plate assembly to the lock case.

All locks are factory set for a 1¾ inch thick door. To adjust for a 1⅜ inch to 1½ inch door, remove the back plate and cylindrical unit assembly from the lock case, and remove the spacer that is positioned between the cylindrical unit and the back plate (FIG. 18-35). Reassemble the cylindrical unit with the shorter (6-32 inch × ¼ inch) screws provided with the unit. Insert the roll pin in the front knob shaft one hole closer to the front knob (from position #2 to #3 in Fig. 18-35). Again, be sure that there are two lockwashers under each screwhead joining the cylindrical unit to the back plate.

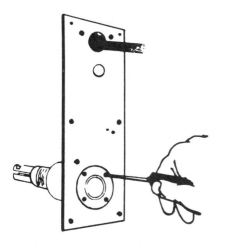

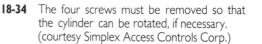

18-34 The four screws must be removed so that the cylinder can be rotated, if necessary. (courtesy Simplex Access Controls Corp.)

To adjust for 2 to 2¼ inch doors, follow the above instructions, but add the extra spacer provided (lip edge to lip edge) to the one already assembled to the lock. Use the longer (#6-32×¾ inch) screws provided, and insert the pin in the shaft in the hole furthest from the front knob (from position #2 to #1 in the shaft in the hole furthest from the front knob (from position #2 to #1 in Fig. 18-35). When remounting the cylindrical unit, be sure it is central to the hole in the back plate and turns freely. Double check to ensure the two lockwashers are under each screwhead.

Hold the latch assembly in the edge of the door and slide the lock case in from the front. Depress the latch slightly to engage the "T" (marked "A" in FIG. 18-36) in the lock retracting shoe (marked "B"). Viewing the unit from the back side of the 2⅛ inch hole, be sure the curved lips of the latch housing slide within the opening of the cylindrical case ("C") in FIG. 18-36. Hold the lock case to the front of the door with one hand, and insert the two top mounting screws from the back. Finger-tighten both screws, to start, to ensure the holes were properly positioned. If you are not able to start the screws into the back of the lock case by finger tightening, the mounting holes were not positioned correctly. Forcing can cause the threads to be stripped. To correct this condition, remove the lock case and file out the holes to ensure proper alignment of each to the nuts on the back plate.

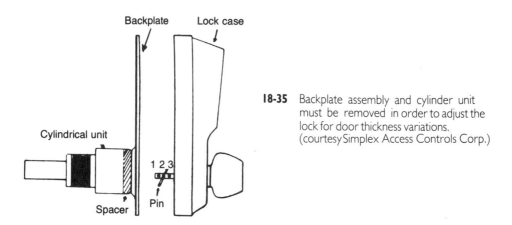

Backplate   Lock case

Cylindrical unit

1 2 3

Spacer   Pin

18-35 Backplate assembly and cylinder unit must be removed in order to adjust the lock for door thickness variations. (courtesy Simplex Access Controls Corp.)

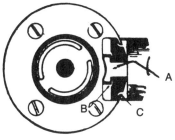

18-36   Latch assembly with critical portions indicated; these three points (A-B-C) must be worked in exact order to ensure proper lock installation. (courtesy Simplex Access Controls Corp.)

Screw on the inside rose and tighten with the spanner wrench. Slide the knob onto the inside sleeve, depressing the knob catch spring by pushing the pointed tip of the wrench through the hole in the side of the knob collar. When the knob is pushed into the stop, remove the tool and make sure the knob is securely fastened.

The trim plate is mounted horizontally over the 1 inch in diameter combination change hole and the two mounting screws (FIG. 18-37). With the center punch, indent the door slightly for the two protrusions on the plate used to prevent rotation. Cut the lock screw to length as required. Insert the control key and keep turning clockwise to thread the lock screw into the combination change sleeve assembly on the back plate of the lock case. Tighten the screw until the plate is snug to the door, but do not overtighten. The key can only be removed in a vertical or horizontal position. Overtightening may cause the lock screw to jam in the change sleeve, requiring the removal of the screw with your needle-nosed pliers.

All locks are factory set for the same combination: buttons #2 and #4 are pushed together, and then #3 individually. For security reasons, you should change the combination immediately, using as few as one button, or as many as five. When the buttons are set for pushing at the same time with other buttons, it is a different combination than if the buttons are set to be pushed individually (i.e., 1, 2, 3 individually pushed buttons is a different combination than if 1, 2, 3 are pushed at the same time). The same button cannot be used more than once in any given combination. If your Unican has a passage set function, be sure it is not in use when the combination code is changed.

To set the combination, the door must be open. Remove the inside trim plate with the control key. Turn the front knob fully clockwise, and then release. Push the existing

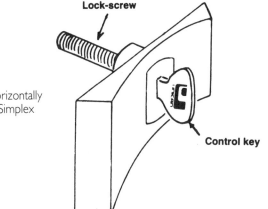

18-37   The trim plate must be positioned horizontally over the hole in the door. (courtesy Simplex Access Controls Corp.)

combination, and then release the buttons. The buttons should never be held depressed during lock opening. Insert the flat blade end of the special tool into the slots of the combination change sleeve and rotate clockwise to stop, but do not force. Then, rotate counterclockwise to stop and remove the tool. The combination change sleeve will rotate easily when the correct combination is pushed. It will not rotate with an incorrect combination, and forcing can damage the mechanism.

Now, turn the front knob fully clockwise to stop and release. Turn only once at this stage of the combination change. Depress the buttons firmly and deliberately in the sequence desired for the new combination. Record the combination at once, so it will not be forgotten. Turn the front knob fully clockwise to stop to set in the new combination. Try the new combination before closing the door and be sure you have properly recorded it. *Note:* If the Unican will open without pushing a code, you are in the "0" combination mode because the front knob was turned more than once, or out of sequence. In such a case, repeat all the steps in the change procedure, except pressing the existing (old) combination. It is very important that the front knob be turned fully *clockwise to stop* when called for in the combination change procedure to ensure that the old code is completely cleared out and the new code is completely set in. Full rotation will be impaired by any of the following:

- Using a latchbolt assembly not supplied by the manufacturer (especially one with a shorter backset or shorter throw).
- A shallow strike box that prevents the full ¾ inch throw of the latch when the door is shut.
- Any binding of the latch assembly in the door or in the strike caused by improper door or frame preparation or by door sag and/or buckling.

Further installation tips:

- M.A.G. type filler plates (FIG. 18-38) in 9 inch and 14 inch lengths are available for covering up existing mortise lock preparation or badly butchered doors. These can be obtained for 1¾ inch doors only. They are handy to use for an installation template as well.
- Do not allow either the inside or the outside knob to hit against the wall. A floor mounted door stop should be used to prevent damage to the factory adjusted clutch mechanism behind the knob. The manufacturer's one year warranties are voided if the knobs are allowed to hit. If there is a door closer, be sure that it is properly adjusted for the Unican.
- Due to its unique design, clockwise rotation of the front knob performs two functions. It clears out any incorrect combination attempts and reactivates any buttons previously pushed, and it withdraws the latch bolt if the combination has been pushed properly.
- When operating the lock, *never* hold the buttons depressed while turning the front knob, since damage to the unit can result. Just depress the buttons until they stop and then release them.
- Proper installation and operation of the Unican will result in many years of satisfied customer usage. Just remember that it starts with a good installation.

Servicing of the Unican 1000 series may be required. In the majority of instances, servicing will be the result of customer mishandling and abuse of the unit. Once in a very great while the unit itself may malfunction for some unknown reason. It is at this time that you require some basic information concerning the various parts of the Unican 1000 unit and illustrative material for the proper assembly or disassembly. Figure 18-39 is an exploded

18-38 M.A.G. filler plate. (courtesy M.A.G. Engineering and Mfg., Inc.)

view of the Unican 100. Figure 18-40 is a parts item description and the assembly number and parts.

All Simplex auxiliary locks are shipped with the following combinations: #2 and #4 pushed at the same time, then #3 pushed singly.

Some people forget how to operate this lock, even though the procedure is very simple:

Turn the control knob to the left to activate the buttons.

Press the correct buttons in the proper order.

Release the last button(s) before turning the control knob. Turn the control knob to the right to open.

To relock, turn the control knob to the left. Model NL locks automatically.

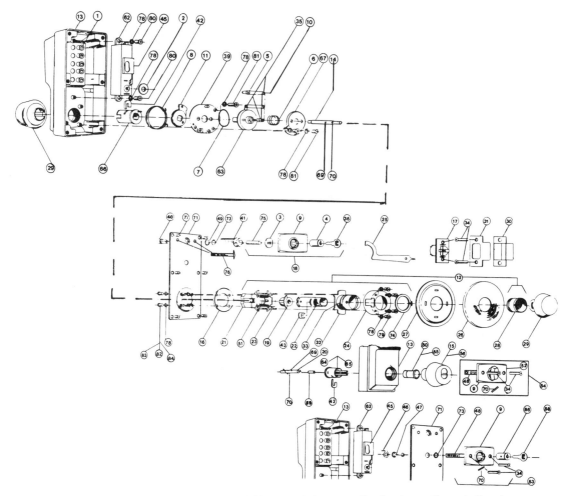

18-39   Exploded view of the Unican 1000 Lock unit. (courtesy Simplex Access Controls Corp.)

Some customers will look at the lock they contemplate purchasing and wonder about the control knob. Since it operates the bolt mechanism, if it is removed, can the lock be opened?

The answer is a resounding no. The front control knob cannot be forced to open the lock since it is connected to the lock housing by a friction clutch. If the knob has been forced, it will be at an angle and can be turned back to the vertical position by hand or with a pair of pliers without damaging the lock.

Preso-Matic

The Preso-Matic keyless pushbutton combination lock comes in either a deadbolt version (FIG. 18-41) or a deadlatch version (FIG. 18-42). This is another completely mechanically operated lock; no electricity is required. From the outside, the customer presses the selected four-digit combination. The deadbolt or deadlatch automatically retracts, opening the door. With the deadlatch model, the lock automatically locks each time the door is closed.

ITEM NO.	PART NUMBER	DESCRIPTION	PER UNIT
1	200024	Push Buttons	5
2	200026	Bushing,Shaft	1
3	200079	Spring, Nut	1
4	200080	See Assembly Chart	1
5	201006	Link, Combination Chamber	1
6	201018	Spring, Balance	1
7	201018	Spring Double Wave	1
8	201019	Knob Return Spring	1
9	201037	Inside Trim Plate Only	1
10	201038	Pin, Linkage	1
11	201040	Stop Plate	1
12	201048	See Assembly Chart	1
13	201049	Front Plate	1
14	201050	See Assembly Chart	1
15	201093	See Assembly Chart	1
16	201102	Spacer Door Thickness	2
17	201144	Latch	1
18	201155	See Assembly Chart	1
19	201218	Drive Insert	1
20	201267	See Assembly Chart	1
21	201272	Shoe	1
22	201275	Inside Sleeve	1
23	201276	Shoe Retainer and Bridge	1
24	201277	Cover for Shoe Housing	1
25	201278	Installation Wrench	1
26	201280	Rose	1
27	201281	Rose Reinforcing Plate	1
28	201282	Thread Ring	1
29	201284	Knob Standard	2
30	201287	Strike Box	1
31	201288	Strike Plate	1
32	201289	Shoe Retainer Cap	1
33	201290	O Ring 1 in. I.D.	1
34	201291	Screw, Philips Combination No. 8	4
35	201293	Clip	3
36	201298	Key DF 59 and Key Ring	2
37	201301	See Assembly Chart	1
38	201303	See Assembly Chart	1
39	201304	Clutch Backing Plate	1
40	201335	Combination Change Stud	1
41	201336	Combination Change Sleeve	1
42	201343	Front Knob Retainer Clip	1
43	201344	Inside Knob Retainer Clip	1
44	201345	See Assembly Chart	1
45	201350	Chamber Release	1
46	201351	Chamber Release Cam	1
47	201352	Bushing, Passage Set	1
48	201358	Connecting Bar	1
49	201362	Washer	1
50	201365	Knob Insert Best/Falcon	1
51	201366	Shoe Spring	2
52	201367	Turn Knob	1
53	201397	See Assembly Chart	1
54	201398	See Assembly Chart	1
55	201414	Knob Insert Russwin/Corbin	1
56	201416	See Assembly Chart	1
57	201417	See Assembly Chart	1
58	201422	Key DF 5 for Passage Control	2
59	201425	Shaft Self-Aligning Bushing 1000-2,4 & 6	1
60	201426	See Assembly Chart	1
61	201427	See Assembly Chart	1
62	201429	Combination Chamber	1
63	201430	See Assembly Chart	1
64	201431	See Assembly Chart	1
65	201432	See Assembly Chart	1
66	201433	See Assembly Chart	1
67	201434	See Assembly Chart	1
68	201435	See Assembly Chart	1
69	201436	Cross Pin .075 Dia.	1
70	201437	Cross Pin .100 Dia.	1
71	201438	Back Plate	1
72	201439	See Assembly Chart	1
73	201440	Retainer Ring 3/8 I.D.	1
74	201441	Retainer Ring 1 1/4 I.D.	1
75	201442	Screw Stud 1/4-20 X 1 5/8	1
76	201443	Screw 8-32 X 2 1/2	2
77	201444	Screw Philips 6-32 X 3/8 F.H.M.S.	6
78	201445	Split Washer No. 8	17
79	201446	Screw Philips 6-32 X 1/4 R.H.M.S.	4
80	201447	Screw Philips 6-32 X 5/16 R.H.M.S.	2
81	201448	Screw Philips 6-32 X 3/8 R.H.M.S.	3
82	201449	Screw Philips 8-32 X 1/4 R.H.M.S.	4
83	201450	Screw Philips 8-32 X 1/2 R.H.M.S.	4
84	201451	Screw Philips 8-32 X 3/4 R.H.M.S.	4
85	201490	See Assembly Chart	1
86	201496	See Assembly Chart	1
87	201497	See Assembly Chart	1

ASSEMBLY CHART

ASS'Y NO.	INCLUDING ITEM NUMBERS	ASSEMBLY DESCRIPTION
200080	4, 36	Control Lock Assembly Combination Change (DF-59)
201048	19, 21, 22, 23, 24, 28, 32, 33, 43, 51, 74, 78, 79	Cylindrical Drive Unit Assembly
201050	14, 69, 70	Shaft Assembly Standard
201093	15, 50	Knob Assembly Key Override for Best/Falcon
201155	3, 4, 9, 36, 75	Inside Trim Plate Assembly
201267	20, 69, 70	Shaft Assembly, Key Override
201301	16, 19, 21, 22, 23, 24, 32, 33, 40, 41, 43, 49, 51, 71, 73, 74, 78, 83	Back Plate Assembly
201303	26, 27, 28	Rose Assembly
201345	22, 32, 33, 43	Inside Drive Sleeve Assembly
201397	9, 34, 48, 58, 70, 86	Key Actuator Assembly for Passage Set
201398	9, 34, 48, 52, 70	Turn Knob Actuator Assembly for Passage Set
201416	55, 56	Knob Assembly Key Override for Corbin/Russwin
201417	20, 42, 55, 56, 59, 65	Conversion Assembly 1000-1 to 1000-2C
201426	1, 2, 8, 13, 14, 29, 42,62,63,66,67,78,80,81	Front Plate Assembly 1000-1
201427	1, 2, 8, 13, 15, 20, 42, 50, 59,62,63,64,67,78,80,81	Front Plate Assembly 1000-2B
201430	6, 7, 63	Clutch Sub-Assembly
201431	11, 64	Sleeve Drive Assembly Key Override for Best/Falcon
201432	11, 65	Sleeve Drive Assembly Key Override for Corbin/Russwin
201433	11, 66	Sleeve Drive Assembly Standard
201434	39, 67	Clutch Cover Assembly
201435	5, 10, 35	Link Assembly
201439	40, 41, 49, 71, 73	Back Plate Sub-Assembly
201490	15,20,42,56,59,64	Conversion Assembly 1000-1 to 1000-2B
201496	58, 86	Control Assembly for Passage Set (DF-5)
201497	1, 2, 8, 13, 15, 20, 42, 55, 56, 59,62,63,65,67,78,80,81	Front Plate Assembly 1000-2C

18-40 Parts identification and assembly chart for the Unican 1000 lock unit. (courtesy Simplex Access Controls Corp.)

18-41 Pres-Matic keyless pushbutton deadbolt combination lock. (courtesy Pres-Matic Lock Co., Inc.)

With several models available, each customer can have a version with the appropriate options to meet his special need. All models come with an instant-exit feature that requires only the depressing of an unlock button on the door inside (FIG. 18-43). Also available is a nightlatch button, which provides an added measure of protection for the home or apartment owner or the business person working late at night. It means that nobody is going to interrupt (or surprise!) him at work or at home.

For businesses that rely upon electric strikes operated from remote locations, the unit can also be used. The Trine model 012 electric strike unit is given a unique configuration by the Preso-Matic factory to accept the deadlatch bolt.

The Preso-Matic lock has several combination possibilities:

- ☐ Four-digit combination.
- ☐ Seven-digit combination.
- ☐ Master combination (for use with the four-digit combination version).

The four-digit combination provides for a maximum of 10,000 possible combinations; the seven-digit combination can give up to ten million possible combinations. The Master Combination (MC) provides for a master combination in addition to the individual four-digit combination for units. In this way, all locks of a given system will open on one combination (the MC) which is a six-digit number, but each unit within the system can still be opened by its own four-digit combination. The MC combination variations allow for security protection and flexibility that are not available with other pushbutton type keyless combination locks. For concerns that require numerous units, the advantages to top management of a master combination are important.

The numbers of the combination for each lock unit are determined by the two "combination slides" (FIG. 18-44) which are inserted into the unit. To change the combina-

18-42 Deadlatch version of the same lock. (courtesy Preso-Matic Lock Co., Inc.)

18-43   Instant exit feature and also locking ability with just a touch of a button. (courtesy Preso-Matic Lock Co., Inc.)

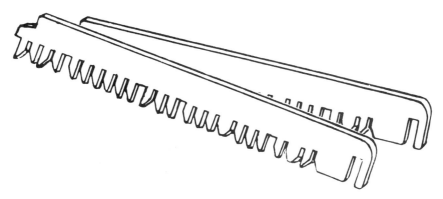

18-44 Combination slides used with the lock unit. (courtesy Preso-Matic Lock Co., Inc.)

18-45 Preso-Matic installation template, 2 3/8″ backset.

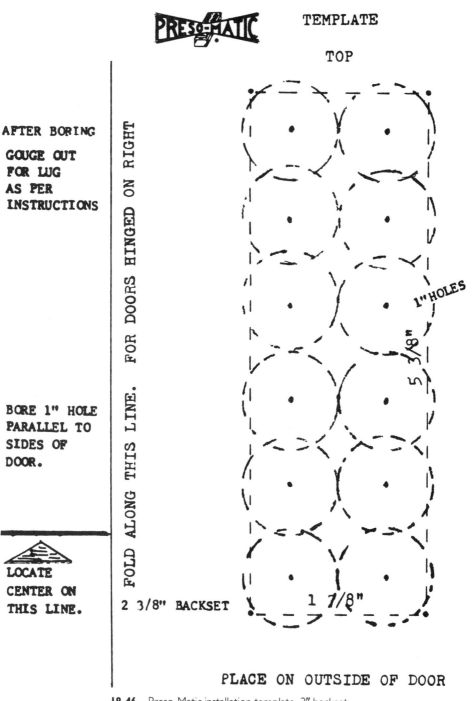

PRESO-MATIC   TEMPLATE

TOP

AFTER BORING

GOUGE OUT
FOR LUG
AS PER
INSTRUCTIONS

FOR DOORS HINGED ON RIGHT

1" HOLES

5 3/8"

BORE 1" HOLE
PARALLEL TO
SIDES OF
DOOR.

FOLD ALONG THIS LINE.

LOCATE
CENTER ON
THIS LINE.

2 3/8" BACKSET

1 7/8"

PLACE ON OUTSIDE OF DOOR

18-46   Preso-Matic installation template, 3" backset.

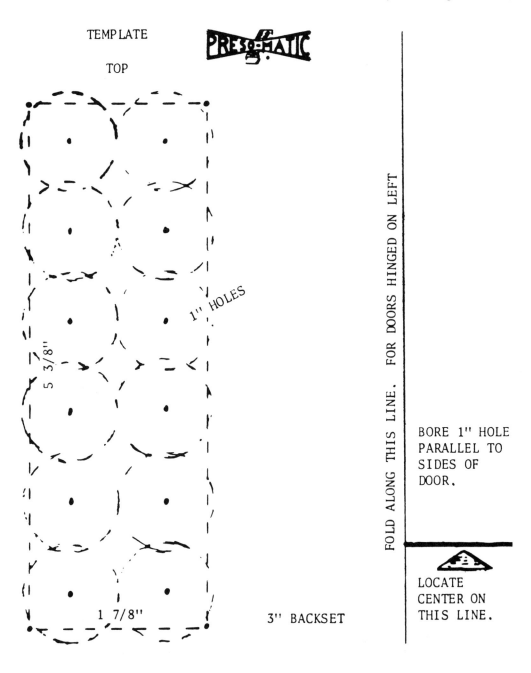

TEMPLATE

TOP

1" HOLES

5 3/8"

1 7/8"

3" BACKSET

FOLD ALONG THIS LINE. FOR DOORS HINGED ON LEFT

BORE 1" HOLE
PARALLEL TO
SIDES OF
DOOR.

LOCATE
CENTER ON
THIS LINE.

PLACE ON OUTSIDE OF DOOR

18-47 Preso-Matic installation, Step I.

18-48   Preso-Matic installation, Step 2.   **18-49**   Preso-Matic installation, Step 3.

MODEL	LOCK TYPE	BOLT PROJECTION	BACKSET	NOTES
8101	Deadbolt	⅝″	2⅜″ (2¾″ also available)	Manual locking. Built-in night latch button. Hardened steel free-turning bolts. Manually locks.
LT8102	Deadbolt	1″	3″	Same
8103	Deadlatch	½″	3″	Night latch button. Locks automatically. Inswinging doors only.
8103A	Deadlatch	½″	3″	Same, except without night latch button.
8200	Deadlatch	½″	3″	Hardened steel latchbolt. Locks automatically. Night latch button.
8200A	Deadlatch	½″	3″	Same, except without night latch button.

18-50   Installation layout variations. (courtesy Preso-Matic Lock Co., Inc.)

OPTIONS:
1. Change of combination slides for a 7 digit code available.
2. Lock shields.
3. Stay open function; keeps the deadlatch in an unlocked position.
4. Door stop strike plate; deadlatch models only. Case hardened; ideal for double door installation.
5. Extra wide strike plate for metal doors (2½″ × 2½″); case hardened.
6. Heavy duty ANSI electric door opener. 24 Vac. for normally locked units.

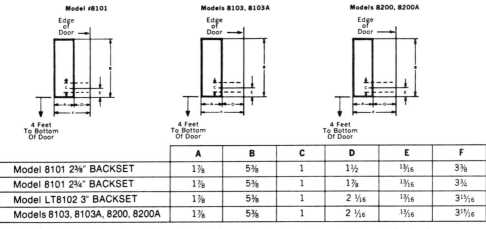

	A	B	C	D	E	F
Model 8101 2⅜″ BACKSET	1⅞	5⅜	1	1½	¹³⁄₁₆	3⅜
Model 8101 2¾″ BACKSET	1⅞	5⅜	1	1⅞	¹³⁄₁₆	3¾
Model LT8102 3″ BACKSET	1⅞	5⅜	1	2¹⁄₁₆	¹³⁄₁₆	3¹⁵⁄₁₆
Models 8103, 8103A, 8200, 8200A	1⅞	5⅜	1	2¹⁄₁₆	¹³⁄₁₆	3¹⁵⁄₁₆

18-51 Overview of the various lock models. (courtesy Preso-Matic Lock Co., Inc.)

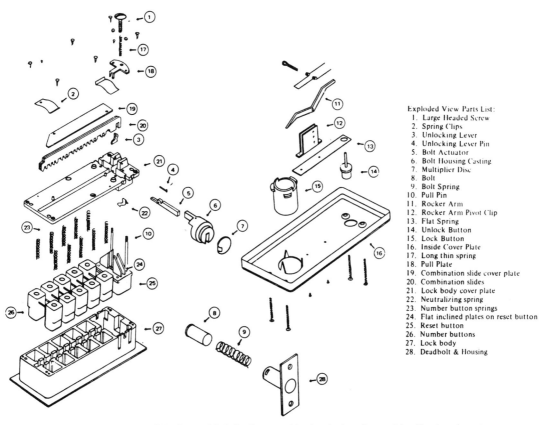

Exploded View Parts List:
1. Large Headed Screw
2. Spring Clips
3. Unlocking Lever
4. Unlocking Lever Pin
5. Bolt Actuator
6. Bolt Housing Casting
7. Multiplier Disc
8. Bolt
9. Bolt Spring
10. Pull Pin
11. Rocker Arm
12. Rocker Arm Pivot Clip
13. Flat Spring
14. Unlock Button
15. Lock Button
16. Inside Cover Plate
17. Long thin spring
18. Pull Plate
19. Combination slide cover plate
20. Combination slides
21. Lock body cover plate
22. Neutralizing spring
23. Number button springs
24. Flat inclined plates on reset button
25. Reset button
26. Number buttons
27. Lock body
28. Deadbolt & Housing

18-52 Exploded view of the Preso-Matic keyless combination lock and parts identification. (courtesy Preso-Matic Lock Co., Inc.)

tion, you simply remove the inside cover plate, turn two spring clips clear of the slot in the lock housing, lift out the combination slides, and slip in new ones.

All the Preso-Matic locks are carefully designed and engineered to aid you in quick, easy, trouble-free installation. Actually, there are only three basic steps:

1. Bore the opening for the lock and latch (FIG. 18-45) using the required template that comes with the unit (FIGS. 18-46, 18-47).

2. Mortise the door edge for the bolt face and insert the bolt housing and tighten the screws (FIG. 18-48).

3. Insert the lock and tighten only one screw, connecting the bolt housing to the lock and making the unit operable (FIG. 18-49). Attach the cover plate on the inside of the door. Installation is complete.

More specific installation procedures will be given further on. It must be noted that for volume installations in large construction projects or numerous prefabricated doors, you should use the Preso-Matic round face drive-in bolt which installs in seconds and saves time.

Three basic installation layouts are available, as shown in FIG. 18-50. Figure 18-51 is a quick summary of the various lock models. Figure 18-52 provides an exploded view of the lock and part identification.

19

Electrical access and exit control systems

*E*lectrically operated release latch strikes and locks are easily installed in place of standard units to provide remote controlled access for a door. These units are not new to business and industry and really shouldn't be that new to the locksmith—but they are. Why? Because, except for a few instances in large cities or industrial regions, the average locksmith doesn't get involved with such units. Since it is "electrical," the job is passed on to an electrician. By so doing, the locksmith loses business, and valuable business at that.

Throughout this chapter, we will be looking at some electrically operated locks and release latch strikes. We will also discuss the circuitry behind the various types of units, where they can be installed, and the potential of such units.

ELECTRIC RELEASE LATCH STRIKES

Electrically actuated release latch strikes are more common than locks, so this subject area will be dealt with first. These units can normally be reversed for either right or left-hand doors. They operate on low voltage and the strikes will fit a hollow jamb channel as shallow as 1.6 inches (1.75 for mortise latches) of they can be, if necessary, installed in wooden jambs. Further on in this discussion we'll provide the information necessary for you to consider in-unit selection.

In this important area, though, especially since it is a very rapidly-growing one in terms of access control (security), it is important that you receive that latest and best information available. We contacted the manufacturer considered the nation's leader in the electrical release latch strike field, Adams Rite, in order to obtain the following information. For Adams Rite's assistance we are extremely grateful.

Electrical release strikes are gaining popularity among small and large business concerns; it is evident that the trend will continue to increase at a greater rate during the coming decade. The electrical release strike comes in a variety of sizes (FIG. 19-1) to meet varying needs and requirements.

There are many electrical strikes available on the market, but while all of them meet minimum standards, you should still want to consider the manufacturer and the purpose for which the units were initially designed and developed prior to your purchase of them.

First and foremost in the field is Adams Rite, whose electrical strikes are architecturally sound and designed to add increased security features to areas requiring a certain amount of

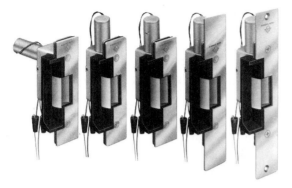

19-1 A sampling of the various types and sizes of electric release latch strikes. (courtesy Adams Rite Mfg. Co.)

traffic control. When properly powered and installed, these electrical release strikes will provide a very long, maintenance-free life.

The fact that electrical power is used for the operation of the strike in no way makes the unit an electrical appliance, as some people tend to think. With a few exceptions, the electrical circuitry unique to these units are designed to meet the needs of the strike, not the other way around. The hardware specifier (that's you, by the way) should select the strike and expect the electrician (which, again, *could* be you) to supply the power at the point of installation. In a few cases an existing circuit can be used, and thus the strike is selected and installed to it. In selecting strikes, two areas of consideration must be followed: electrical and hardware.

ELECTRICAL CONSIDERATIONS

The first electrical factor is to determine the duty of the unit. Is the operation to be intermittent or continuous? If the door is normally locked and released only momentarily from time to time, it is intermittent. If it is the rare case where the strike is activated (unlocked) for long periods, the duty becomes continuous. A still rarer requirement is reverse action in which the strike is locked only when its current is switched on.

Intermittent duty. For a normal intermittent application, the electrical strike will have to be 24 volts ac. This alternating current call-out gives enough power for almost any entrance, even one with a wind-load situation. Yet the low voltage range is below that requiring U.L. or building code supervision. At this voltage, good, reliable transformers are available.

Continuous duty. When a continuous duty application is required, additional components are added to the unit at the factory.

Reverse action. A long period of unlocking can also be obtained by using a reverse action strike. This might be required to provide the same service as a continuous duty strike, but preserves current because it is on battery dc power. It could also be used to provide a "fail-safe" unlocked door in case of a power failure or building emergency (such as a fire) where quick, guaranteed unlocked doors are required. *Note:* Reverse action strikes, when ordered by the locksmith for installation, must specify the hand of the door on the order. These types are not reversible, as are some other strikes.

Monitoring. If a visual or other signal is required to tell the operator that the electric strike is doing its job properly, then a monitoring strike will be required. This will mean that two sensor/switches are added. One is activated by the latchbolt's penetration of the strike, and the other by the solenoid plunger that blocks the strike's release.

Transformer. Low voltage is necessary for proper electric strike operation, and this is obtained by the use of transformers which are stepped down from the normal ac power voltage to 12, 16, 24, or other lower voltage. Whether or not you perform the electrical side of the installation (when wires are required to be run) three electrical specifiers are necessary:

□ Input voltage (110—standard household current).

□ Output voltage (12, 16, 24, etc.—from the transformer unit).

□ The capacity of the transformer, called volt amps (output voltage times the output amperes).

Skimping on the capacity of the transformer to save a few dollars will underpower the door release and is likely to bring in complaints of "porr hardware." Intermittent duty electric strikes draw from 1½ to 3 amps; a continuous duty model draws less than 1 amp.

In any application where the current draw of the strike must be restricted to less than 1 amp, regardless of the duty, a current limiter which stores electrical energy to relieve high use periods is available.

Wiring. The wiring used must carry the electrical power from the transformer through the actuating switch to the door release. It must be large enough to minimize "frictional" line losses and deliver most of the input from the transformer to the door release. A small diameter garden hose won't provide a full flow of water from the nozzle, particularly if it's a long run; neither will an undersized wire carry the full current!

Electrical troubleshooting. When insufficient electrical power is suspected in a "weak" door release, a simple check can be made. Measure the voltage at the door release while the unit is activated. If the voltage is below that specified on the hardware schedule, the problem is in the circuit—probably an under-capacity transformer, if the current length is short. A long run may mean both a transformer and a wire problem.

The second consideration concerns the hardware itself, and has several points:

Face shape (sectional): With two exceptions, electrical strikes from Adams Rite have a flat face. These exceptions are the two having a radius-type to match the nose shape of a paired narrow stile glass door.

Face size: The basic Adams Rite size conforms to ANSI standards for strike preparation: 1¼ inches × 4⅞ inches. However, two other sizes are offered to fill or cover existing jamb (or opposing stile) preparation from a previous installation of another unit, such as an MS deadlock strike or a discontinued series electric strike.

Face corners: Strikes are available with two types of corners: round for installation in narrow stile aluminum where preparation is usually done by router, and square corners for punched hollow metal ANSI preparation or for wood mortise installation.

Jamb material: Strikes with vertically mounted solenoids are designed to slip into hollow metal stile sections as shallow as 1.6 inches. Horizontal solenoids for wood jambs requires an easily bored mortise of 3⅛ inch depth.

Previous strike preparation: If the jamb was previously fitted with another, now discontinued strike, such as a Series 002 electrical strike (for example), the new strike unit to be installed—say an Adams Rite 7510—will fit in with only minor alterations. In this example, it would mean increasing the lip cutout ¼ inch at each end. In the case where a hollow jamb was originally prepared to receive the bolt from another type deadlock, another latch such as the 4510 series would have to be substituted. In such a case, the radius or flat jamb would cover the old strike cutout.

Lip length: the standard lip on all basic Adams Rite electric strikes accommodates a 1¾ inch thick door which closes flush with the jamb. Where the door/jamb relationship is different, a long lip can be added.

Compatible latch: Certain electric strikes will mate with all latches in a given series. The key-in-knob latches and mortise latches are also made to mate properly with the various electric strikes available.

Handing of the strike unit: All standard operation strikes are unhanded; thus, they can be installed for either a right or a left-hand door. However, reverse action (locked only when the current is on) units require a specification of the hand when the order for the unit is placed.

To repeat: understand that electrical strikes are not just electrical appliances; they are hardware. In the case of Adams Rite strikes, they are made by hardware people, because they know that the hardware has to be carefully matched to its application.

Also realize that when a unit is ordered for installation, check with the local electrical codes to ensure that you can actually install it. In some jurisdictions, an electrician may be required to do the job.

INSTALLATION AND PROCEDURES

Let's move on to discuss and illustrate some of the procedures required for a variety of strike units.

Series 750 for an aluminum jam installation:

Prepare the door jamb per FIG. 19-2.

Install the mounting clips to the jamb using 8-32 × ⅜ screws and pressed metal nut. Leave the screws slightly loose to permit easy alignment of the base assembly and clip.

Spacers are provided to assure flush final assembly of the face plate and jamb. Add one or more spacers between the jamb and mounting clip when the face plate extends beyond the jamb. When the face plate sets inside the jamb, spacers must be added between the mounting clip and the electric strike case. See detail "A" in FIG. 19-2. To attach spacers to the mounting clip, remove the protective coating from the spacer and press to the desired mounting clip surface. Make sure the clearance hole in the spacer aligns properly with the hole in the mounting clip.

Using the wire nuts, connect the wires coming from the unit to the wires coming from the low voltage side of the transformer.

Insert the electrical release into the jamb by tipping the solenoid coil into place behind the mounting clip, then drop the unit onto the clips.

Attach the unit with the #10 screws and lock washers provided.

Attach the face plate using the screws furnished (8-32 × ¼ inch).

Secure the screws holding the mounting clips to the jamb.

Figure 19-2 also shows the dimensions of the electric door release that would be installed in the aluminum jamb.

Replacing an obsolete unit with an electrical strike model 7510 wood jamb:

Prepare the jamb to the dimensions shown in FIG. 19-3. Care should be taken to mortise cut, to clear the solenoid.

Using wire nuts provided, connect the wires from the 7510 unit to the wires from the low voltage side of the transformer.

Insert the electric release into the jamb. Attach this with #10 flat head wood screws.

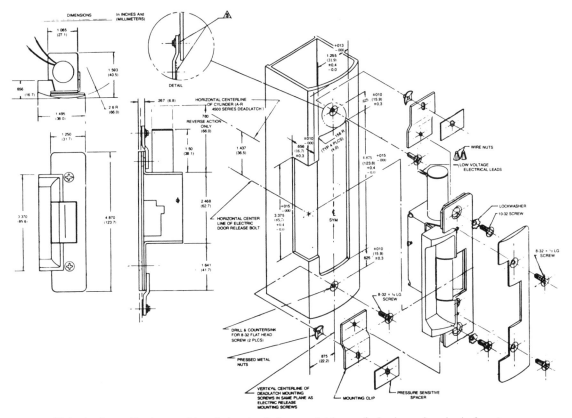

19-2 Jamb specifications and installation details for a radial (curved) aluminum door jamb. (courtesy Adams Rite Mfg. Co.)

Replacing the obsolete unit (002 as an example):

Remove the 002 unit and disconnect the low voltage wires. Do not remove the existing mounting clips.

Enlarge the jamb cutout by removing .250 inch from each side.

Using the wire nuts, connect the 7510 wires to the low voltage side of the transformer.

For a flush installation, it will be necessary to provide a .062 spacer between the mounting surface and the face plate.

Insert the electric door release into the jamb, and attach with screws.

Figure 19-3 also shows the dimensions of the 7510 electrical door release unit.

The next installation is for an electric strike, model 7520, to meet ANSI specifications. Note the 'molded' strip down the center of the door frame in the illustration. Installation is as follows:

Prepare the metal jamb to the dimensions shown in FIG. 19-4.

Spacers are provided with the unit to assure the flush mounting of the face plate to the jamb. To use the spacer(s), remove the protective cover and press them to the mounting bracket. Make sure the clearance hole in the spacer aligns with the tapped hole in the bracket.

Connect the wires from the unit to the transformer wires.

Insert the electric release into the jamb by tipping the solenoid coil into place behind the mounting bracket. Then drop the unit onto the mounting brackets. Attach the unit with the screws furnished.

Attach the face plate.

This installation is for an electric strike for key-in-knob latch, set to ANSI specifications. Here we look at an Adams Rite 7540 model Unit:

Prepare the door jamb as shown in FIG. 19-5.

Install the mounting clips using the screws and pressed metal nut. Leave the screws slightly loose to permit easy alignment of the case assembly and clip.

Use the provided spacers as necessary.

Connect the wires from the 7540 unit to the low voltage wires from the transformer.

Insert the electric release into the jamb and attach with the #10 screws provided.

Attach the face plate and secure the screws holding the mounting clip to the jamb.

This installation is for an electric strike into a wooden jamb:

Prepare the wood door jamb to the dimensions indicated in FIG. 19-6.

Connect the wires from the unit to the transformer wires.

Insert the strike into the jamb and attach with the #10 wood screws.

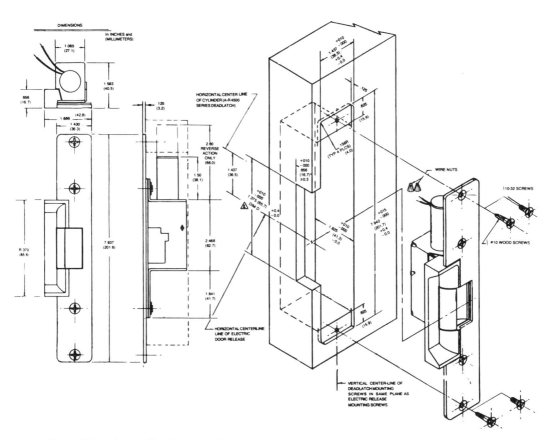

19-3 Dimensions and installation details for unit to be mounted in a wood frame. (courtesy Adams Rite Mfg. Co.)

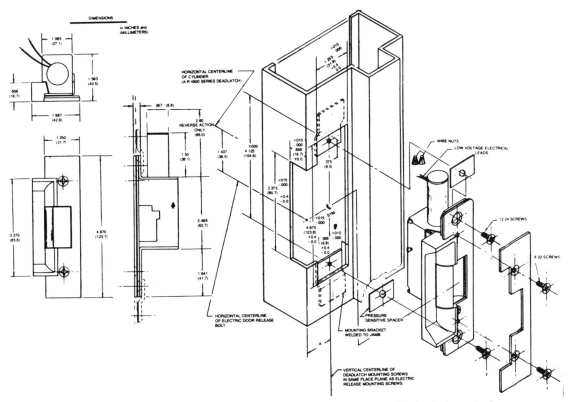

19-4 Jamb preparation and dimension data for a metal frame that has a molded strip down the frame center. (courtesy Adams Rite Mfg. Co.)

Attach the face plate.

Figure 19-7 provides the technical dimensions of the strike unit to be installed.

CIRCUITRY

Figure 19-8 illustrates the typical wiring diagram for an audible (ac) operating strike unit. Figure 19-9 is the same unit, but with a rectifier used to provide for silent (dc) operation of the strike unit.

You might have an instance where the installation requires a monitoring signal operation. In this case, FIG. 19-10 provides such a typical wiring diagram. Many possible monitoring arrangements can be had: the green light indicates that the latch bolt is in the strike *and* that the solenoid plunger is blocking the strike. The yellow light indicates the latch bolt is in the strike but is unblocked, allowing access. The red light signals that the latch bolt is out of the strike.

You have seen several of the more common among a dozen-plus types of electrical strike release units. Let's consider that you have installed one or more of these types, and that for one reason or another, you have some trouble—after installation—with the unit. What do you do?

Troubleshooting Adams-Rite electric strike latch units

Accurately checking an electrical circuit, just as any other job, requires specific tools to perform. Make sure you have a *good* 20,000 volts per ohm volt-ohm-milliammeter. The Simpson model #261, one of several available on today's market, will cost about $60, but it is money well spent, especially when you will be installing and maintaining electric strike latch units.

First, read the volt-ohm-milliammeter (vom) instructions and make some practice runs on simple low voltage circuits to ensure that you properly understand the operating instructions and can correctly read the results on the meter.

1. *How to check voltage.*
    a. Zero the pointer.
    b. Be sure the power is turned off to the circuit being measured.
    c. Set the function switch to the correct voltage to be measured (+dc or ac).
    d. Plug the black test lead into the common (−) jack; plug the red test lead into the (+) jack.
    e. Set the range selector to the proper voltage scale. *Caution:* it is very important that the selector be positioned to the nearest scale voltage to be measured.
    f. Connect the black test lead to the negative side of the circuit and the red lead to the positive side. This is applicable to the dc circuit only. Turn the power on to the

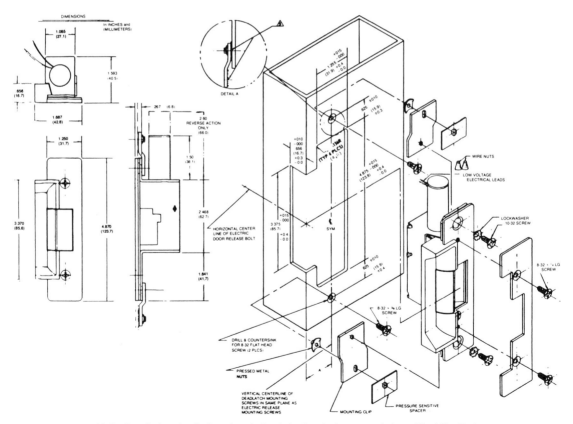

19-5   Installation details for a key-in-knob latch unit. (courtesy Adams Rite Mfg. Co.)

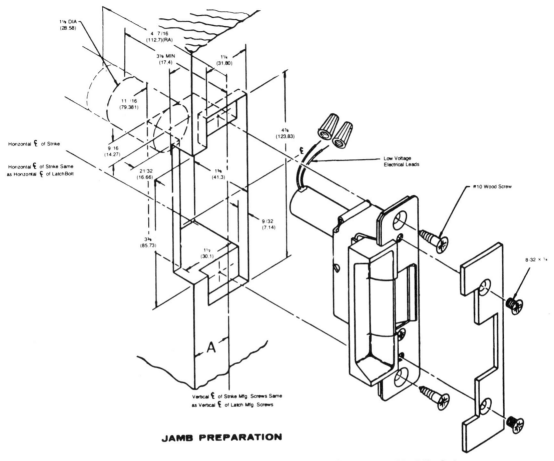

JAMB PREPARATION

19-6 Another wooden door jamb preparation. (courtesy Adams Rite Mfg. Co.)

circuit being tested. If the pointer on the vom moves to the left, the polarity is wrong. Turn the switch function to −dc and turn the power back on. The pointer should now swing to the right for the proper reading on the dc scale. For an ac circuit, the connections are the same except as noted in c above. You don't have to worry about polarity in the ac circuit.

    g. Be sure to turn the circuit off before disconnecting the vom.

2. *How to check current (dc).*

    a. Zero the pointer.

    b. Be sure the power is turned off to the circuit being tested.

    c. Connect the black test lead to the −10A jack and the red test lead to the +10A jack.

    d. Set the range selector to 10 amps.

    e. Open the circuit to be measured by disconnecting the wire that goes to one side of the solenoid. Connect the meter in series—that is, hook the black lead to one of the disconnected wires and the red lead to the other wire.

    f. Turn the power on to the circuit and observe the meter. If the pointer moves to the left, reverse the leads in the jacks.

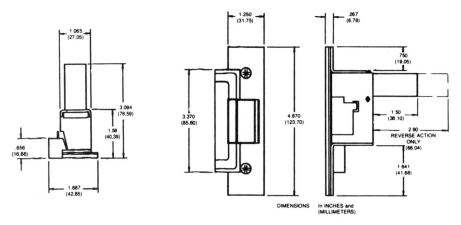

19-7   Strike unit dimensions for wood door jamb installations. (courtesy Adams Rite Mfg. Co.)

    g. Turn the power back on to the circuit and read amperage on the dc scale.

    h. Turn the vom off before disconnecting.

3. *Checking for line drop.* Measure the line drop by comparing voltage readings at the source (the transformer's secondary or input side) with the reading at the electric strike connection.

4. *How to find circuit shorts in a hurry.* Again the vom is the most reliable instrument for detecting a circuit short. This is accomplished by setting up the vom to measure resistance.

    a. Set the range switch at position R×1.

    b. Set the function switch at +dc.

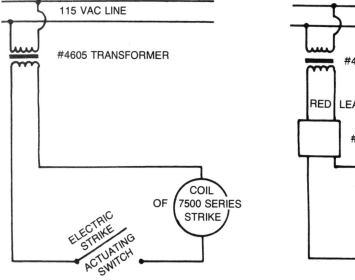

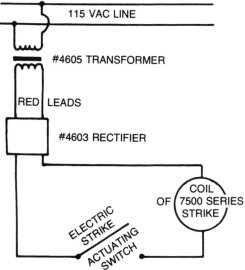

19-8   Typical wiring diagram for an audible (ac) operation. (courtesy Adams Rite Mfg. Co.)

19-9   Same circuitry, but with a rectifier added for silent (dc) operation. (courtesy Adams Rite Mfg. Co.)

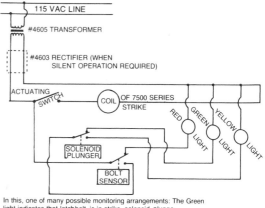

19-10 Wiring for a monitor signal operation. (courtesy Adams Rite Mfg. Co.)

In this, one of many possible monitoring arrangements: The Green light indicates that latchbolt is in strike solenoid plunge is blocking strike. Yellow light indicates latchbolt is in strike but strike is unblocked, allowing access. Red light signals that latch bolt is out of strike.

    c. Connect the black test lead to the common (−) jack and the red lead to the + jack.

    d. Zero the pointer by shorting the test leads together.

    e. Connect the other ends of the test leads across the resistance to be measured—in the case of a solenoid, one of the test leads to one coil terminal, and the other test lead to the other terminal.

    f. Watch the meter. If there is no movement of the pointer, the resistance being measured is open. If the pointer moves to the peg on the right hand side of the scale, the resistance being measured is shorted closed. If you get a reading in between these two extremes, it is very likely the solenoid is okay.

What to do when:

1. The strike will not activate after installation:
   a. Check the fuse or circuit breaker supplying power to the system.
   b. Check to make sure that all wiring connections are securely made. When wire nuts are used, ensure that both wires are firmly twisted together in order for good electrical transfer between the wires.
   c. Check the solenoid coil rated voltage (as shown on the coil label) to make sure that it corresponds to the output side of the transformer within ±10%.
   d. Using the vom, check the voltage at the secondary (output) side of the transformer.
   e. Using the vom, check the voltage at the solenoid. This would assure you that there are no broken wires, bad rectifiers, or bad connections.
   f. Check the coil for a possible short.

2. Transformer overheats:
   a. Make sure that the rated voltage of the transformer and the rated voltage of the coil correspond within ±10 percent.
   b. Make sure the VA rating is adequate. For all Adams-Rite units, it is recommended you see 40 VA. 20 VA is the absolute minimum, and you could experience transformer heating in even moderate use applications.

3. Rectifier overheats:
    a. The rectifier is wired wrong. This means the overheating is of a temporary nature (a few milliseconds) and then it's burned out; or,
    b. There are too many solenoids being supplied by a single rectifier and you are pulling more current through it than the system diodes are rated for.
4. Solenoid overheats:
    a. The coils used in an electric strike latch, when used as rated (continuously or intermittently by pulsing once per second) gave a coil temperature rise rating of 65 degrees C (149 degrees F) above ambient. That means 149 degrees F plus 72 degrees F, for example, equals 221 degrees F. The coil insulation is rated at 130 degrees C, which is 266 degrees F. Regardless of whether the coil goes to 221 degrees or 266 degrees, we're talking about something too hot to keep your fingers on!
    b. With the above in mind, I think it is fair to say the vast majority of intermittent duty units never see that kind of use. If you have a coil that gets extremely hot on very short pulses at two or three second intervals, you either have the wrong coil or the wrong transformer output. The same is basically true for continuous duty coils. If the coil temperature exceeds the ratings, it has to be because the coil voltage or the transformer are improperly coordinated.
    c. If you set the meter up as if testing for a short (see above) and obtain the exact resistance, you can compare to the specifications found in FIG. 19-11; this becomes another way of knowing you have the correct coil. Figure 19-11 also provides all the power data for the various electric strikes from Adams Rite that you might ever be required to install on various premises.

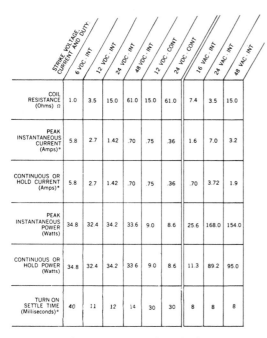

STRIKE VOLTAGE, CURRENT AND DUTY	6 VDC. INT	12 VDC. INT	24 VDC. INT	48 VDC. INT	12 VDC. CONT	24 VDC. CONT	16 VAC. INT	24 VAC. INT	48 VAC. INT
COIL RESISTANCE (Ohms) Ω	1.0	3.5	15.0	61.0	15.0	61.0	7.4	3.5	15.0
PEAK INSTANTANEOUS CURRENT (Amps)*	5.8	2.7	1.42	.70	.75	.36	1.6	7.0	3.2
CONTINUOUS OR HOLD CURRENT (Amps)*	5.8	2.7	1.42	.70	.75	.36	.70	3.72	1.9
PEAK INSTANTANEOUS POWER (Watts)	34.8	32.4	34.2	33.6	9.0	8.6	25.6	168.0	154.0
CONTINUOUS OR HOLD POWER (Watts)	34.8	32.4	34.2	33 6	9.0	8.6	11.3	89.2	95.0
TURN ON SETTLE TIME (Milliseconds)*	40	11	12	14	30	30	8	8	8

* Readings taken from oscilloscope. Photographs made.

19-11 Power data for electric strike units. (courtesy Adams Rite Mfg. Co.)

The last topic to cover concerns the correct determination of the proper wire size to use for electric strikes. Some people—especially those not trained in the various phases of proper electricity—assume that a low voltage means you can use the smallest and cheapest wire size available. This is certainly not the case. The size of the wire is dependent upon the use, the amount (voltage) of electricity going through the wire, the distance of the wire run, and also the number of strikes to be used on the wire run.

To determine the correct wire size, obtain the wire size factor from Table 1 of FIG. 19-12 and multiply this number by the distance to the strike (in feet). Take the resulting number in circular mills (cm) to determine the wire size required from Table 2 of the figure.

The following two examples will show you what we mean:

Example 1: A 24 volt ac intermittant (24 Vac-int) is to be mounted 60 feet from the transformer. The wire size factor for the unit is 34. Multiply 34 times 60 and this equals 2040. 2040 on table 2 corresponds to a wire size of 16 gauge.

Example 2: What is the maximum distance that you could mount the same 24 Vac int using a 20 gauge wire? From Table 2, the largest CM number corresponding to 20 gauge is 2000. This, divided by the wire size factor for 24 Vac-int, which is 34, produces the number 29; 29 feet is the maximum distance.

In the case when two or more strikes are to be actuated simultaneously on the same circuit, simply multiply the wire size factor by the number of strikes. Then, multiply by the distance to the farthest strike to obtain the cm number, and thus the wire gauge required for the installation.

ELECTRIC DOOR OPENERS

Another major supplier of electric door openers in the Trine Consumer Products Division of the Square "D" Company. Trine units are available for standard wood and metal doors, in addition to the narrow stile doors, such as those obtainable from Adams-Rite (FIG. 19-13). Fourteen different models are currently available for such diverse application as entrance doors to apartments, banks, institutions, industrial buildings and complexes, offices, and interlocking door control.

Installed in the door jamb in place of the lock strike or in conjunction with various other locksets, these units offer the ability to lock or unlock a door from a remote control

STRIKE USED	WIRE SIZE FACTOR
6 VDC — INT	128
12 VDC — INT	50
24 VDC — INT	13
48 VDC — INT	3.5
12 VDC — CONT	14
24 VDC — CONT	3.5
16 VAC — INT	10
24 VAC — INT	34
48 VAC — INT	9

CM	WIRE GAUGE
1000 AND BELOW	20
1001 — 1600	18
1601 — 2550	16
2551 — 4100	14
4101 — 6500	12

19-12 Tables provide data for determining what size required based on type of electric strike unit used. (courtesy Adams Rite Mfg. Co.)

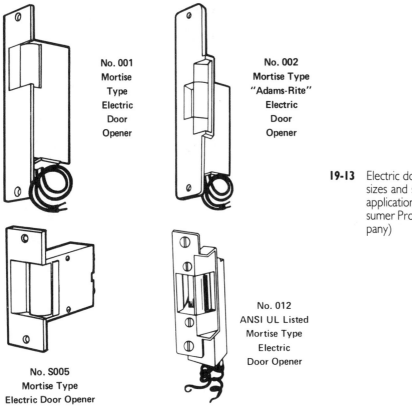

No. 001
Mortise
Type
Electric
Door
Opener

No. 002
Mortise Type
"Adams-Rite"
Electric
Door
Opener

No. S005
Mortise Type
Electric Door Opener

No. 012
ANSI UL Listed
Mortise Type
Electric
Door Opener

19-13 Electric door openers come in a variety of sizes and shapes to fit different installation application needs. (courtesy Trine Consumer Products Division, Square "D" Company)

station. This variety of electric strike can be used with most standard brands of locksets in both mortise and surface mount types.

Remember that electric strike units have two systems of operation. For unlocking doors (normally in the locked position), it opens when energized. For locking doors (normally unlocked) the door is locked when the unit is energized.

For unlocking doors, the units come in standard voltages: 3 to 6 Vdc and 8 to 16 Vac. These units are electromagnetic coil types. For locking doors they come in standard voltage of 24 Vac, with a 47 ohm solenoid, and 24 Vac 119 ohm solenoid. These units are suitable for constant duty operation without overheating or burnout.

Both types of units are fail-safe, in that they will automatically go to the unlock position in the event of power failure, which assures an unlocked passageway in case of emergency, such as a fire on the premises.

Figure 19-14 provides the electrical operating characteristics for both the electromagnet and solenoid types of units. Figures 19-15 through 19-17 illustrate several models that are commonly installed by locksmiths in the course of work.

At FIG. 19-18 and 19-19 are specification guide details relating to the actual full-size opening of the various types of Trine electric door opener latch cavities. Remember that when you install one, check the lock latch width and throw to the cavity size of the door opener selected, allowing for a minimum ⅟₃₂-inch clearance at all points. For lock latch height for the unit installed, always be sure to check the detailed drawing provided with the lock latch to ensure that you have properly cut the cavity and not over-or under-cut. For the

ELECTRICAL OPERATING CHARACTERISTICS
Average Current Ratings

ELECTRO-MAGNET TYPE
(normally LOCKED . . . UNLOCKED when energized)

AC Voltage	*Coil-Ohms Resistance	Current Draw	DC Voltage	*Coil-Ohms Resistance	Current Draw
10	2.0	1.95	6	2.0	3.0
16	2.0	2.0			
**24	15.0	.66	**24	15.0	2.0
48	73.0	.18	48	73.0	.66

* Plus or minus 10%.
** Suitable for CONSTANT DUTY OPERATION.

SOLENOID TYPE
Available (normally LOCKED) and (normally UNLOCKED)

AC Voltage	*Coil-Ohms Resistance	Current Draw	DC Voltage	*Coil-Ohms Resistance	Current Draw
		FOR INTERMITTENT USE			
24	47.0	.42	24	119.0	.26
			48	200.0	.10
		FOR CONSTANT DUTY OPERATION			
24	119.0	.26	24	119.0	.26

*Plus or minus 10%

19-14 Electrical operation characteristics for Trine units. (courtesy Trine consumer Products Division, Square "D" Company.)

lock latch heights, see the detail drawings in FIG. 19-20 and 19-21. A typical installation detail, for example, using the Trine Model 001, is shown at FIG. 19-22. Jamb preparation for the installation of an electric strike in a metal door frame is at FIG. 19-23.

During jamb preparation, a neat cutting job is required for good appearance, so special care must be taken to ensure that the dimension of the height is adequate for a loose fit insertion in the cutout. Too tight a fit here will cause the binding of the keeper in the case and result in poor performance.

19-15 Trine model 014 for key-in-knob locksets. (courtesy Trine Consumer Products Division, Square "D" Company)

*No. 014
For ANSI - Key In-Knob Lock Sets
Std. 115.3
2-7/8" high

*Patent applied for

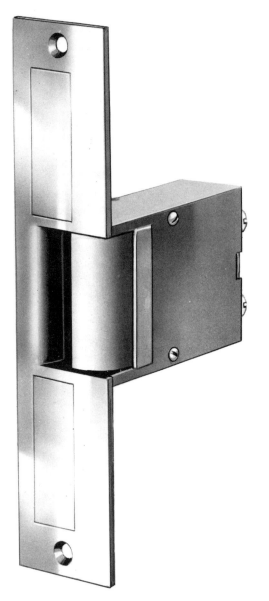

19-16   Solenoid coil unit for high intensity use. (courtesy Trine Consumer Products Division, Square "D" Company)

The dimension indicated by an asterisk must be determined by the installer or the jamb fabricator; it is dependent on the lockset used. This dimension establishes the relationship between the keeper and the bolt. Incorrect location of the strike will prevent the door from shutting tightly, or cause excessive binding between the keeper and the bolt. Leave approximately ⅟₃₂ inch free space between the keeper and the bolt.

While it may seem simple and easy to do (which it is), the two strike wires must be attached securely by wires nuts to the supplied voltage lines. Be careful not to damage the wires when inserting the strike in the jamb cutout. This can be accomplished by hooking the

19-17 Model 003 with a magnet coil for normal use; this unit is used with rim type locksets. (courtesy Trine Consumer Products Division, square "D" Company)

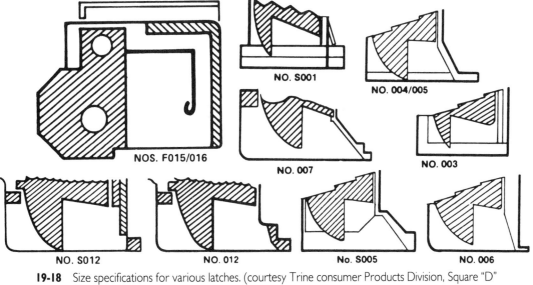

19-18 Size specifications for various latches. (courtesy Trine consumer Products Division, Square "D" Company)

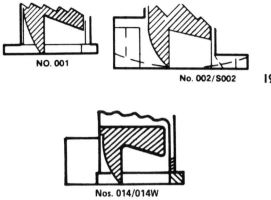

NO. 001

No. 002/S002

Nos. 014/014W

19-19 More latch cavity specifications. (courtesy Trine Consumer Products Division, Square "D" Company)

wire outlet side under the jamb edge first, and then pushing the other side of the strike in proper position.

LOW VOLTAGE ELECTRIC DOOR OPENER CIRCUITRY

Figure 19-24 illustrates and provides specifics for audible (ac) operation, while FIG. 19-25 is for silent (dc) operation.

While it seems impracticable, some businesses even use the automatic electric interlock type of circuitry for rooms that serve more than one consumer, such as a bathroom between two offices, or separate businesses located in the same building. Figure 19-26 shows how the electric interlock would work in such a situation. Such interlocks can also be used

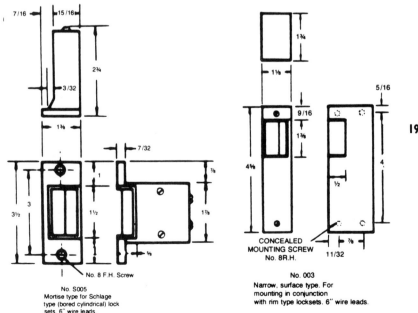

No. 8 F.H. Screw

No. S005
Mortise type for Schlage type (bored cylindrical) lock sets. 6" wire leads.

CONCEALED MOUNTING SCREW No. 8R.H.

No. 003
Narrow, surface type. For mounting in conjunction with rim type locksets. 6" wire leads.

19-20 Detail specification drawings for magnet coil units. (courtesy Trine Consumer Products Division, Square "D" Company)

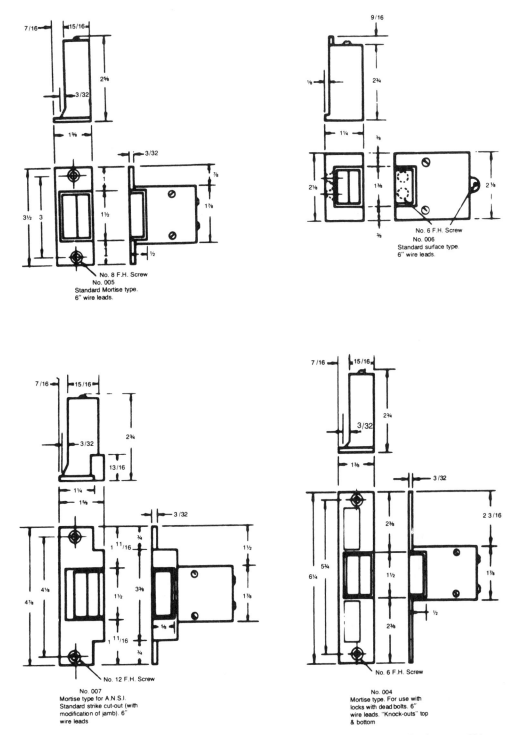

19-21 Specification details for high-intensity solenoid coil electric door opener units. (courtesy Trine Consumer Products Division, Square "D" Company)

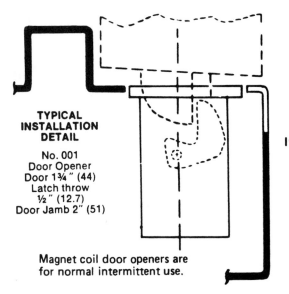

**TYPICAL
INSTALLATION
DETAIL**

No. 001
Door Opener
Door 1¾" (44)
Latch throw
½" (12.7)
Door Jamb 2" (51)

**Magnet coil door openers are
for normal intermittent use.**

19-22 Typical installation detail for model 001 magnet coil unit. (courtesy Trine Consumer Products Division, Square "D" Company)

for air shower rooms, money counting rooms, x-ray and photographic rooms, radiation and biological labs, and other such type installations.

In such a case as shown in FIG. 19-26 the door opener is normally unlocked, but it becomes locked when energized. The door indicator lights (DL-1/DL-2) can be used if a visual signal is required. The opening of either door operates the door switch (a mortise type, Trine #340 or surface type, Trine model #316). It also activates the opposite unlocked door opener, locking the door. When the door shuts, the opposite door returns to the unlocked position.

Returning now to the use of the unit in a communicating bathroom locking situation, upon entering the bathroom the occupant turns the outer knob, unlocking the lockset. To lock the bathroom's doors, the occupant closes both doors if open, and operates the wall

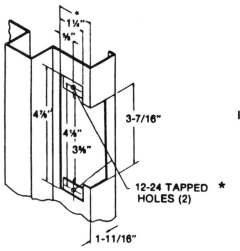

1¼"
⅝"
4⅞"
4¼"
3⅜"
3-7/16"
12-24 TAPPED ✳
HOLES (2)
1-11/16"

19-23 Typical procedural details for a mounted door frame installation. (courtesy Trine Consumer Products Division, Square "D" Company)

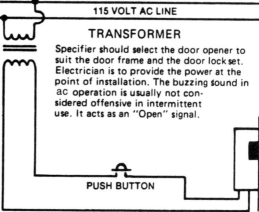

19-24 Low voltage electrical circuit for an audible operation. (courtesy Trine Consumer Products Division, Square "D" Company)

115 VOLT AC LINE

TRANSFORMER

Specifier should select the door opener to suit the door frame and the door lock set. Electrician is to provide the power at the point of installation. The buzzing sound in ac operation is usually not considered offensive in intermittent use. It acts as an "Open" signal.

PUSH BUTTON

DOOR OPENER

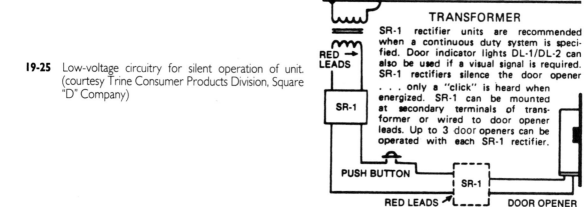

19-25 Low-voltage circuitry for silent operation of unit. (courtesy Trine Consumer Products Division, Square "D" Company)

115 VOLT AC LINE

TRANSFORMER

RED → LEADS

SR-1

SR-1 rectifier units are recommended when a continuous duty system is specified. Door indicator lights DL-1/DL-2 can also be used if a visual signal is required. SR-1 rectifiers silence the door opener . . . only a "click" is heard when energized. SR-1 can be mounted at secondary terminals of transformer or wired to door opener leads. Up to 3 door openers can be operated with each SR-1 rectifier.

PUSH BUTTON

SR-1

RED LEADS   DOOR OPENER

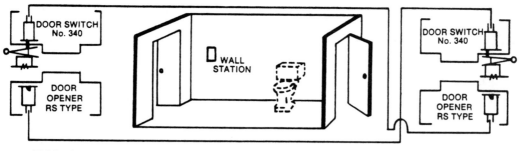

DOOR SWITCH No. 340

WALL STATION

DOOR SWITCH No. 340

DOOR OPENER RS TYPE

DOOR OPENER RS TYPE

19-26 Automatic electric interlock for the operation of opposite side doors in an area servicing more than one user. (courtesy Trine Consumer Products Division, Square "D" Company)

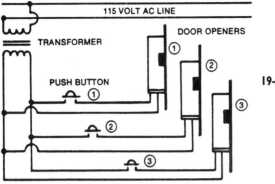

19-27   Circuitry for multiple wiring of door openers. (courtesy Trine Consumer Products Division, Square "D" Company)

switch to locked position, illuminating both DL-2 door indicator lights and locking both doors. Operating the wall switch to the unlocked position turns off both door indicator lights and unlocks both the doors. After leaving the bathroom, the occupant presses the pushbutton in the lockset, preventing anyone from the adjoining room from entering the occupant's room. An emergency release is obtained by installing a push button at each door, which will unlock its own door.

While this is only one type of situation where more than one door opener (always an RS type unit in such interlocking cases) is used, there are other instances where multiple door openers are required, with each having its own pushbutton control switch; further, each one is operated off the same power line. Figure 19-27 illustrates how the wiring of multiple door openers is accomplished.

Electrical ratings for solenoids for fail-safe (continuous duty) and standard (intermittent duty only) units are at FIG. 19-28.

Due to the remote control applications allowed for electrical strike units, there can be a considerable distance between the strike location and the actual control for that particular strike. At FIG. 19-29 are the *minimum* wire sizes required for standard and fail-safe electrical strike unit voltage at various distances.

Troubleshooting trine electrical strikes

Earlier we discussed troubleshooting procedures, but no matter how carefully units are tested and retested at the point of manufacture, problems can still arise. These troubleshoot-

ELECTRICAL RATINGS FOR SOLENOIDS	FAIL SAFE UNITS (CONT. DUTY)		STANDARD UNITS* (INTERMITTENT DUTY ONLY)		
	24 VDC	12 VDC	24 VDC	12 VDC	48 VDC
RESISTANCE IN OHMS	115	33	16	4	90
WATTS SEATED	4.8	4.4	36	36	25
AMPS SEATED	.2	.37	1.5	3.0	.53

*Bridge rectifier may be supplied with any unit for ac operation.

19-28   Electrical ratings for intermittent use and constant duty operation for solenoid coil door openers. (courtesy Trine Consumer Products Division, square "D" Company)

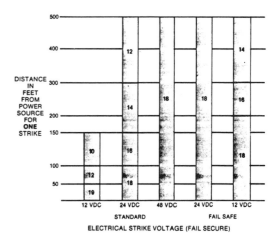

19-29 Electrical strike voltage (fail-secure mode of operation). (courtesy Trine Consumer Products Division, Square "D" Company)

ing procedures concern the unit after installation, and they can relate to any electrical strike unit in use.

After installation, the strike hums but does not operate properly when the power is applied.

Probable cause: ac voltage has been supplied, but the rectifier has not been installed between the transformer and the strike.

Correction: Install the rectifier per the wiring diagram at FIG. 19-30.

After installation, the strike does not operate even if the power is left on for one minute (maximum time) and the strike is noticeably warm.

Probable cause: (1) Static loading against the door is excessive due to weather stripping, a warped door, wind loading on the door, or improper jamb preparation. (2) Insufficient voltage is reaching the strike due to an inadequate supply wire size.

Correction: (1a) Adjust the strike bolt relationship, so that some slight gap exists in the locked position. (1b) Balance air conditioning pressures on opposite sides of the door. (2) Provide sufficient sized service wire (FIG. 19-29).

The keeper moves to fully open on power application, but does not return to the locked position when the door closes.

Probable cause: If the deadbolt is extended, but it does not pull the keeper to the fully locked position on door closure, the return spring may need adjusting.

Correction: Deform the return spring slightly by forcing it toward the keeper.

The keeper binds in the strike case on both opening and closing.

Probable cause: The cutout in the jamb is too narrow and it is forcing the strike case walls inward toward the keeper.

Correction: Increase the vertical dimension until the keeper operates freely.

19-30 Circuitry installation of a rectifier to alleviate a strike hum. (courtesy Trine Consumer Products Division, Square "D" Company)

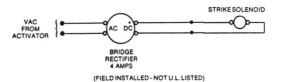

It should be noted that electric door openers are also available and work extremely well for deadbolts. They can be used with just about any manufacturer's lockset. They include the following types:

☐ Mortise locks with deadbolt.

☐ Mortise lock with deadlatch and guarded latch.

☐ Individual deadbolt lock.

☐ Panic exit devices.

☐ Maximum security deadbolts (¾″ throw).

☐ Release up to 1″ throw deadbolts.

☐ With optional adapters, they may be used with cylindrical locks, alarm locks or with Detex bolts.

These units will also allow for the strike operation of mortise locks, with both latchbolt and deadbolt release by the action of a single keeper. The keeper is held open after the door is opened if the deadbolt was in an extended position at the time of the door opening. The keeper automatically returns to the locked condition when the deadbolt reenters the strike. The door may also be closed with deadbolts extended. In strikes with options H, J and K (FIG. 19-31), the keeper is forced open by the extending latchbolt and springs closed to the locked position to accommodate the returning latchbolt. Other options are also shown at FIG. 19-31.

Figure 19-32 illustrates two Trine unit model (015 and 016) specifications. Along with the basic unit for installation, the necessity of a heavy duty transformer (FIG. 19-33) is necessary to convert the primary voltage to the operating voltage.

ELECTRIFIED MORTISE LOCKS

The electrified mortise lock provides for a wider number of additional installation sales and servicing possibilities, especially when access security controls for your customer are

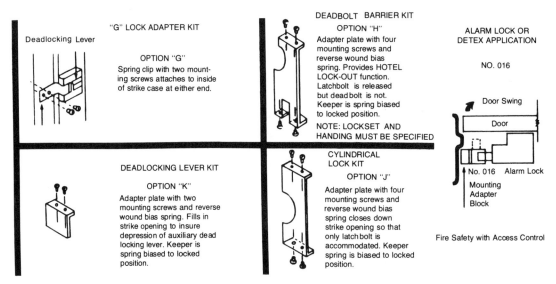

19-31   Various strike options for electric door openers used with deadbolt locks. (courtesy Trine Consumer Products Division, Square "D" Company)

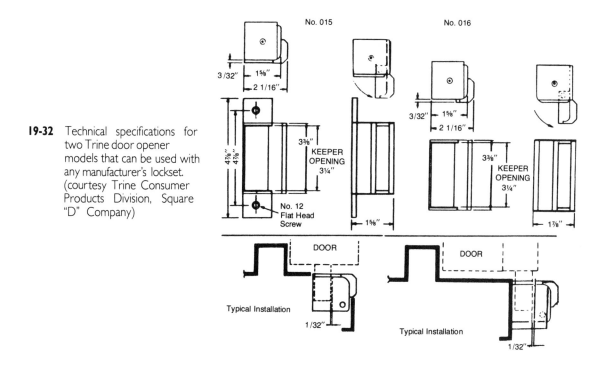

19-32 Technical specifications for two Trine door opener models that can be used with any manufacturer's lockset. (courtesy Trine Consumer Products Division, Square "D" Company)

concerned. The electrified mortise lock (FIG. 19-34) is another excellent, very durable product that has minimal installation procedures, yet offers long life and reliability for your customer.

This lock unit has the standard rugged lock case construction of stainless steel, plus a normal mortise strike. The inside knob is always free, while the outside knob is rigid except when the 24 Vdc internal solenoid is energized, which releases the outside knob, allowing for latchbolt retractment. Like any other mortise lock which is standard, it is reversible, allowing for different hands.

Installation of the electrified mortise lock:

Mark the door and jamb (FIG. 19-35). Mark the horizontal centerline of the lock on both sides and the edge of the door. Mark the vertical centerline of the lock at the door edge and the vertical centerline on both sides of the door as measured from the door edge.

19-33 Heavy-duty transformer for voltage conversion. (courtesy Trine Consumer Products Division, Square "D" Company)

19-34   Electrified knob lock. (courtesy Alarm Lock Corporation)

Mortise the door for the lock case and front (FIG. 19-36). Refer to the installation template that comes with the lock unit for the specific dimensions.

As indicated on the template and as shown in FIG. 19-36, drill the required holes. *Caution:* Check the lock for function, hand, and bevel before drilling. Mark and drill only those holes required for your function.

Remove the front plate and mount the lock in a mortised opening. Apply the trim (four trim variations are possible; see FIGS. 19-38 through 19-41 at the end of the installation instructions). Apply the cylinder (using a minimum of four turns); then, fasten with the cylinder clamp screw. Replace the front plate. *Caution:* Check the cylinder and lock for proper function before closing the door.

Install the strike (FIG. 19-37). Position the strike on the door jamb as shown. Mortise the jamb and apply the strike. When a box strike is used, mortise deep enough to allow the latchbolt and deadbolt to fully extend.

Reversing the lock and bevel:

At times you might be required to replace or remove and reinstall and lock on another door.
Remove the cover from the lock.
Reverse the latchbolt and the auxiliary bolt.
Reverse the hubs.

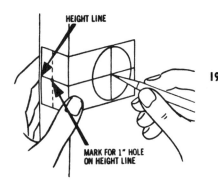

19-35   Marking the door for installation. (courtesy Alarm Lock Corporation)

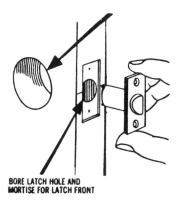

19-36   Bore and Mortise for latch front. (courtesy Alarm
Lock Corporation)

**BORE LATCH HOLE AND
MORTISE FOR LATCH FRONT**

To change the bevel, loosen the two screws on the top and bottom of the case. Adjust the proper bevel and tighten the screws.

Replace the cover. Be certain that all parts are in their proper place before fastening the exterior screws. *Caution:* When the hand of the lock is changed, make sure that the proper hand strike is also used. A left-hand lock requires a left-hand strike; a right-hand lock requires a right-hand strike.

Figure 19-38 through 19-41 show the various trims available for this unit. Figure 19-38, the heavy duty screwless trim; FIG. 19-39, the lever handle trim; FIG. 19-40 is the medium duty trim with a Glenwood escutcheon; and FIG. 19-41 is the lever handle trim with a Utica escutcheon.

ELECTRIFIED KNOB LOCK

The electrified knob lock (FIG. 19-42) by the Alarm Lock Corp. is a heavy-duty lockset with a key bypass in the outside knob. It has a deadbolting latchbolt and stainless steel trim for long-lasting exterior or interior use against the elements. The inside knob is always free for

19-37   Strike installation details. (courtesy
Alarm Lock Corporation)

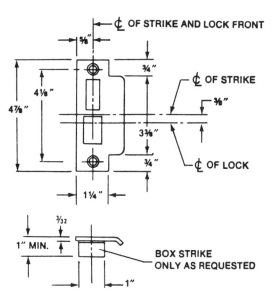

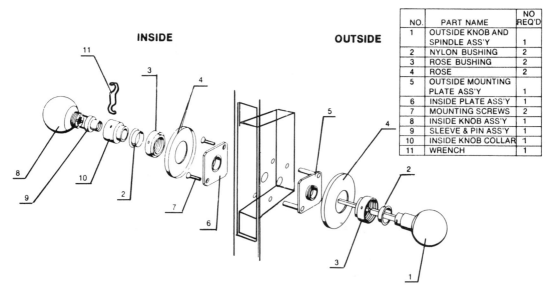

INSIDE   **OUTSIDE**

NO.	PART NAME	NO REQ'D
1	OUTSIDE KNOB AND SPINDLE ASS'Y	1
2	NYLON BUSHING	2
3	ROSE BUSHING	2
4	ROSE	2
5	OUTSIDE MOUNTING PLATE ASS'Y	1
6	INSIDE PLATE ASS'Y	1
7	MOUNTING SCREWS	2
8	INSIDE KNOB ASS'Y	1
9	SLEEVE & PIN ASS'Y	1
10	INSIDE KNOB COLLAR	1
11	WRENCH	1

19-38   Heavy-duty screwless trim. (courtesy Alarm Lock Corporation)

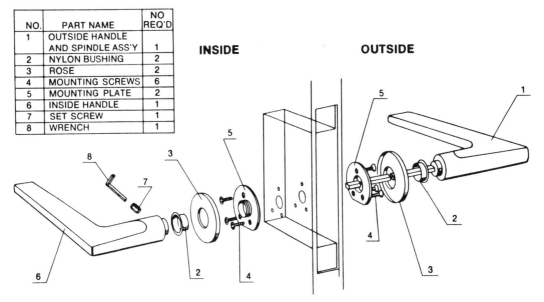

NO.	PART NAME	NO REQ'D
1	OUTSIDE HANDLE AND SPINDLE ASS'Y	1
2	NYLON BUSHING	2
3	ROSE	2
4	MOUNTING SCREWS	6
5	MOUNTING PLATE	2
6	INSIDE HANDLE	1
7	SET SCREW	1
8	WRENCH	1

INSIDE   **OUTSIDE**

19-39   Lever handle trim. (courtesy Alarm Lock Corporation)

INSIDE   OUTSIDE

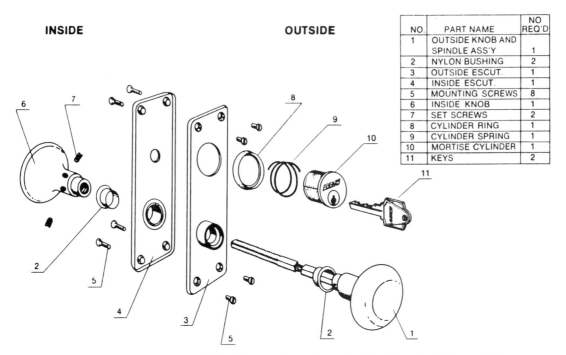

NO.	PART NAME	NO REQ'D
1	OUTSIDE KNOB AND SPINDLE ASS'Y	1
2	NYLON BUSHING	2
3	OUTSIDE ESCUT.	1
4	INSIDE ESCUT.	1
5	MOUNTING SCREWS	8
6	INSIDE KNOB	1
7	SET SCREWS	2
8	CYLINDER RING	1
9	CYLINDER SPRING	1
10	MORTISE CYLINDER	1
11	KEYS	2

19-40   Medium-duty trim with the Glenwood escutcheon. (courtesy Alarm Lock Corporation)

INSIDE   OUTSIDE

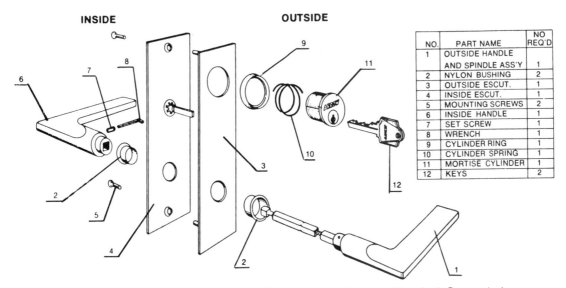

NO.	PART NAME	NO REQ'D
1	OUTSIDE HANDLE AND SPINDLE ASS'Y	1
2	NYLON BUSHING	2
3	OUTSIDE ESCUT.	1
4	INSIDE ESCUT.	1
5	MOUNTING SCREWS	2
6	INSIDE HANDLE	1
7	SET SCREW	1
8	WRENCH	1
9	CYLINDER RING	1
10	CYLINDER SPRING	1
11	MORTISE CYLINDER	1
12	KEYS	2

19-41   Lever handle trim using the popular Utica escutcheon. (courtesy Alarm Lock Corporation)

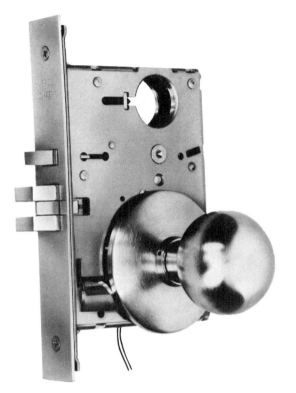

19-42   Model #121 electrified mortise lock. (courtesy Alarm Lock Corporation)

exit. The outside knob is rigid to prevent entry except when the internal solenoid is energized, which releases the outside knob to retract the latchbolt and open the door. The electrified knob lock has a continuous duty solenoid that uses 24 Vac power supply (360 milliamps).

Installation procedures:

1. Mark the door. Mark the height line on the edge of the door 38 inches from the floor (or other height that may be required by the customer; this is the suggested height, although it may vary according to the individual needs). Position the centerline of the template that comes with the lock on the height line, and mark the center point of the door thickness and the center point for the 2⅛-inch hole through the door (FIG. 19-43).

2. In boring the holes, bore the 2⅛-inch hole at the point marked from both sides of the door. Bore the 1-inch latch unit hole straight into the edge of the door from the center point of the height line. Mortise the latch front and install the latch unit (FIG. 19-44).

3. Removing the lock from the box, remove the inside trim. This is accomplished by inserting the small end of the spanner wrench through the hole in the knob bearing sleeve. This hole faces the latch retractor. Press the knob catch to slide the knob off the spindle and remove the rose from the threaded sleeve (FIG. 19-45).

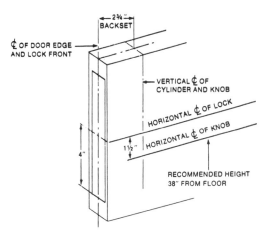

19-43 Electrified knob lock installation, Step 1.

4. The door thickness adjustment is accomplished by adjusting the outside rose either in or out until the indicator on the lock shows the proper thickness for the door. Make sure the rose clicks into place (FIG. 19-46). The lock will fit on all doors 1⅜ inch to 2-inches thick. (If the lock you require is for a door of greater thickness, it will have to be obtained as a special request to the company.)

5. Next, install the main unit (FIG. 19-47). The main housing must engage with the latch prongs and retractor with the latch tailpiece as shown in FIG. 19-48.

6. Secure the trim by replacing the rose and screwing it on to the threaded spindle with your fingers as far as it will go. Then tighten the rose with the spanner wrench (FIG. 19-49).

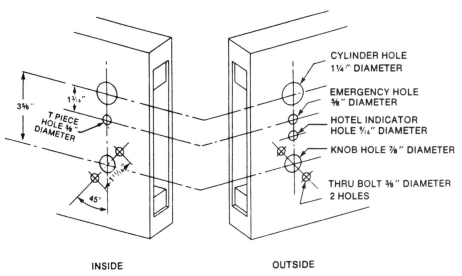

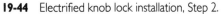

19-44 Electrified knob lock installation, Step 2.

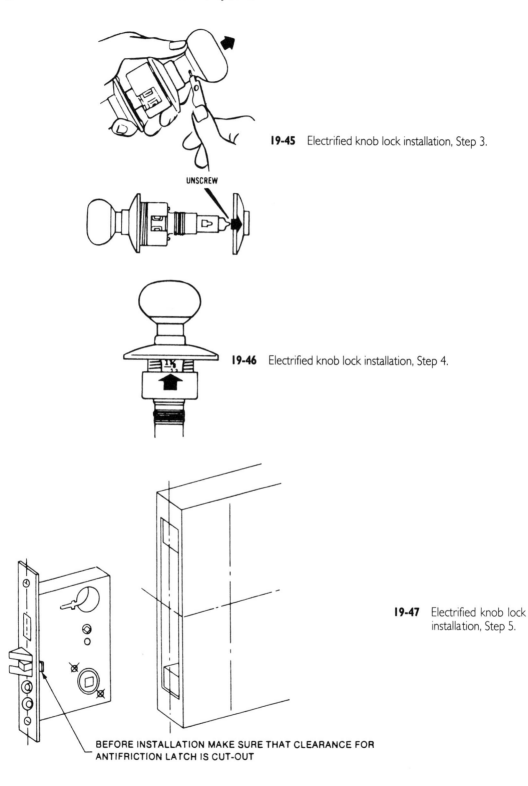

19-45   Electrified knob lock installation, Step 3.

UNSCREW

19-46   Electrified knob lock installation, Step 4.

19-47   Electrified knob lock installation, Step 5.

BEFORE INSTALLATION MAKE SURE THAT CLEARANCE FOR
ANTIFRICTION LATCH IS CUT-OUT

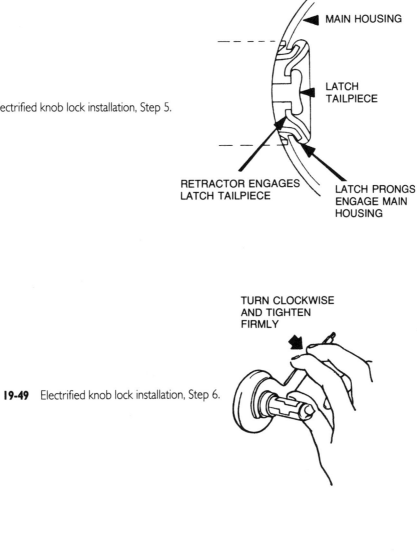

19-48   Electrified knob lock installation, Step 5.

MAIN HOUSING

LATCH
TAILPIECE

RETRACTOR ENGAGES
LATCH TAILPIECE

LATCH PRONGS
ENGAGE MAIN
HOUSING

TURN CLOCKWISE
AND TIGHTEN
FIRMLY

19-49   Electrified knob lock installation, Step 6.

19-50   Electrified knob lock installation, Step 7.

DEPRESS KNOB CATCH

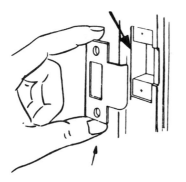

19-51   Electrified knob lock installation, Step 8.

7. Replace the knob by lining up the keyway in the knob with the slot in the spindle. Then push the knob straight on until the knob hits the knob catch. Depress the knob catch and push the knob straight on until the knob catch clicks into place. Test both knobs to make sure they are securely fastened to the spindle (FIG. 19-50).

8. Install the strike by making a shallow mortise in the door jamb to align with the latch face and install the strike as in FIG. 19-51.

If you must reverse the cylinder of the lock, the following applies:

Insert the key into the cylinder, turn the key clockwise as far as it will go, and at the same time, depress the knob catch retainer. When the retainer moves down, pull the knob off.

Turn the knob 180 degrees, and replace it as follows: Remove the key and push the knob on the spindle as far as it will go. Insert the key and turn it clockwise to its most extreme position while depressing the knob catch retainer. When the retainer moves down, push the knob onto the spindle, allowing the retainer to click into position.

If you must remove the cylinder, push the cylinder tailpiece sideways with your finger until the tailpiece unsnaps from the plug. Then turn the knob over and hit the shank of the knob with a sharp blow on a flat surface. This will release the knob ferrule and will permit the removal of the cylinder. Rekeying of the cylinder is conventional.

Now that you understand the specifics of the lock installation itself, consider the electrical considerations related to the lock. The electrical power for the lock can either be concealed within the door by use of a current transfer unit (FIG. 19-52) or left exposed on the interior side. The use of a current transfer unit is a unique means of bringing the electrical current from the lock side of the door to the door. The face and strike must be thoroughly matched and the door closed for power to flow through. With this in place, from a remote switch activation, power will flow through allowing the door to be opened.

The other method, leaving the wiring exposed, is not really what it seems. The use of an armored door loop is used and it provides a much easier means for bringing electrical

19-52   Current transfer unit. (courtesy Alarm Lock Corporation)

current from the hinge side of the door frame to the lock in the door. It consists of an 18-inch flexible armored cable which is attached to the frame on one end, and just below the lock on the inside of the door. The lock wires can be passed through the door wall via a small hole and connected to the loop wire ends.

ELECTROMAGNETIC LOCKS

While electric locks are available and used for many applications, there is also the need for electromagnetic locks. Such locks can be used in hospitals, convalescent homes, banks, universities, museums, libraries, factories, laboratories, penal institutions, and airports, among others.

The electromagnetic lock, while operating at less than 12 Vdc, can exert a holding force of 1500 pounds. This is fantastic for a lock that consists of two components—the lock itself, and its armature.

Imagine a lock that mounts rigidly to a door frame and an armature that mounts to the door (FIG. 19-53). When the door is closed, the two make contact; upon activation of the lock, the two are attracted to each other. The door, for all practical purposes, is bonded solidly to the frame.

Ten years ago, such a lock would have been priced out of reach, but today it is possible for almost every business or institution that requires such a locking device to afford one. As a locksmith, you must be prepared and familiar with this "unconventional" (in the traditional sense) lock.

Security Engineering, the leader and major proponent of this lock, has indicated the major features as follows:

☐ A high holding (1500 pounds)—effectively seals the door from both sides, keeping intruders out and/or valuables safely in.

☐ Conserves energy—the efficient low power consumption design (2 amps) ensures a long life and economical operating cost for the lock.

☐ High reliability—recommended for high trafficked openings with extreme usage, where conventional locks, locks with moving parts, or electromagnetic locks with a lesser holding force may not be capable of withstanding the abuse.

19-53 Model 3900 electromagnetic lock. (courtesy Security Engineering, Inc.)

□ Fully fail safe—locking and unlocking is positive and instantaneous, with no residual magnetism or moving parts that might stick, jam, bind, wear out, and/or need replacement.

□ Extremely low maintenance—no maintenance or adjustments are required after initial installation.

□ Exclusive mounting design—this permits rapid efficient installation in new or existing buildings and automatically compensates for normal door wear, sag, warpage, or misalignment.

□ Self-contained unit—it is completely factory wired with all options for the unit already built in, fully concealed, and tamper-resistant. There is only one external wire connection to be made.

□ Four operating voltages—12 Vdc; standard 12 Vac; 24 Vdc; and 24 Vac.

Figure 19-54 illustrates a typical installation and the various features (standard and optional) that can come with a unit.

Series 3900 electromagnetic lock mounting:

It is important that prior to actually attempting to install the electromagnetic lock, that you handle the equipment with care, since damaging the mating surfaces of the armature of the

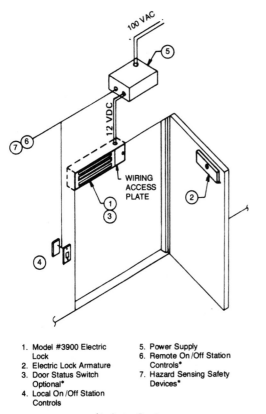

19-54 Model 3900 installation layout. (courtesy Security Engineering, Inc.)

1. Model #3900 Electric Lock
2. Electric Lock Armature
3. Door Status Switch Optional*
4. Local On /Off Station Controls

5. Power Supply
6. Remote On /Off Station Controls*
7. Hazard Sensing Safety Devices*

*As System Requires

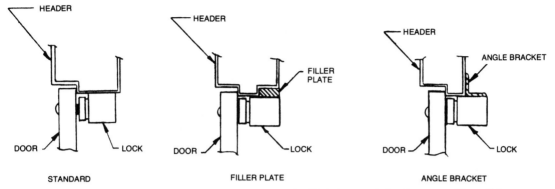

19-55 Door frame header determination tells you if a filler plate, angle bracket, or nothing will be required. (courtesy Security Engineering, Inc.)

electromagnetic may reduce its locking efficiency. The electromagnet mounts rigidly to the door frame header. The armature mounts to the door with special hardware provided that allows it to pivot about its center to compensate for door wear and misalignment. All template use must take place with the door in its normal closed position.

The lock face is covered with a rust inhibiting film. This should not be removed. If the film is accidentally removed, you should clean the lock face with a clean dry soft cloth (do not touch the lock face with your hands) and reapply a rust inhibitor such as M1 (manufactured by Starret) or LPS3 (manufactured by LPS Laboratories); these rust inhibitors are readily available in most hardware stores. Then apply a light film of rust inhibitor generously to the armature face.

Note the type of door frame header and install the filler place or angle bracket as required (FIG. 19-55).

Fold the template (FIG. 19-56) where indicated to form a 90 degree angle. For a swinging door, place the template against the door header and door opposite hinge side of the door jamb. For a pair of swinging doors, place the template against the door and door header at the center of the door opening. Transfer the hole locations to the door and door frame header.

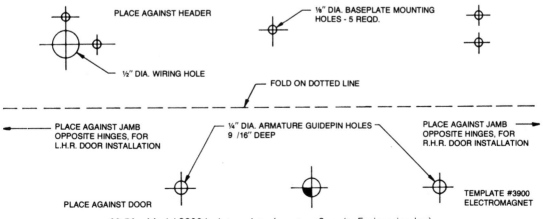

19-56 Model 3900 lock template. (courtesy Security Engineering, Inc.)

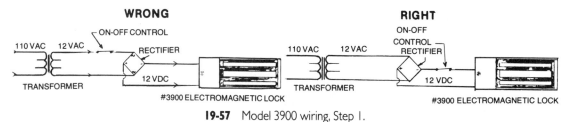

19-57   Model 3900 wiring, Step 1.

Follow the template instructions for the specific hole sizes required. See Fig. 19-56 to determine the proper armature mounting hole preparation. The hole is designated as the mark ⊕ in Fig. 19-56.

Mount the armature to the door with the hardware provided as shown in the figures.

Install the mounting plate to the header with the five #10 screws.

Fasten the model 3900 lock to the mounting plate with the two socket screws and compensate for any misalignment by adding or subtracting washers at the armature mounting screws.

Finally, firmly tighten all the screws. Install the anti-tamper plugs into the holes over each socket head mounting screw. Use a soft hammer to avoid damaging the lock case.

Wiring instructions:

In some instances, you may be using a prefabricated power supply for the lock unit. But suppose you are not. What then? To achieve the best possible electromagnet lock installation, note the following suggestions (both right and wrong ways illustrated for each procedure).

The proper placement of the on-off station control in the electrical circuit will ensure instantaneous door release and avoid any delay in complete unlocking (FIG. 19-57).

Care must be taken when adding any auxiliary controls or indicators to the electrical circuit. Improper placement may cause a slight delay in the door release (FIG. 19-58).

Solid state on-off devices must be properly protected from the back voltage generated from the collapsing electromagnetic lock coil when the unit switched off. This may be accomplished by installing a metal oxide varistor (MOV) across the lock coil lead wires (FIG. 19-59). The MOV is available from Security Engineering; see the address in the Appendix B.

Installation of the MOV is for arc suppression in ultra high usage installations to prevent switch contact deterioration. In any situation, though, the installation of a diode across the leads is not recommended (see FIG. 19-58).

Another lock to consider is the surface-mounted electrically actuated lock, the series 2200 from Security Engineering. This lock (FIG. 19-60) has been designed to offer the

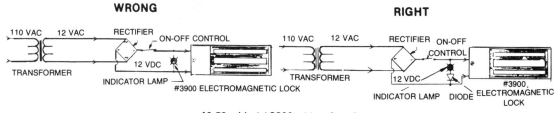

19-58   Model 3900 wiring, Step 2.

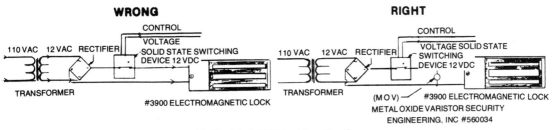

WRONG   **RIGHT**

19-59 Model 3900 wiring, Step 3.

advantages of remote electric locking, coupled with a versatility limited only by the user's imagination.

For basic operation, the lock mounts to a door frame or a door. When energized, it locks or unlocks the door when the solenoid-operated deadbolt projects from the unit and makes contact with a strike.

Standard features include the following:

☐ Fail-safe operation—if the power is interrupted, the unit automatically unlocks.

☐ Fail-secure operation—if the power is interrupted, the unit automatically unlocks.

☐ Easy installation—the unit is not handed and thus, can be mounted on any plane.

☐ Versatility—it can be used with inswinging, outswinging, sliding, overhead or rollup doors.

☐ Maintenance free—unit parts are ruggedly built and self-lubricating.

☐ Current draw—.5 amps at 24 volts.

☐ Low voltage—24 Vac standard; 24 Vdc, 12 Vac, and 12 Vdc available.

☐ Long and silent operation—the unit has a continuous duty solenoid that will function for millions of cycles.

☐ Compact size—2″ × 2″ × 6¾.″

19-60 Model 2200 series surface-mounted lock. (courtesy Security Engineering, Inc.)

The unit has optional features, such as a choice of one or two signal lights; or a built-in bolt position switch to indicate whether or not the bolt is extended or retracted for interfacing with automatic door equipment or signaling the lock status to a remote location unit. *Note:* The fail secure locks are not recommended where life safety may be compromised and is also not recommended where panic crash bar hardware is the only means of emergency egress.

Lock mounting instructions (refer to Fig. 19-61)

Remove the four screws from the unit; remove the housing cover.

Manually extend the solenoid bolt by pushing the BPS pin forward (if equipped) or inserting a wire into the rear of the solenoid.

Engage the extended solenoid bolt and strike and locate as shown in the figure on the door and door frame.

Using the mounting holes in the lock base and strike as templates, mount the solenoid lock and strike with the screws provided. (Several strike types are available and can be used, depending upon the situation required. See FIG. 19-62.)

Check for the proper alignment and see that the lock is free, having no binding in operation.

Connect the electrical wiring per the system option or requirements.

Check the operation of the lock for any final adjustments; if any, adjust as required.

Reinstall the housing cover and secure with screws.

A more compact but just as effective lock is the 1300 series mortise solenoid lock (FIG. 19-63). This mortise solenoid activated deadbolt is designed primarily for installation in new construction, with emphasis placed on tamper resistance and concealment of the unit.

As for operation, upon activation of the solenoid, the deadbolt projects ½ inch from the faceplace or retracts ½ inch to flush with the faceplate, depending upon the solenoid selected and used.

Among the standard features of this lock unit are as follows:

☐ Fail safe and fail secure operation—interrupted power and the unit automatically unlocks.

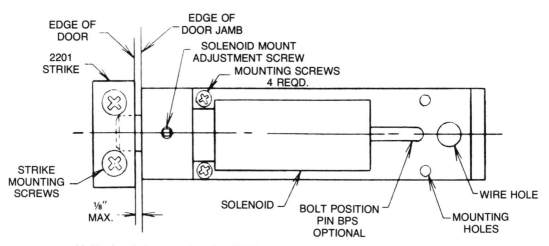

19-61   Installation procedure for #2200 series lock. (courtesy Security Engineering, Inc.)

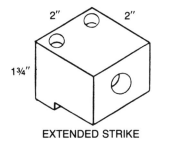

EXTENDED STRIKE

ANGLE STRIKE

19-62 Strikes. (courtesy Security Engineering, Inc.)

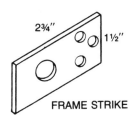

FRAME STRIKE

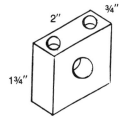

NARROW STRIKE

19-63 Series 1300 mortise solenoid lock. (courtesy Security Engineering, Inc.)

□ Compact size—1⅜ inch in diameter × 4½ inches long.

□ Voltage and current draw are the same as the 2200 series.

ELECTRICAL KEYLESS LOCKS

Simplex security locks have also kept pace with the electrically operated locks and latch unit. The Simplex keyless entrance control is a wall-mounted unit control for electrical door locks (FIG. 19-64) and has a built-in adjustable time delay feature not found in other currently available marketed units. The elimination of key control problems in high-risk areas, for home or business, is a distinct and great advantage. The Simplex entrance control is an easy, yet very secure installation. The unit is secured by two rugged draw bolts that pass completely through the wall. The bolts can be cut to any desired length to permit installation in walls from 4 inches to 11-inches thick (FIG. 19-65). The adjustable timer delay of 15 to 20 seconds permits locating the control at any convenient position near or by the door.

Two styles of front mounting plates are included to meet varying installation needs. One is 3¼-inches wide × 9⅝ inches high; the smaller is 1¾-inches wide × 8½ inches. Both front mounting plates are of stainless steel. Some security features for this unit include the following:

□ A special visual shield for each entrance control unit to protect against unauthorized observation of the button sequence operation.

□ If the control knob is forced by pliers or other tools, it will spin free without activating the electrical circuit.

19-64 Wall-mounted electrical control unit for door lock or latch. (courtesy Simplex Access Controls Corp.)

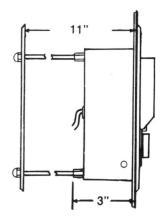

19-65 Lock unit can be mounted on any thickness wall up to 11". (courtesy Simplex Access Controls Corp.)

☐ The electrical terminals and wire leads are protected against a short circuit attach by a heavy-duty metal back cover that completely encloses the entire electro-mechanical device inside the wall.

Also in demand by many businesses is the Simplex/Unican 2000-15 model (FIG. 19-66) which can be used with many leading brands of surface-mounted panic exit devices. Another popular unit is the model TM pushbutton lock (FIG. 19-67) which can be used in both business and home.

A sophisticated digital access control system that does not require a console board or monitoring station is available which the locksmith can easily install. The Memorilok system has two basic units, an access control keyboard (FIG. 19-68) and a program unit (FIG. 19-69). With these two units, 10,000 combinations are available to the customer. Now there is a third optional piece of equipment, the card reader (FIG. 19-70), which provides a substantially higher degree of security integrity for the area to which access is desired.

The Memorilok system incorporates a solid state electronic memory for the storage of the selected code combination. The entry code combination is electronically stored. Unlike other access control systems, there are no wheels, dials, or other visual indicators to betray the code combination.

Standard unit features include:

☐ *Automatic address.* A pre-set timer is activated when the first digit of the code combination is entered. The remainder of the code entry must be completed before the expiration of this time, or the entry will be canceled and the programmer will activate the penalty time.

☐ *Automatic penalty.* A penalty time of approximately five seconds is incurred if an incorrect key is depressed during the code entry, or if two or more keys are pressed simultaneously. This feature defeats random button manipulation.

☐ *Code bypass.* There is provision within the access control unit for the connection of an external (remote) switch or pushbutton to operate the lock via the programmer unit without the use of the entry keyboard. This normally would be sited within the protected access area and be used for giving access to personnel or visitors who do not possess the code combination, but for whom access is warranted and authorized by the customer.

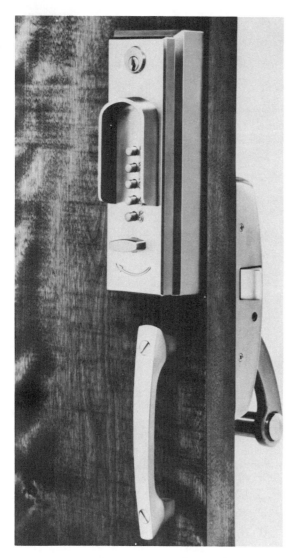

19-66 Unican 2000-15 unit used with an emergency exit panic bar. (courtesy Simplex Access Controls Corp.)

□ *Memory Protection.* The Memorilok keeps the code combination stored and undamaged in case of a main power failure. This is due to its constantly charged nickel cadium battery that is contained within the unit.

Optional extra features include the following:

□ *Duress/error alarm.* The duress/error alarm option is available as an integral part of the Memorilok unit, and if desired by the customer, must be ordered as original equipment. The duress alarm is activated by depressing a factory fixed last digit (from 0 to 9) instead of the last digit of the selected code combination. If entry has to be made under duress, the first three digits of the programmed code combination plus the predetermined last digit—the Duress Digit—is entered on the keyboard. This

19-67 Model TM lock installed in an office cabinet. (courtesy Simplex Access Controls Corp.)

19-69 Programmer unit.

19-68 Pushbutton control unit.

19-70 Pushbutton unit with card reader for increased security.

automatically operates a set of Form C (Changeover) dry contacts which, in turn, triggers any desired alarm system in effect: audible, silent, remote, or local.

The error alarm circuit will react upon any number of selected unsuccessful attempts at entry. As soon as the predetermined number of incorrect attempts (from one to eight) have been made, the alarm is activated, warning that the system is being tampered with.

☐ *Flush mounting.* To flush mount the programmer, keyboard, or even the battery standby, an oversized front panel is used. The panels, 8 inches × 9 inches are affixed to the standard front panels in order to conceal the roughout opening around the Memorilok unit.

The optional card reader fits into the front of the keyboard panel. It provides higher levels of security integrity because the coded card *must* be used by the individual seeking entry. It must be entered into the Memorilok before the code combination will be accepted by the electronic memory circuit.

Wiring details are in FIG. 19-71. The following notes apply, as referenced in the figure:

These connections are not used unless the remote entry/exit is required. Use the open switch.

All contacts are shown (NO and NC) without the lock being energized, and without being in the alarmed state.

On installations where there is a long wire run between the programmer and the lock or other device being operated, carefully compute the minimum requirements of the lock or device and the voltage drop which will occur in the wire.

WIRING INSTRUCTIONS FOR MODEL 400 MEMORILOK®

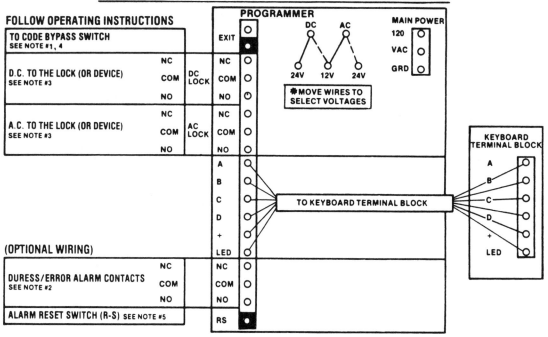

19-71   Memorilok wiring instructions (refer to text for "notes"). (courtesy Alarm Lock Corp.)

SPECIFICATIONS:

Dimensions:	
Keyboard Unit	5 ¼" (128 mm) Wide 6 ¼" (153 mm) High 2 ⅝" (66 mm) Projection
Programmer Unit	Same as Keyboard
Shipping Weight:	Complete Memorilok - 6 pounds Battery Standby Unit - 6 pounds
Main Power Input:	120 Vac or 240 Vac 50 /60 HZ. Specify voltage required.
*Output to Lock:	12 or 24 Vac or Vdc at 1 Amp. Voltage is user selectable.
Program Key Switch:	Contacts are closed in the "ON" position for code programming.
Wiring Information:	Includes ten feet of six conductor 22 gauge wire.

19-72 Memorilok #400 unit specifications. (courtesy Alarm Lock Corp.)

Code bypass switch requires a momentary switch with normally open (NO) switch contacts.

The alarm Reset Switch (R-S) requires a momentary switch with normal switch contacts. Thus should be wired between terminal #2 of the exit and terminal R-S indicated on the diagram as "blacked out block" on the programmer wiring hookup. Use this only for duress/error entry.

To program the Memorilok, the programmer key is turned to the *on* position. Enter the desired four digit code combination on the keyboard. A green LED indicator light will glow to indicate that the code is properly programmed into the memory system. Return the programmer key to the off position. Reinsert the code to test the unit for proper operation. Now the green LED light will indicate acceptance of the code combination and supply of power to the lock or other device used.

The adjustable time delay for the Memorilok affects the lock open time (time delay for the unit) and it can be varied from one to ten seconds by turning an adjusting screw.

Unit specifications are at FIG. 19-72. Specification installation details, template, and other specific information concerning conductor wire are provided with the unit from the factory.

19-73 Micro Kaba rotary switch. (courtesy Lori Corp.)

KEY ENTRY		No. of Switch elements
 Spring Return		1 2 3 4
One Key Entry 9h		1 2 3 4
One Key Entry 6h		1 2 3 4
Two Key Entry		1 2 3 4

19-74 Four different key positions provide an opportunity for different control features. (courtesy Lori Corp.)

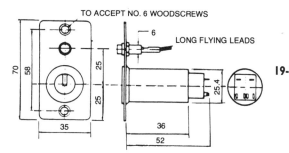

KS 50S

19-75 Two-pole switch. (courtesy Lori Corp.)

TO ACCEPT NO. 6 WOODSCREWS

6

LONG FLYING LEADS

70   58   25   25   25   25.4   35   36   52

19-76 Specifications for the two-pole switch. (courtesy Lori Corp.)

19-77 Surface-mounted remote control station with a waterproof key cover. (courtesy Lori Corp.)

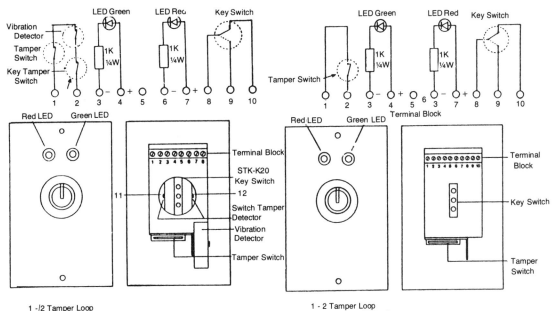

1 -/2 Tamper Loop
3 - 4 12 Vdc LED "on"
6 7 8 ON/OFF Changeover Switch. KS64

1 - 2 Tamper Loop
3 - 4 12 Vdc LED "on"
6 7 8 ON/OFF Changeover Switch KS64

WIRING DIAGRAM FOR REMOTE
CONTROL STATIONS KS-RCS/1

WIRING DIAGRAM FOR REMOTE
CONTROL STATIONS KS-RCS/2

19-78 RCS/1 and RCS/2 wiring diagrams. (courtesy Lori Corp.)

KEY-ACTUATED SWITCHES

The full-service locksmith should have available for customers, especially those in business, a variety of key actuated switches. Figure 19-73 shows a MicroKaba rotary switch. This miniature switch is operated by the 12-mm diameter cylinder, which has over 10,000 different possible key combinations. Four key entry variations are available, each having up to four switch elements (FIG. 19-74).

19-79 The narrow remote control switch allows for many different installation applications. (courtesy Lori Corp.)

19-80 Remote control station housing mounting box. (courtesy Lori Corp.)

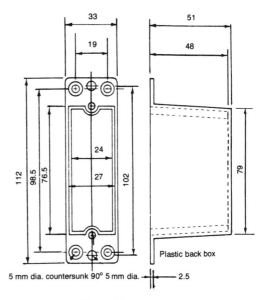

19-81 Mounting box specification data. (courtesy Lori Corp.)

Plastic back box

5 mm dia. countersunk 90° 5 mm dia. ——|——— 2.5

Cut out dimensions 83 mm × 32 mm

19-82 Electrical specifications for the remote control station. (courtesy Lori Corp.)

Rated Voltage (V)	Non-Inductive Load		Inductive Load	
	Resistive (A)	Lamp (A)	Inductive (A)	Motor (A)
ac 110 to 125	5	1.5	3	2.5
220 to 250	5	1.0	2	1.5
dc 6 to 8	5	2	5	3
110 to 125	0.4	0.05	0.4	0.05
220 to 250	0.2	0.03	0.2	0.03

Two-pole switches (FIG. 19-75) are also available, which provide for 90 degree movement with two key withdrawal positions, thus allowing the electronic unit to be held in either the on or off position. The unit can take up to 250 volts ac, at 4 amps. Dimensions of the unit are at FIG. 19-76.

The surface-mounted remote control stations (KS RCS/1 and /2) are die-cast aluminum with a stainless steel front plate, tempered against opening, and with a red and green LED (FIG. 19-77). The waterproof version ensures unit use during all types of weather conditions. Figure 19-78 provides basic wiring information for the RCS/1 and /2 units.

Another remote control station with an anti-tamper switch is at FIG. 19-79. Like the other units, it uses the Mini-Kaba or the Kaba 20 key/switch. A remote control station mounting box (FIG. 19-80) is available and used for mounting the unit. Mounting box dimensions are at FIG. 19-81.

Remote control switches are UL approved. Electrical specifications are at FIG. 19-82.

20

The business and law of locksmithing

- ☐ *Police clearance*—Many jurisdictions require licensing and fingerprinting.
- ☐ *Professional qualifications*—Some cities insist on tests to determine professional competence.
- ☐ *Basic tools and equipment*—You should ensure that you have all you need to tackle any locksmithing job.
- ☐ *Supply stock*—It's possible to work out special arrangements with some local supply houses.
- ☐ *Financing*—Setting up shop involves an initial cash outlay.
- ☐ *Bookkeeping*—Records must be kept of all transactions.
- ☐ *Vehicles*—Necessary for out-of-shop calls.
- ☐ *Advertising*—Sometimes it takes more than just locksmithing competence to bring in the business Good advertising usually includes signs, business cards, window displays, display racks, display materials, etc.
- ☐ *Literature*—Locksmith supply house catalogs, manufacturers' literature, reference books, business forms, etc.
- ☐ *Contacts*—It pays to know other locksmiths, factory representatives, etc. They can help you from time to time with locksmithing problems.

Locksmithing is a profession involving skill, competence, and public trust. It is a career that can be highly rewarding. This chapter will give you the background necessary to use the basics of what you have already learned, and apply it in a business.

The Occupational Outlook Handbook indicates that in 1972 there were approximately 8000 full-time locksmiths in the United States. Many of these 8000 individuals operated their own locksmithing business, but others were in a shop which employed one to three persons. The situation is very much the same today, meaning that in this country, the demand for competent locksmiths is overwhelming.

Fifty to 75 million keys are lost and must be replaced each year. Every household, automobile, and business has two to five locks. In a medium-sized city of 75,000 to 100,000 persons, there are easily over 250,000 locks. This means there are literally billions of locks and keys in this country—plenty of business for a good locksmith. Statistics show that locksmiths make good money, too.

BUSINESS CONSIDERATIONS

If you're thinking about launching a career in locksmithing, there are several things you should consider:

SELECTION OF THE BUSINESS SITE

Selecting the business site is a major decision, and one that will affect the business throughout its infancy.

In determining the location, consider the types of businesses you will be near and what effect they will have upon your customers. Shopping malls are ideal for locksmithing shops because of the walk-in business generated by people who need one or two keys duplicated. Should you undertake home servicing, you would want a location that enables you to get to your customers quickly.

The type of building must also be considered. It would be unwise to obtain a building that is in need of massive repairs. The building should also be large enough for your needs. Thought should be given to the building's potential for future expansion.

Demographic and economic factors of the surrounding area should be considered too. Make a study of population density, population growth, projected economic growth, the tax structure, etc. Examine any potential competition with other locksmiths.

SHOP LAYOUT

Window displays are a good way to attract customers. The display itself should be simple and effective, neat, and clean. Window displays cannot be cluttered; you must be discriminating and orderly in developing them. The display should tell the viewer something—something interesting and appealing about your products or services.

The front counter area should be carefully organized; this is where sales are made. Ensure that the key machines are near the counter—visible from the street. Spread out your keys on a key board; such a display stays in your customers' minds longer than a pile of keys on a table. Code books usually are kept near the key machines.

The workbench usually should be out of the customers' sight, preferably in the back room. If it is out front, it should be clean at all times.

Lighting of the work and customer service areas is most important. It should be adequate but not overlighted. The lights should be of the non-glare variety. The work area can have additional small lights placed several feet above the bench.

MERCHANDISING THROUGH ADVERTISING

Advertising takes many forms, from radio, television, and handbills to putting your name on each key duplicated. There are no hard and fast rules. Good advertising is advertising that gets results.

The selection of the proper advertising can be important for a beginning locksmith. Consider the population; do they listen to the radio, watch television, or read the paper more than in other areas, or do they look for gimmicks? What are the costs of the various advertising media? You can approach the advertising staffs of the local media to determine these answers.

Once an advertising scheme is carried out, evaluate the results. Did the advertising increase business? Was the advertising worth the investment?

Ultimately, the locksmith himself is the most effective advertising. Responsible locksmithing, a neat appearance, and a little courtesy go a long way. In spite of all the sales talk and advertising hoopla, the customer is only interested in one thing—quality service.

Whether your work as a locksmith is out of your home, your vehicle or out of a formal locksmithing shop, you need to have some form of 'advertising' at your shop. To have a sign outside that merely states "Locksmith" isn't really enough. Whenever possible, you should have and use the small stickers that can be affixed to windows, the side of a van truck, or whatever, as they can increase your business.

Figure 20-1 shows a variety of such aids, including decals, window banners, and a sign. Whenever possible, you should be using these types of aids. They promote business and tell your potential customers exactly what services you can or cannot provide. Such signs also save you time in answering general inquiries from walk-ins off the street who are not sure if you can provide a specific service, such as cutting a circular ("Ace" type) key. In this instance, having a sign that indicates that these keys are duplicated in your shop can increase business.

A hanging sign (FIG. 20-2) with a slightly different slant to it—"Keys Cut While you Watch"—is another visual aid to customers. Again, by using the properly aligned key machines, you can ensure that the phrase "guaranteed fit" is accurate.

Decals for your office or on the service van also aid in promotion of your business and security in general (FIG. 20-3).

KEY BLANKS AND THEIR SELECTION

As a locksmith, it is important to carry a variety of key blanks to meet the continuing needs of your customers. More importantly, you must be able to determine what *blanks* are most used and popular in your area. For you to purchase one or two boxes of every type of key blanks imaginable that *might* possibly be used during the course of a career in locksmithing is ridiculous and extremely costly. Purchasing such items randomly upon just any key catalog that comes across your desk shows poor business judgment. The large companies that supply all kinds of locks and security devices will, naturally, also have the widest possible variety of key blanks available for your purchase and use. That assortment doesn't mean that you really need such an extensive variety of key blanks. In almost every case, you do *not*!

20-1 Various sales aids for use in the locksmith shop. At upper left is a keysign; others are window banners. (courtesy Dominion Lock Co.)

20-2   Advertising sign to hang in window of locksmithing shop. (courtesy Dominion Lock Co.)

To consider what you *should* have, let's view the key determination process from the viewpoint of a key blank manufacturer. The following information comes from Herbert Stein, president of Star Key & Lock Manufacturing, Co., Inc., of New York. This company has been in the business of carefully foreseeing the key blank needs of locksmiths for quite a number of years.

The first step is to consult a number of the largest competitors in different parts of the country for their opinion of the relative popularity of a particular blank and the anticipated annual demand. (Some blanks may be extremely popular in certain areas of the country, but

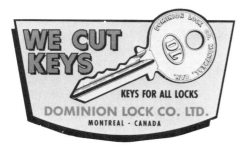

20-3   Decals for use in office or on side of van. (courtesy Dominion Lock Co.)

very slow in other parts.) On the basis of the private poll, a decision will be made. This decision is also influenced considerably by such factors as whether the blank's lock is made by a major company or not, and whether it is a big volume lock or a small one (for example, if the blank is for an automobile, is it for the major line of one of the top domestic or foreign automobile makers, or is it just for one of the small volume imported cars?). Naturally, if a deluge of requests come in from across the nation requesting a specific type blank, the priority for putting that one into immediate production will go to the top of the list.

Now, knowing how a key blank is selected for production helps, but how do you go about finding out what the most common blanks are in your area? Here is where your regional representative of the manufacturer, or direct contact with the factory, can help you. Knowing what keys are most used in your area is part of their job. Through various national and regional trends, regional and area market studies, and previous sales to your area, the manufacturer can determine with great accuracy what the most common and popularly requested blanks will be for any given part of the country.

While manufacturers keep such records, you might feel that you are imposing on them with a "trifling request." Not so, but even then, some locksmiths just look through a catalog, especially when starting out in a business, and contact a local locksmith distributor/supply house in the area. The supply house can assist you, but many sales of key blanks don't go through the supply house; they are purchases from other area distributors or directly from the manufacturer. So the supply house, while in touch with your area, isn't always as fully cognizant as the manufacturer in this regard.

At the supply house you will probably be shown one or several catalogs with a myriad of key blanks with various ones indicated that you should have on your shelf. These may or may not meet your requirements. Again, Star Key & Lock has thought of this; their catalog lists the most commonly used key blanks for just about every situation. This is where dealing directly with the manufacturer is to your advantage. By purchasing assortments of different types, you reduce your initial key blank costs, reduce your stockroom inventory, have more space for other products, and know exactly what your customers will expect. In addition, you don't have several hundred or more blanks to look through for any given type, but perhaps only a hundred and fifty or so.

By purchasing in an assortment mode, you have the advantage of getting the keys in display mounts that also provide a subtle buying hint to the customers who walk into your shop. These displays hold the *most* popular key blanks (usually between 90 and 150 different) that will meet just about all of your needs. The few oddball keys that come in for duplication are also the ones that give every locksmith a headache, because it may be only once every two to five years that a key of that specific type is ever needed.

A little footwork at your end helps you assess the needs of your shop. Check out the local general supply houses and see what blanks are the most popular with them. Also ask them about the most commonly purchased locksets in your area, and get to know these locks, the different keyways possible and available for them, and also what *were* the big sellers within the past ten years—not just the big sellers in today's market.

From all of this information you should have learned:

- ☐ The most popular key blanks in your area.
- ☐ The volume of blanks used in the past several years for different types of locks.
- ☐ The top dozen locks, by manufacturer, selling currently in your area.
- ☐ The top dozen locks, again by manufacturer, that were big sellers during the past decade.
- ☐ The various keyways necessary for these locksets.

This knowledge (coupled with the size of your shop and projected business) gives you a very good idea of which key blanks will be the most in demand, which ones will require a medium stock, and which ones you have to consider only once every two or three years. The minimal stock blanks will last for a long time, but the more popular ones may need frequent restocking—possibly every month.

With all this research done, you can now properly select the blanks that you feel are necessary. Remember that it is probably best to consider the general assortment of key blanks when first starting out. Get the catalogs that provide these types of blanks and consider their assortments. Later on, as your business expands, you can move on to include other key blank manufacturers for other, less common blanks that you see a need to stock.

After completing your key blank research, you should now know the types of locks required, as well as exactly what types of key reference catalogs for key codes your shop should carry. You can also start listing the specific cylinders, pin sizes, and dimensions required, specific repair parts kits, key machines that may be necessary (say, for cutting Ace keys), and some other specifics that will be necessary to perform as a locksmith.

The professional locksmith knows his (or her) key blanks and doesn't need to constantly refer to charts and catalogs for this information. In the beginning, every locksmith has trouble identifying one key among hundreds that may be on the shop wall. Star Key & Lock has come up with a nifty idea that is extremely helpful (and something that I wish that all manufacturers of key blanks would do): they put an "instant" identity on the key blank. This new feature has put Star far out in front of other manufacturers. What it amounts to is "stamping the make of the lock" on the key blanks in addition to the comparative numbers. This achieves immediate identification of the key blank by the locksmith, saving time and trouble, and is a valuable learning tool for the student and apprentice locksmith. The make of lock stamped directly on the key also helps convince the customer that he is receiving the correct key for his lock, which aids in establishing the locksmith's credibility.

An added feature that you as a locksmith can supply to customers who require something special or an identifying logo is embossing. This means imprinting the bow of the key blank with anything from a fancy logo design to several lines of advertising copy. Imprinting can be done on any number of the brass keys available from Star. I mention this because it is not a common or standard service, though some other manufacturers offer it as well.

Special embossing usually consists of three lines or less of standard type copy. This is based on a maximum of 14 figures on the first line, 12 on the second line, and 9 on the third line. The manufacturer must make back the costs of embossing; as a minimum, consider an order of 1000 embossed keys of assorted key blank numbers in increments of at least 50 pieces per number ordered.

The embossing is an excellent way to advertise your business. It costs less than average advertising, and still gets your message across to the public that needs and uses your products and services (FIG. 20-4). The chief advantages of special embossing are excellent advertising, increased locksmith shop prestige, and indisputable proof that the keys were made by the name embossed thereon.

In closing out this discussion of key blanks, we must comment that the keys in use today, for the most part, are of brass (all brass); this is important. All-brass key blanks are best for impressing because they have structure which will show distinct impression marks. The blanks cut very easily on key machines or by hand filing, but the myth that the brass is soft because of this is not true. The all-brass key retains maximum strength (77,000 p.s.i.); the special brass alloy which is used by Star Key & Locks (and other manufacturers) gives the best combination of desirable properties of impression-sensitive structure, free-cutting duplication, and the maximum strength necessary for a key blank.

20-4 Excellent advertising increases the locksmith shop prestige and also provides indisputable proof that the keys were made by you. (courtesy Star Key & Lock Co.)

Both as a locksmith and as an individual who requires keys in your own day-to-day movement (home, auto, office), you will see through experience that certain key blanks are better than others. Use the products that are of the best quality, not those that cost the least!

A key machine for the office or for use on the job in remote locations is shown at FIG. 20-5. This compact highly dependable cylinder key cutter, the "Companion," will go wherever your locksmithing work may be. It has dual voltage (110 Vac or 12 Vdc) capability with no extra wiring necessary; for traveling situations, it comes in a sturdy carrying case and weighs only 25 lbs.

The machine itself has four-way jaws to grip any cylinder key; no special adapters are required. The precision alignment of the cutter ensures a correctly duplicated key

20-5 The "Companion" model portable key machine operates off 110 Vac or 12 Vdc current. (courtesy Dominion Lock Co.)

everytime. (Of course, always check the cutter alignment periodically to ensure that the precision alignment is true. Rough handling and occasional bumps may throw the alignment out slightly. Be safe, be sure, and cut a true duplicate key every time.)

The key display and machine is a visual selling aid in the locksmith shop. Usually located near the counter or cash register, it promotes sales subtly by showing your customers that you have the commonly used keys in stock and can quickly duplicate a key for them while they wait. In the case of the display at FIG. 20-6, it is important that you know exactly what is available. This particular key assortment and cutting machine is a self-standing revolving counter rack and covers both U.S. and foreign automotive keys in addition to keys for business and residential locks.

The jaws on this key machine can be rotated to present four different gripping surfaces to cut four different types of keys without using any adapters. The jaws simply lift up and rotate, eliminating the fumbling necessary with adapters for the key machine. The "standard" station of the cutter jaws holds regular cylinder keys, while the "wide" jaws grip Ford double-sided keys. The "A" station jaws hold the Schlage wafer key (SC 6) and the "W" station jaws are for the Schlage water key (SC22). Other keys, such as Chicago and double-sided Datsun keys, can be cut with the "A" station jaws.

KEY MACHINE CUTTERS

After continued use, the cutters for your key machine(s) will become dull. This, in turn, means that each key you cut will not have the precision cuts required. Any such key may operate roughly in the lock, or not at all. To alleviate this, you should have at least one replacement cutter on hand of each type you normally use.

The basic cutters are shown at FIG. 20-7, and include the file, slotting, milling and the side milling slotter. The cutter specifications and key type uses are as follows:

Material	Used For	Cutter Type
File	High carbon steel	Cylinder keys
Slotting	High speed steel	Flat keys with .045" cut*
Milling	Tungsten-chrome alloy steel	Cylinder keys
Side Milling Slotter	High speed steel	Cylinder and flat keys (.045" cut)

*Two different slotting cutters are available with different thicknesses for .030 inches and .045-inch cuts on flat steel keys.

DISPLAYS

Shown in FIG. 20-8 is the color key carousel. Because colored keys are becoming more popular, some customers will prefer to have another color key when having their key duplicated. Color keys assist the customer in rapid identification of specific keys. In some instances, they will not accept a key that is not colored.

A key comparison board allows you to rapidly determine which blanks, by manufacturer, are the same in their component makeup (i.e., the specifics of the key identification). A key comparison board is a great advantage to the locksmith, and even more so for the apprentice or locksmith trainee.

In the case of these two items (as with others that you will find in various catalogs), the maker isn't just hyping a product in hopes that you will buy it; the manufacturer genuinely wants to provide you, as a locksmith, with items that will assist you in the locksmithing trade.

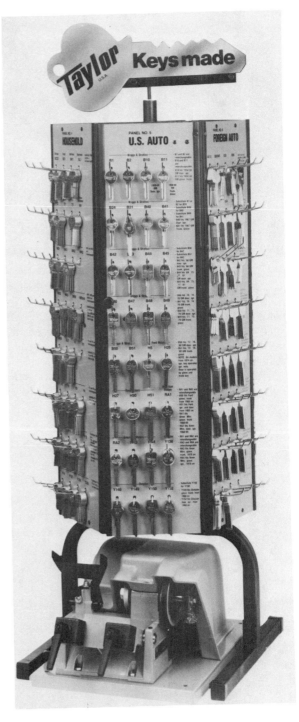

20-6 Key machine with the most commonly cut key blanks available for immediate duplication for customers in need. (courtesy Taylor Lock Company)

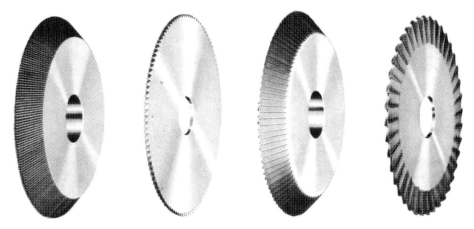

20-7   A set of the basic replacement key machine cutters: file, slotting, milling, and a side-milling slotter. (courtesy Taylor Lock Company)

By helping you to do your job a little better, easier, faster, and more professionally, and by making available the supplementary business items required, the manufacturer is helping everyone: you, your customer, and himself. When you have problems, questions, or need assistance, the manufacturer or his representative is there to help you. Likewise, the manufacturer is looking ahead to your varying needs and has a variety of items to help you to help yourself.

Another example of a manufacturer's concern for assistance to the locksmithing business is at FIG. 20-9. Lock displays like this unit are available for setup and demonstration

20-8   Color key carousel from manufacturer to assist you in the locksmithing business. (courtesy Star Key & Lock Co.)

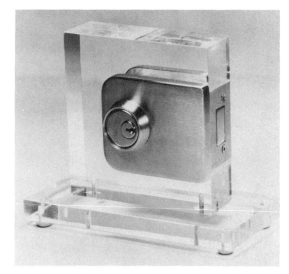

20-9 Manufacturer-prepared lock display for customer consideration. Every working shop should have a variety of such displays for the customer. (courtesy M.A.G. Engineering and Mfg., Inc.)

of specific products within your shop. The customer is interested in a specific type of product—in this case, a lock. With a variety of these type of displays available, the customer can try out each one, and as the choice of locks is narrowed down, you provide specifics concerning the remaining locks. In this particular case you might point out the very salient and positive features of the Ultra 800, the Medeco cylinder, the strength and positive locking action of the mechanism, and other specifics.

For narrow stile door installation, Adams Rite has created a specific installation kit. The advantages of this kit should be obvious: extra parts and detailed information for each stile installed. For the locksmith who plans continued installation, repair, and/or replacement of lock units in narrow stile doors, this kit is a must. It allows for a smoother, more rapid installation of numerous units, saving a tremendous amount of time (FIG. 20-10).

THE LOCKSMITH AND THE LAW

The locksmith, because of the uniqueness of his profession, must have a better understanding of various laws than most people. In this regard, when setting up a locksmithing business it is always prudent to consult with your attorney regarding all laws that concern you as a locksmith. In many jurisdictions there are laws covering licensing, control of locksmithing tools, and the registration of code books. Some local laws regulate the conduct of certain locksmithing business practices, such as duplicating master keys, making bank deposit box keys, opening automobiles, etc.

When first entering the business, aside from consulting with your lawyer and possibly the police, you should also contact your area locksmithing association, if one exists. Information can be obtained from the national locksmithing organizations too, through their newsletters and publications. These are excellent sources of information about your legal responsibilities.

Your legal responsibilities demand that you be very careful about the jobs you accept. You must be certain that you are not breaking the law by complying with a customer's wishes. Authorization statements from supervisors, such as from a bank or post office, for duplicating a key should always be doublechecked with the main offices. Verification of such written statements is a must; they should also be filed with the job order.

When jobs of this type arise, the following information should be included in your files:

☐ Name of the person who brought the job in.
☐ Identification (Social Security card or driver's license).
☐ Address of the person who brought the job in.
☐ Business telephone (call to doublecheck).

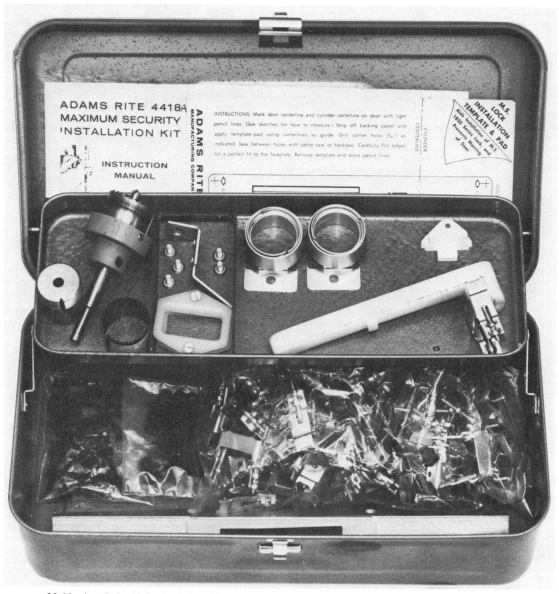

20-10   Installation kit for the Adams Rite narrow stile door units, a must item involved in this type of lock installation. (courtesy Adams Rite Mfg. Co.)

☐ Name of the firm.

☐ Business address.

☐ Type of service performed.

☐ Type of payment (if a check, it should be a business check, not a personal check).

☐ Automobile make, model, serial number, license number, state it is registered in.

You may also have some forms printed up for the individual to fill out. A single form can be used to cover all the situations you may run into. The form should be kept along with the work order.

In some states or cities, laws require that you take an examination before your locksmithing peers to show that you are a competent locksmith meeting professional standards. Besides this, police checks of your reputation, qualifications, background, and previous employers may be made to ascertain that you are of good moral character.

Various laws have been enacted to protect the public from unscrupulous individuals posing as locksmiths and to protect the locksmithing trade itself from such individuals. The following is extracted from the Los Angeles Code, Ordinances No. 83,128, as an example of a local law regulating the locksmithing profession:

SEC 27.11 LOCKSMITH—REGULATING APPLICABLE TO

A. Definitions

"Locksmith" shall mean any person whose trade or occupation, in whole or in part, is the making or fashioning of keys for locks, or similar devices, or who constructs, reconstructs or repairs or adjusts locks, or who opens or closes locks for others by mechanical means other than with the regular keys furnished for the purpose by the manufacturers of the locks.

B. Trade of locksmith—permit required

No person shall engage in the business of locksmith, or practice or follow the trade or occupation of locksmith without a permit therefore from the Board of Police Commissioners.

C. Permit—application

Such permit shall be issued only upon the verified application of the individual seeking the permit. The application shall be upon a form prescribed by the Board and shall set forth the proposed location of the applicant's place of business, the names and address of five character references, and such other things as the Board may require to determine the character, honesty, and trustworthiness of the applicant. Specimen fingerprints of the applicant shall be furnished with the application.

D. Permits—fees—expiration

Each application for a permit shall be accompanied by a fee of $10 and each application for the annual renewal thereof, by a fee of $5. Each permit, unless sooner revoked, shall expire on December 31st, following the date of issuance. Each permit shall bear a serial number.

E. Permits—issuance and denial

The Board shall cause an investigation to be made upon each application, and if the Board finds that the applicant's reputation for honesty is good, that he has not used his skill or knowledge as a locksmith to commit or aid in the commission of burglaries, larcenies, thefts, or other crimes, that he intends honestly and fairly to practice the trade of locksmith in a lawful manner, and that he has not been convicted of a felony, then the permit shall issue. Otherwise it shall be denied.

F. Permittee—must keep record

Each permittee must keep a book, which shall be open to inspection by any police officer at all times, in which the following must be entered:

1. The name and address of every person who whom a key is made by code or number.
2. The name and address of every person for whom a locked automobile, building, structure, house, or store, whether vacant or occupied, is opened, or a key fitted thereto.

G. Keys to be stamped

It shall be unlawful for any locksmith to fail to stamp the serial number of his permit upon any key made, repaired, sold, or given away by him.

H. Signs to be displayed

Every locksmith shall display in a conspicuous manner in the place where he is carrying on such business, trade, or occupation, a sign of a style, size, and color to be prescribed by said Board, reading, "Licensed Locksmith," together with the official permit number.

I. Permits—revocation

The Board, upon proceedings had as prescribed in Section 22.02, may revoke or suspend any permit issued hereunder upon any of the following grounds:

1. Misrepresentation in obtaining such permit.
2. Violation of any provision of this section.
3. That the permittee has committed or aided in the commission of or in the preparation for the commission of any crime by the use of his skill or knowledge as a locksmith or by using or letting the use of his tools, equipment, facilities, or supplies.

21

Collecting locks
and keys

Within the past few years, a lot of people have come to realize how much fun and profitable lock and key collecting can be.

During the writing of the first edition of this book, lock and key collecting was an obscure hobby enjoyed by a handful of people. Although unusual and intricately designed locks and keys have always been fascinating to people, very few knew of places to acquire the items and fewer people knew how to price them.

It's difficult to begin a collection of items you know little about and can't find. The general public's ignorance of locks was largely due to the secretive nature of locksmithing. Early locksmiths (then called blacksmiths) jealously guarded information about locks from both the general public and from each other. Within about the last hundred years, locksmiths became a little less secretive, and began sharing more information with one another through writings.

In recent times, books such as the first edition of this handbook helped many more people learn about locks.

Today there are lock and key museums, lock and key shows, and national organizations that promote lock and key collecting. Information about the hobby is easier than ever to obtain. No longer is it necessary to be a locksmith in order to begin a substantial lock and key collection.

PADLOCKS

The vast majority of locks collected today (perhaps as many as 90 percent) are padlocks. Padlocks are so popular because they're among the oldest types of locks; they're easily obtainable; and most of them are small and easy to display or store. Figures 21-1 and 21-2 show some collectible padlocks.

Railroad locks are one of the most popular types of padlocks among collectors. Perhaps older collectors see railroad locks as nostalgic reminders of childhood days. Within the last few years, many railroad locks have rapidly increased in value.

21-1   Wyeth Padlock. (courtesy Donald Edwin Stewart.)

MASTERPIECE LOCKS

During the Renaissance period, apprentice locksmiths qualified to become master lock-smiths by designing and making locks by hand that were acceptable to the locksmiths guild. These locks are called masterpiece locks. Masterpiece locks often took several years to make.

After a masterpiece lock was accepted by the guild, it was used as a display of the locksmith's ability. Many of those locks are still available. Prices for them vary, depending on their quality and whether or not they're available with original keys.

21-2   Miller Cut-away Padlock. (courtesy Donald Edwin Stewart.)

21-3 Folger Adam Jail Lock & Key. (courtesy Donald Edwin Stewart.)

JAIL LOCKS AND KEYS

Many collectors specialize in collecting jail locks and keys. Figure 21-3 shows a jail lock and its key. Those keys are often misrepresented by antique dealers and other sellers. Sellers often try to pass off imported gate keys as old jail keys because few people outside of a jail system know what the keys look like. Most of us rely on ficticious television portrayals for our "knowledge" of what such keys look like.

Genuine old jail keys are for lever tumbler locks, and are accurately cut (see FIG. 21-4). Many imported gate keys, on the other hand, are crudely cut bit keys. Figure 21-5 shows an old bit key lock. The fake "jail" keys often come from France, Mexico, or Spain.

KEYS

A lock collection is much more valuable when it includes keys for the locks—which is why most lock collectors also collect keys. But some people are more interested in collecting keys than in collecting locks. Keys are relatively inexpensive, easy to acquire, and easy to store.

EGYPTIAN AND ROMAN KEYS

Few early egyptian keys are around today, because they were made of wood. But later models, from about 60 to 300 years old, are available patterned in basically the same design as the early models.

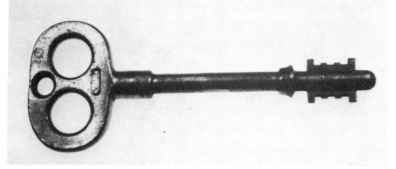

21-4 Yale Spike Jail Key. (courtesy Donald Edwin Stewart.)

21-5   Old Rim Lock (Cottage brand). (courtesy Donald Edwin Stewart.)

Early Roman keys were made of bronze or iron, and many can be found today. "Finger Ring" keys are the most common. They are decorative and were designed to be worn as rings. The most valuable finger ring keys are made of precious metals and are decorated with gems.

CAR KEYS

Original keys for antique and classic cars are popular among many key collectors. While few people can afford a Packard or Model T, most people can afford to own the keys.

21-6   Car Crest Keys (1950s/60s). (courtesy Donald Edwin Stewart.)

21-7 Ford Model "T" Key (Briggs mint mark). (courtesy Donald Edwin Stewart.)

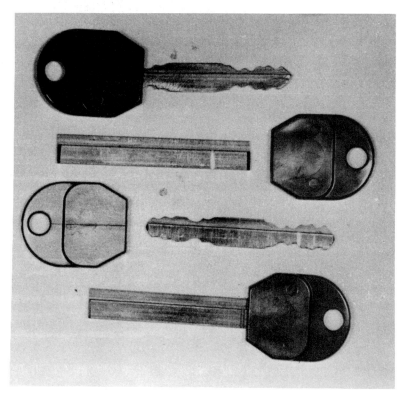

21-8 Ford Double End Car Keys and Blanks. (courtesy Donald Edwin Stewart.)

21-9   Steel Folding Door Keys. (courtesy Donald Edwin Stewart.)

Door locks weren't used on the earliest automobiles. Apparently, people saw no need to lock the doors on their "Horseless Carriages." In his memoirs, Henry Ford mentioned that he carried a chain and padlock with him so he could chain his personal vehicle to telegraph poles.

Although they didn't have door locks, many early cars had other locks. They had locks for battery boxes, hoods, spare tires, radiator caps, etc.

Among the most valuable car keys for collectors are the following: Packard's 50th Anniversary keys; early gold plated-keys; Crest keys (see FIG. 21-5); ".45 keys"; and early Ford keys (especially Model T keys). Figures 21-6 and 21-7 show some early Ford keys.

FOLDING KEYS

Folding keys are interesting collectibles because few are being made today. Figure 21-9 shows folding keys. A lot were made between 1700 and 1900, and come in a variety of shapes and sizes. Most were made of bronze, iron, or brass.

21-10   Ship Keys and Tags. (courtesy Donald Edwin Stewart.)

21-11 Skate Keys. (courtesy Donald Edwin Stewart.)

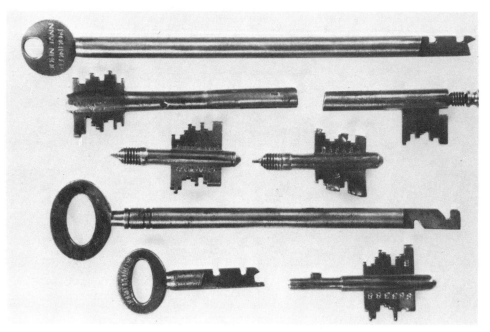

21-12 Two-Piece Safe Keys. (courtesy Donald Edwin Stewart.)

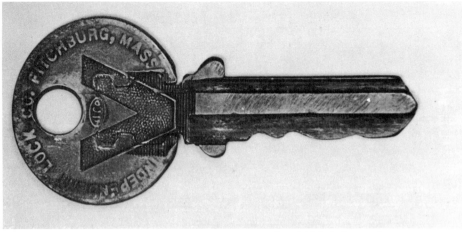

21-13   Ilco WW II Victory Key. (courtesy Donald Edwin Stewart.)

OTHER KEYS

The number of collectible types of keys is limited only by a collector's imagination. People collect ship keys (see Fig. 21-10), casket keys, watch keys, skate keys (see Fig. 21-11), safe deposit keys (see Fig. 21-12), hotel keys, Keys to the City, etc. Two popular keys among general key collectors include the Ilco WW II Victory key (see Fig. 21-13), and the Ilco Bottle Opener key (see Fig. 21-14).

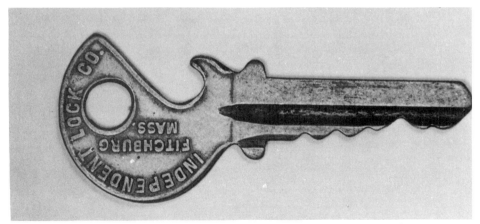

21-14   Ilco Bottle Opener Key (1931). (courtesy Donald Edwin Stewart.)

22

Key duplicating machines

One of the most important investments you'll make in locksmithing equipment is a key duplicating machine. A wide variety of them are available, with prices ranging from a few hundred dollars to several thousands of dollars. Figure 22-1 shows a few types of key duplicating machines.

A wide difference in price doesn't necessarily indicate a substantial difference in functions and quality between key duplicating machines. This chapter will help you better understand how the machines work, and how to find the best one you can afford.

A key duplicating machine consists of four basic parts:

Two vises move in unison. One vise holds the key being duplicated; the other holds the key blank.

A key guide traces the profile of the key.

A cutter wheel cuts the key blank in accordance with the profile traced by the key guide.

Most inexpensive key duplicating machines are designed to only duplicating cylinder keys; but a few inexpensive models can also duplicate flat keys. More sophisticated models are designed to duplicate other types of keys. The most expensive ones can duplicate bit keys, tubular lock keys, angle keys (such as Medeco), and dimple keys. You can save money by purchasing a machine that duplicates only the types of keys you're most interested in duplicating.

CRITICAL DESIGN FACTORS

When evaluating a key duplicating machine, look carefully at the pivot mechanism. The key should meet the cutter wheel squarely on a dead parallel with the axis of the wheel. A slight angle is enough to upset the dimension of the duplicated key. (See FIG. 22-2.)

Some machines don't have pivots, and arrange matters so that the vises move laterally into the cutter. As long as the bearings are true, this ensures that the duplicated key is a faithful copy of the original.

The jaws of the vises must be carefully engineered to ensure that both the key and key blank do not shift during the duplication process and that the key and key blank are held squarely against the cutter.

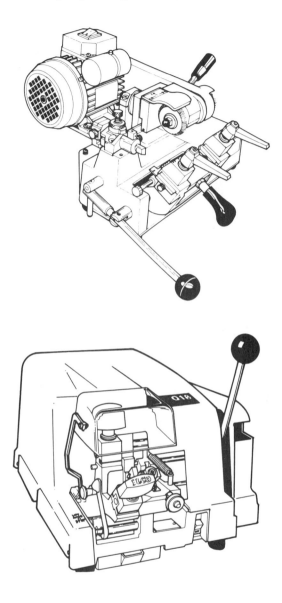

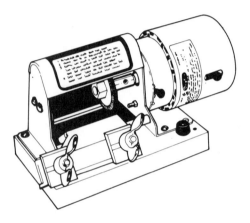

22-1   Key cutting machines. (courtesy Ilco Unican Corp.)

CUTTERS

There are three basic types of cutters. *Filing* cutters and *milling* cutters are both used to duplicate cylinder keys. A *slotter* cutter is used for square-ended keys (such as bit keys and flat keys). Figure 22-3 shows some examples.

The diameter of the cutter is important. Large diameter cutters leave a very slight concave on the newly cut key. Smaller cutters make a deeper concave. The most useful key duplicating machines allow you to use various sizes and types of cutters.

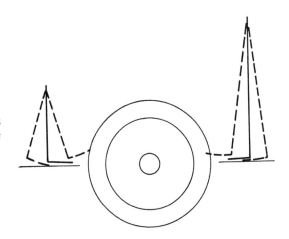

22-2 A key machine should be constructed so the key is almost parallel to the cutter. A long pivot (shown on the right) reduces the angle.

KEY GUIDE

The key guide should be checked regularly. In the absence of manufacturers' instructions for doing so, check the guide by mounting two identical key blanks in the vises; then, lift the vises up to the key guide and cutter wheel. While keeping the vises in place, slowly rotate the cutter wheel. The cutter wheel should barely scrape one of the key blanks. If the cutter wheel doesn't touch a key blank, or if it digs deeply into a key blank, the machine is out of alignment and needs to be adjusted. Figure 22-4 illustrates how to check a key guide.

FRAMON'S DBM-1 FLAT KEY MACHINE

Framon Manufacturing Company's DBM-1 is a high-quality flat key duplicating machine. It's designed to duplicate a wide variety of flat and corrugated keys (See FIG. 22-5.) By understanding how it works, you'll be able to operate many similar machines.

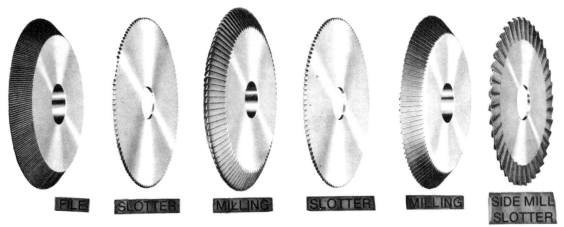

22-3 Various cutter wheels.

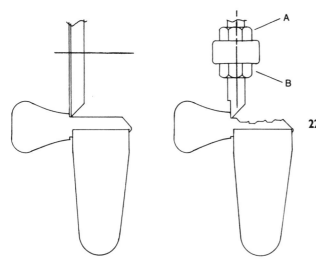

22-4 The cutter wheel should barely touch the key blank (left). (courtesy Keynote Engineering.)

Cutting procedure

All keys should be set from the tip for spacing. Insert the pattern key in the right hand vise with the tip of the blank protruding slightly beyond the left side of the vise (see FIG. 22-6). The reason for this position is to allow cutting of the tip guide on the blank if the blank tip is slightly different than the pattern key. In this position the tip can be cut without the cutting vise.

Push the guide shaft rearward and lock into this position by tightening the locking knob (see Fig. 22-7). This relieves spring pressure so tip setting is easier. Lift the yoke, and set the tip of the pattern key against the right hand side of the guide. While holding this position, insert the blank, and set the tip of the blank against the right hand side of the cutter (see FIG. 22-8). This procedure assures proper spacing.

22-5 Framon's DBM-1 Flat Key Machine. (courtesy Framon Manufacturing Co.)

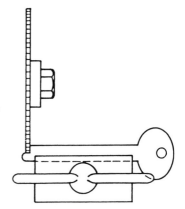

22-6 Set key from tip for spacing. (courtesy Framon Manufacturing Co.)

22-7 Tighten locking knob to lock guide shaft into position. (courtesy Framon Manufacturing Co.)

22-8 Set tip of blank against right hand side of cutter. (courtesy Framon Manufacturing Co.)

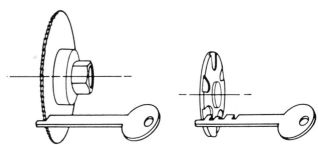

Release the guide by loosening the locking knob; the key is now ready to be cut.

See the cut in the pattern key against the guide, and lift the yoke into the cutter to make the cut (see FIG. 22-9); lower the yoke and repeat for the next cut. Follow this procedure until all cuts (including the throat cut) are made and the key is complete. All cuts should be made with a straight in motion. This will ensure clean, square cuts.

You will notice that there is no side play in the guide assembly, so all cuts on duplicate keys will be the same width as pattern keys.

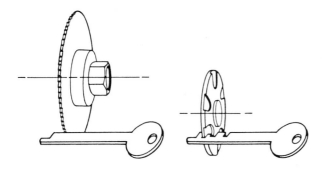

22-9 Set cut in pattern key against guide. (courtesy Framon Manufacturing Co.)

Adjustments

With tip setting for spacing there is no problem with improper spacing.

As to depth setting, this can be checked by using two blanks that are the same; check the depth by drawing the blank against the cutter guide. If depth adjustments are needed, simply loosen the set screw on the depth ring and adjust the ring until the cutter barely touches the blank. To make cuts deeper, rotate the depth ring counterclockwise. To make cuts shallower, rotate the depth ring clockwise.

Tighten the set screw after adjustment is made, but do not overtighten.

Another way to check depth is to make one cut on the duplicate key; check the depth cut on both the pattern key and duplicate. As an example, if the cut on the duplicate key is .003 deeper than cut on the pattern key, loosen the set screw on depth ring, rotate the ring .003 clockwise, and tighten the set screw. *Note:* Calibrations on the depth ring are in increments of .001.

Check the guide setting; the guide must be set to the same width of the cutter used. The DBM-1 is supplied with one .045 width cutter. This is the best width for general work. Cutter widths of .035, .055, .066, and .088 will be available if needed. Cutter width of .100 (LeFebure) can be obtained by using an .045 and an .055 at the same time. All of these cutters are solid carbide.

To set the guide, simply loosen the cap screw (DBM-42). Rotate the guide to cutter width, and tighten the cap screw. The Detent screw in the guide shaft will align the guide. No adjustment is required when changing guide settings.

Maintenance

Yoke rod (DBM-09), guide shaft (DBM-09) and vise studs (DBM-18) should be lubricated using very fine oil sparingly (Do Not Use Motor Oil).

Be sure to wipe off all excess oil. To lubricate the guide shaft, unscrew the locking knob (DBM-22) and put one or two drops of oil in the opening. Replace the knob.

Other than these parts, cleanliness is the best maintenance.

THE KD50A

Ilco Unican Corporations' KD50A is a high-quality machine for duplicating cylinder keys. By understanding how it works, you'll know the basics of operating most other cylinder key duplicating machines. The following information about the KD50A came from an instruction manual published by the manufacturer.

General operating sequence

The KD50A has a constant power switch which must be turned on. However, the machine motor will not operate until activated by the carriage assembly. (See FIG. 22-10.)

After both key and blank are properly clamped and aligned, pull down on the carriage handle. Use your thumb to depress the carriage release knob; the key setting gauge will

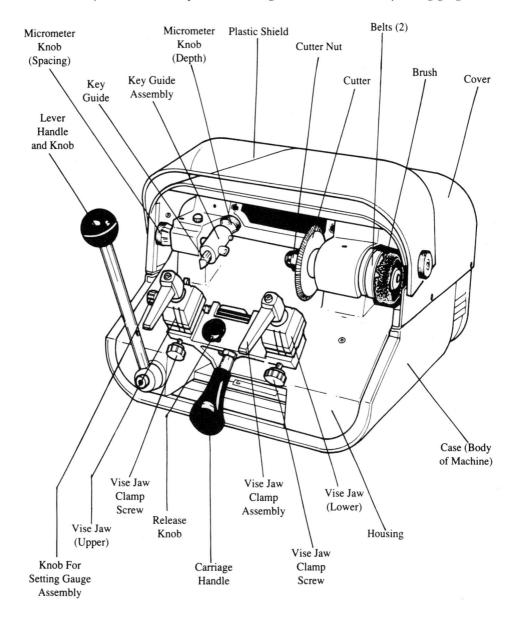

NOTE: On/Off power switch (KD50A-15) is not shown, but is visible on the left side of the machine.

22-10 Ilco Unican's KD50A Key Duplicating Machine. (courtesy Ilco Unican Corp.)

automatically spring away. Spring tension will raise the carriage, and the motor will automatically start.

Move the lever handle sideways, so that the original key touches the key guide in an area between the shoulder and the first cut. Do not let the shoulders touch either the key guide or cutter wheel. Using the lever handle, slide the carriage left and then right to complete the cutting operation. Lower the carriage until it locks into the original position, which will automatically stop the motor and the cutter. Remove the new key and deburr with the brush; do not overbrush or run the key into belts.

Adjusting for proper depth of cut

Remove the wire plug from its electrical socket for safety. Clamp the two service bars into the vise jaws as shown in FIG. 22-11, making certain that both bars rest flat against the bottom of the vise and that they are butting against the edge of each vise jaw. Lift the carriage toward the key guide and cutter until a flat portion of the service bar rests against the key guide. (To lift the KD50A carriage, pull down and press the carriage release button between the vise jaws.)

Turn the cutter by hand. The machine is correctly adjusted if the cutter barely grazes the top of the right service bar. If the cutter is stopped from turning or turns freely without contacting the service bar, the cutting depth must be adjusted, as follows:

Loosen the Allen screw that holds the key guide.

Turn the cutting depth micrometer adjusting knob behind the guide, either left or right. This will move the key guide in or out. Do this until the cutter just grazes the top of the right service bar when the left service bar is resting against the key guide. Turn the cutter by hand; adjust to the high spot of the cutter.

Tighten the key guide Allen screw.

Note: This adjustment must be made if the cutter is replaced, or whenever a test fails to work, indicating that the cutter may have worn down somewhat, resulting in cuts that are too shallow.

Adjusting for proper lateral distance (spacing)

Key cutting accuracy also depends upon the spacing of the key and blank key to be the same as the distance between the key guide and cutter. To assure that the lateral distance adjustment is correct, refer to FIGS. 22-11 to 22-14 and proceed as follows:

Insert the service bars into the vise jaws making sure that each service bar is butting against the edge of each vise jaw. *This is critical!*

Rotate the key setting gauge up, and make certain that both setting gauge shoulders rest *exactly* against the service bar stops as shown in FIG. 22-12. If there is a discrepancy, loosen the right setting gauge Allen screw and adjust so that both gauge shoulders rest exactly against both service bar stops.

Lift the carriage to the key guide and cutter. Insert the key guide and cutter into the V shaped grooves in the service bars as shown in FIG. 22-13. Both the key guide and the tip of a cutter wheel tooth must fit exactly into their V grooves or the setting will not be accurate (make certain that you do not seat the space between two cutter teeth into the V groove).

If the guide and cutter do not seat exactly into the V grooves, the distance between the cutter and guide must be altered. Loosen the Allen screw in the key guide assembly, and turn the micrometer adjusting knob fore or aft. This action will shift the position of the key guide assembly to the left or right. Continue until the key guide and the cutter both drop into the V notches of the service bars.

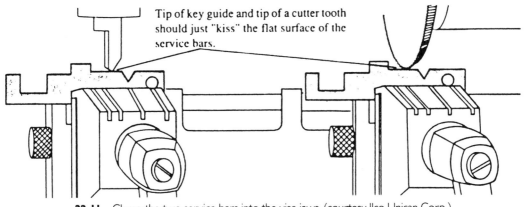

Tip of key guide and tip of a cutter tooth should just "kiss" the flat surface of the service bars.

22-11 Clamp the two service bars into the vise jaws. (courtesy Ilco Unican Corp.)

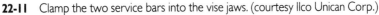

Both setting gauge shoulders should butt exactly against both service bar stops.

22-12 Make certain both setting gauge shoulders rest against the service bar stops. (courtesy Ilco Unican Corp.)

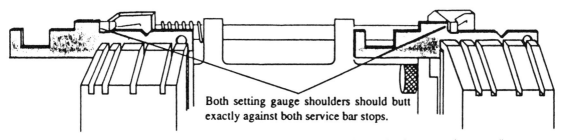

Both the key guide tip and the tip of a cutter wheel tooth must fit exactly into the V-groove in the service bars.

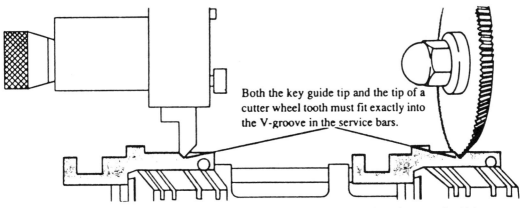

22-13 Insert key guide and cutter into the V shaped grooves in the service bars. (courtesy Ilco Unican Corp.)

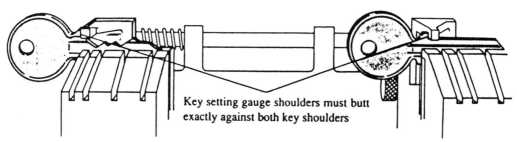

Key setting gauge shoulders must butt exactly against both key shoulders

22-14   The key and key blank are now spaced the correct distance apart for cutting. (courtesy Ilco Unican Corp.)

Insert the pattern key, left to right, into the left vise. Rotate the key setting gauge upward and set its left shoulder against the shoulder of the pattern key. Be sure the key is lying flat along the bottom of the vise. Secure the key by turning the clamp assembly clockwise.

Insert the key blank in the same manner, into the right vise, and secure. Make sure that the key setting gauge is exactly against both key shoulders. The key and key blank now are spaced the correct distance apart, and they are ready for cutting. See FIG. 22-14.

Aligning keys without shoulder (ford and best)

On keys without shoulders, the key setting gauge cannot be used. It is necessary to use the service bar to correctly position the key and the blank.

The vise jaws have a series of slots; and slot can be used for the service bar. Also note the key head rest (KD50A-60), which prevents the key from tilting as the vise jaw is tightened. The key head rest can be moved to properly support the key. See FIG. 22-15.

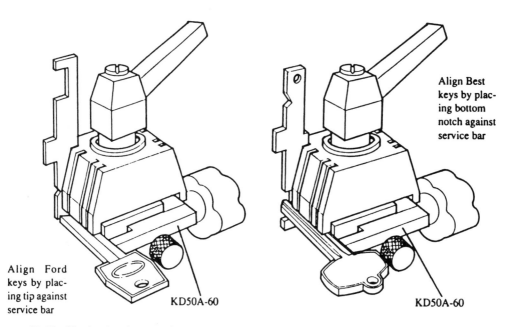

Align Best keys by placing bottom notch against service bar

Align Ford keys by placing tip against service bar

KD50A-60

KD50A-60

22-15   The key head rest can be moved to properly support the key. (courtesy Ilco Unican Corp.)

Aligning narrow blade cylinder keys

Some keys have a very narrow blade; therefore they sit deep in the vise jaws with only part of the cuts showing above the vise. This makes it necessary to use the service pins to raise the key for proper cutting.

Insert an equal size pin under each key and blank on the bottom of the vise jaws. This will raise both the key and blank to allow the correct depth of cut to be made. See FIG. 22-16. Do not cut into the vise jaw!

Aligning double sided cylinder keys

Before cutting this style of key, examine the key to see if there is a milled groove on either side. If so, then reverse the vise jaw and clamp the key, using the V jaws. The key will be held securely when only the top or bottom V jaw fits into a milled groove. When there is no V groove on either side of the key, then use the flat vise jaw.

If the cuts are not the same on both sides of the key, make the shallow cuts first. In this way, when you turn the key over to cut the second side, there will be enough metal to grip the key securely during the actual cutting. To reverse the vise jaw, loosen the retaining screws at the base of the vise jaws. Raise, rotate and reseat both vise jaws; then retighten their retaining screws. Note the V shape of the jaws. Insert the key between the jaws, with a milling groove resting in the point of the V. This will hold the blank securely. Align for spacing and proceed out.

Aligning carriage to prevent vise jaw damage

This machine is equipped with a carriage stop that prevents the carriage from moving all the way up to the cutter. When properly adjusted, it stops the cutter from grinding into the vise jaw. Such a condition could occur when reaching the tip of the cut key, and the carriage lever continues to move the carriage.

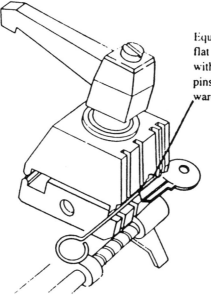

Equal size service pins must rest flat along bottom of each vise with keys resting flat on top of pins so that keys are raised toward key guide and cutter

22-16 Insert pin under key and key blank to allow correct depth of cut to be made. (courtesy Ilco Unican Corp.)

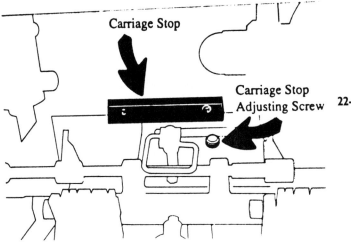

22-17 The carriage stop is a U-shaped channel secured to the housing by set screws. (courtesy Ilco Unican Corp.)

The carriage stop (Part No. KD50A-144) is a U-shaped channel secured to the housing by set screws. It is possible to span the travel of the carriage during the cutting cycle; normally, this position does not change. In addition, there is a carriage stop adjusting screw that is installed in the carriage; this screw controls the distance between the cutter and vise jaw. (See FIG. 22-17.)

The carriage stop adjusting screw is set at the factory to create a clearance of .005 inch between the cutter and the vise jaw. (See FIG. 22-18.) This distance is not critical and can be set without measuring instruments. Just loosen the lock nut and turn the screw in or out, so the cutter does not touch the vise jaw. The machine should be off. When an ordinary business card can slide between the cutter and vise jaw, the adjustment is correct and the accuracy of key cutting will not be affected. CAUTION: Do not make this clearance too wide; key cutting could be affected on some keys having deep cuts.

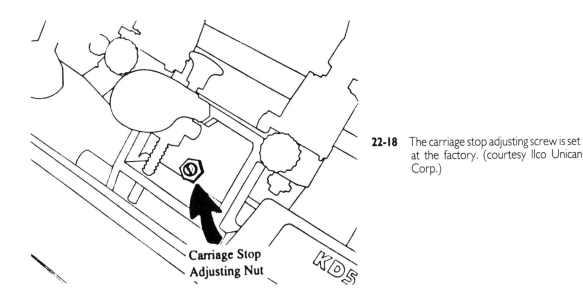

22-18 The carriage stop adjusting screw is set at the factory. (courtesy Ilco Unican Corp.)

It's a good idea to check the clearance on a regular basis, especially when a large quantity of keys are cut. If the cutter is allowed to strike the vise jaw, the edges of the cutter will be dulled immediately, causing a reduction in the life of the cutter.

Some important tips

Clean vise jaws regularly, so that no metal chips lie under the keys. It is essential that both keys lie flat across the entire width of both vise jaws. Neither key should be tilted.

Do not use pliers or other tools to tighten the vise jaws. Firm hand pressure is sufficient.

Keep the carriage shaft free of metal chips. A thin film of oil can be applied to it. The carriage should travel smoothly along its shaft.

Never touch the shoulder of a key to the side of the key guide; this will cause the shoulder of a key blank to touch the side of the cutting wheel. When this happens, some of the metal will be cut away from the shoulder of the key guide. If the resulting duplicating key is duplicating two, three, or four times over, an error will accumulate and cause a non-operating key. Do not grind away the shoulder.

Don't run the cutter into the vise jaw; this will only dull the cutter, and reduce cutter efficiency.

Keep the cutter clean. Don't let any foreign objects or instruments blunt it. This cutter is a precise cutting tool. Handle with care.

Lubrication of moving parts is important. Oil cups are provided to keep the cutter shaft bearings well lubricated. The carriage spindle should be lubricated with a thin film of oil.

Locksmithing Related Associations

AMERICAN LOCK COLLECTORS ASSOCIATION
36076 Grennada
Livonia, MI 48154

AMERICAN SOCIETY FOR INDUSTRIAL SECURITY
1655 N. Ft. Myer Dr., Suite 1200
Arlington, VA 22209

ASSOCIATED LOCKSMITHS OF AMERICA, INC.
3003 Live Oak St.
Dallas, TX 75204

DOOR AND HARDWARE INSTITUTE
7711 Old Springhouse Rd.
McLean, VA 22102-3474

THE LOCKSMITH GUILD
850 Busse Highway
Park Ridge, IL 60068

INSTITUTIONAL LOCKSMITHS' ASSOCIATION
P . O. Box 108
Woodville, MA 01784-0108

KEY COLLECTORS INTERNATIONAL
P. O. Box 9397
Phoenix, AZ 85068

LOCK MUSEUM OF AMERICA, INC.
130 Main St., P. O. Box 104
Terryville, CT 06786-0104

SAFE AND VAULT TECHNICIAN'S ASSOCIATION
5083 Danville Rd.
Nicholasville, KY 40356

SAFECRACKERS INTERNATIONAL/
NATIONAL ANTIQUE SAFE ASSOCIATION
P. O. Box 110099, 1142 Nucla St.
Aurora, CO 80011

VEHICLE SECURITY ASSOCIATION
5100 Forbes Blvd.
Lanham, MD 20706

Appendix **B**

Manufacturers and Distributors of Locksmithing Equipment/ Supplies

Note: Some manufacturers and distributors do business only with professional locksmiths. A letterhead, business card, photocopy of locksmithing license, photocopy of locksmithing bond card, or photocopy of your advertisements from a telephone book's "yellow pages" are all usually acceptable as proof of your professional status.

When requesting product information, price quotes, or a current catalog, be sure to send one form of proof of your professional status. Also, be sure to mention this book as your reference source.

AAA-LOCKSMITH SUPPLY CO.
P. O. Box 6520
Erie, PA 16512

A-1 SECURITY MFG. CORP.
3528 Mayland Court
Richmond, VA 23233

ABUS LOCK CO.
218 W. Cummings Park, P. O. Box 2367
Woburn, MA 01888

ADAMS RITE MFG. CO.
4040 S. Capitol Ave.
City of Industry, CA 91749

ALARM LOCK SYSTEMS
P. O. Box 2001
Pine Brook, NJ 07058

ARROW LOCK CORP.
555 Long Wharf Dr., Suite 12
New Haven, CT 06511

BELWITH INTERNATIONAL
18071 Arenth Ave., P. O. Box 8430
City of Industry, CA 91748

DOM SECURITY LOCKS
225 Episcopal Rd.
Berlin, CT 06037

FRAMON MANUFACTURING CO., INC.
909 Washington Ave.
Alpena, MI 49707

ILCO UNICAN CORP.
400 Jeffreys Rd.
Rocky Mount, NC 27801

M.A.G. ENG. & MFG. CO., INC.
15261 Transistor Lane
Huntington Beach, CA 92649

MEDECO SECURITY LOCKS, INC.
P. O. Box 3075
Salem, VA 24153

MRL INC.
7640 Fullerton Rd.
Springfield, VA 22153

NATIONAL SCHOOL OF LOCKSMITHING AND ALARMS
1466 Broadway, 2nd Floor
New York, NY 10036

PRESO-MATIC LOCK CO., INC.
3048 Industrial 33rd St.
Ft. Pierce, FL 34946-8694

R & D TOOL CO.
7705 R.C. Gorman Ave. NE
Albuquerque, NM 87122-2738

ROFU INTERNATIONAL CORP.
3725 Old Conejo Rd.
Newbury Park, CA 91320

SCHLAGE LOCK CO.
P. O. Box 3324
San Francisco, CA 94119

SECURITECH GROUP, INC.
54-45 44th St.
Maspeth, NY 11378

SECURITRON MAGNALOCK CORP.
1815 W. 205th St., #105
Torrance, CA 90501

SIMPLEX ACCESS CONTROLS CORP.
P. O. Box 4114, 2941 Indiana Ave.
Winston Salem, NC 27115-4114

SLIDE LOCK TOOL CO.
P. O. Box 386
Louisville, TN 37777

Plug Follower and Holder Diameters for Popular Locks

Manufacturer	Diameter (in inches)			
	0.395	.495	0.500	0.550
Acrolock		X		
Corbin		X		
Corbin oversize				X
Corbin small pin	X			
Eagle		X		
Eagle small pin	X			
ILCO		X		
ILCO 4019 rim cylinder			X	
ILCO "peanut" cylinder	X			
Keil			X	
Kwikset		X		
Lockwood		X		
National (Rockford)		X		
National (Ozone Park)		X		
Russwin oversize				X
Sargent		X		
Segal		X		
Taylor		X		
Weslock (except KNK)		X		
Yale		X		
Yale small pin	X			

GLOSSARY

ac abb. Alternating current. For practical purposes, alternating current is the type of electricity that flows throughout a person's house and is accessed by the use of wall sockets.

Ace lock A term sometimes used to refer to any tubular-key lock; but the term is more properly used to refer to *the Chicago Ace Lock*, the first brand name for a tubular key lock.

AHC abb. Architectural Hardware Consultant (as certified by DHI).

ALOA abb. Associated Locksmiths of America.

ampere (or amp) A unit of electrical current.

ANSI abb. American National Standards Institute.

annunciator A device, often used in an alarm system, that flashes lights, makes noises, or otherwise attracts attention.

anti-pick latch A spring latch fitted with a parallel bar that is depressed by the strike when the door is closed. When the bar is depressed, it prevents the latch from responding to external pressure from tools such as a shove-knife or an ice pick.

armored front A plate covering the bolts or set screws holding a cylinder to its lock. These bolts are normally accessible when the door is ajar.

ASIS abb. American Society for Industrial Security.

backplate (rim cylinder) A small plate applied to the inside of a door through which the cylinder connecting screws and bar is passed.

backset (of a lock) The horizontal distance from the edge of a door to the center of an installed lock cylinder, keyhole, or knob hub. On locks with rabbeted fronts, it is measured from the upper step at the center of the lockface.

barrel key A key with a bit projecting from a hollow cylindrical shank. The holllow fits over a pin in a lock keyway and helps keep the key aligned. The key is also known as a hollow post key or pipe key.

BHMA abb. Builders Hardware Manufacturers Association.

bicentric cylinder A lock cylinder having two plugs, and a set of tumblers in each plug. The correct key can be inserted into either plug to open the lock. This permits master keying without reducing the security of the locking mechanism.

bit That part of a key that is cut to operate a lock.

bit key   A key with a bit projecting from a solid cylindrical shank. The key is sometimes referred to as a "skeleton key." A bit key is used to operate a bit key lock.

bit key lock   A lock operated by a bit key lock.

bitting   A cut, or series of cuts, on the bit or blade of a key.

blank   A key that has not been cut or shaped to operate a specific locking mechanism.

bow   That portion of a key that is held between the fingers when the key is being used.

burglar resistant   A term used to describe locks or doors capable of resisting attack by prowlers and thieves for a limited time.

cam   A piece attached to the back of a plug that rotates in conjunction with the plug. In many types of locks, the cam operates the lock bolt. In some cases, such as with many drawer locks, the cam is the lock bolt. Unlike a tang or tailpiece, a cam has no play and operates around a central axis.

case (of a cylinder)   (often called the *housing*; sometimes called the *shell*) The part of a cylinder that houses the plug.

case (of a lock)   The box containing the lock-operating mechanism.

case ward   A ward or obstruction integral to the case of a warded lock.

Cdc abb.   Certified Door Consultant (as certified by DHI).

change key   A key that operates only one lock or a group of keyed-alike locks in a series. In a master keyed system, change keys are the lowest level keys.

CML abb.   Certified Master Locksmith (as certified by ALOA).

combination lock   A lock that may or may not be operated with a key, but can be operated by inserting a combination of numbers, letters, or other symbols by rotating a dial (or dials), or by pushing buttons.

connecting bar   A flat bar attached to the rear of the plug in a rim lock to operate the locking bar mechanism.

core   The term is sometimes used as a synonym for *plug*; but *core* is also used to refer to the figure 8 shaped unit that can be removed and replaced in interchangeable core cylinders.

corrugated key   A key with pressed longitudinal corrugations in its shank to correspond to a compatibly shaped keyway.

CPL abb.   Certified Professional Locksmith (as certified by ALOA).

CPP abb.   Certified Protection Professional (as certified by ASIS)

current   The flow of electricity. Current is measured in amperes.

cuts   The indentations, notches, and cutouts made in a key blank in order to fit the key to a lock.

cylinder   A unit housed in a case (or shell) and usually containing a plug and tumblers.

cylinder guard   A device designed to protect a cylinder from wrenching; it is usually installed around the rim of a cylinder. Free-spinning cylinder guards spin freely (independent of the cylinder) when being turned.

cylinder lock   A lock having mechanisms operated by a cylinder.

cylinder ring   A collar or washer used under the head of a cylinder.

cylinder screw   The set screw in the front of a cylinder lock which prevents the cylinder from being turned after installation.

dc abb.   Direct current. For practical purposes, direct current is the type of electricity obtained from batteries.

deadbolt   A lockbolt having no spring action, usually rectangular and actuated by a key or turn knob.

deadlatch   A lock with a beveled latchbolt that can be automatically or manually locked against end pressure when projected.

deadlock   A lock with a deadbolt only.

depth key   A special key that enables a locksmith to cut blanks made for a special lock according to a code.

derivative code   A special numerical code that relates a lock's tumbler arrangement to the depth of the key cuts necessary to operate the lock.

DHI abb.   Door and Hardware Institute.

dimple key   A key with cuts drilled or milled into the surface of its blade.

disc tumbler   A flat circular or oval-shaped disc with a rectangular hole and one or more side projections. A number of these are used side by side in a disc tumbler lock.

door check   A device consisting of a heavy spring and arm coupled to an air or oil cylinder that controls the speed in which a door closes. Also called a door closer.

double-bitted key   A key with cuts on both sides of its blade.

drill pin   A round pin projecting from the inside of a lock case to receive a barrel key. Also called a barrel post.

drivers   The upper set of spring-activated pins in a pin tumbler cylinder lock. Drivers are usually cylindrical and flat on both ends.

dummy cylinder   A non-locking device that looks like a cylinder, and is used to cover up a cylinder hole.

emergency key   A key capable of opening a privacy function lockset even though the door may be locked from inside a room.

escutcheon   A plate, either protective or ornamental, containing openings for the controlling members of a lock such as knob, handle, cylinder, keyhole, etc.

fence   (also called *the stump*) A piece rigidly attached to the bolt in a lever tumbler lock and, when properly aligned, enters the lever gates to open the lock.

finished key   A key that has been cut to fit a lock.

gate   The opening in lever tumblers that allows them to pass the post, or fence. Also called gating.

heel (of a padlock shackle)   The end of a padlock shackle that is not removable from the case.

jamb   The inside vertical face of a doorway or window frame.

key blank   An uncut key. Once cut, a key blank becomes a finished key.

key caliper   Any small caliper capable of measuring the height of a cut on a key.

key extractor   A device used to remove broken key pieces that have broken off in a lock keyway.

key plate   A small plate or escutcheon having a keyhole only.

keyway   The opening in a lock mechanism into which the key is inserted.

latch   A door fastening device, usually with a sliding or spring bolt.

latchbolt   A beveled spring bolt, usually operated by a knob, lever handle, or thumb piece.

layout board   A board with a number of parallel grooves used to hold pin tumbler parts in order when the locksmith is working on the lock.

LED abb.   light-emitting diode. A device that emits light.

lever handle   A horizontal handle for operating the latchbolt of a lock.

lockface   A plate that shows in the edge of a door after lock installation.

locking dog (of a padlock)   The part that engages the shackle and holds it in the locked position.

master key   A key capable of operating several different locks, each lock being operated by its own change key.

mortise   An opening made in a door to receive a lock or other hardware. Also the act of making such an opening.

mushroom pins   Mushroom shaped drivers.

NSLA abb.   National Locksmith Suppliers Association.

night latch   An auxiliary lock with a spring latchbolt and functioning independently of, and providing additional security to, the regular lock on the door.

ohm A unit of resistance to electrical current.

panic bolt A type of lock fitted with a long bar placed horizontally across the inside of a door. Pressure on the bar releases the lock. Used in theaters, schools, and other public buildings, it is also called a panic bar.

pass key A master key or skeleton key.

pick Any tool or device (other than a key), either from a locksmith supply house or homemade, that is used to manipulate the tumblers in a cylinder into a locked or unlocked position.

pin tumblers Small sliding pins in a lock cylinder that work against coil springs and prevent the cylinder plug from rotating until the correct key is inserted.

pin tweezers A pair of tweezers specially designed to handle pin tumblers.

plug (of a lock cylinder) The cylindrical mechanism in a lock cylinder that houses the keyway.

plug follower (also called a following tool) A tool used to prevent the drivers and springs from coming out of the upper pin chambers while the plug is being removed from the cylinder case.

plug holder A fixture used to hold a plug (or plugs) while rekeying or repairing a lock.

plug retainer A device that secures the plug in a lock cylinder.

post (of a key) With respect to bit keys, the post is the part of the key that the bit is attached to.

power The product of current and voltage.

resistance Opposition to electrical current flow.

resistor A component that resists electrical current flow in a dc circuit.

reversible lock A lock in which the latchbolt can be turned over, so as to adapt it to doors of either hand, opening in or out. This does not apply to beveled front locks and certain cylinder mortise locks.

RL abb. Registered Locksmith (as certified by ALOA).

SAVTA abb. Safe and Vault Technicians Association.

shackle The curved portion of a padlock that passes through the hasp.

shackle spring The spring inside a padlock that projects the shackle from the padlock case when unlocked.

shank (of a key) With respect to bit keys, a shank is the part between the shoulder and the bow.

shear line The space between the case and the plug of a lock cylinder.

skeleton key A warded lock key cut especially thin to bypass the wards in several warded locks, so the locks can be opened.

strike (sometimes called a keeper) The part of a locking arrangement that receives the bolt, latch, or fastener. The strike is usually recessed in the door frame.

top pin (of a padlock) A pin that retains the shackle in a padlock case when unlocked.

tension wrench An obsolute term used to refer to a torque wrench.

torque wrench A device used to apply pressure on a lock while its tumblers are being manipulated by a pick. It's also used to turn the plug to the unlocked position after the lock has been picked.

tubular key A key with a hollow cylindrical-shaped blade that has indentations (usually seven or eight) around the rim of the blade. The key is used to operate tubular key locks.

tubular key lock A type of lock that has its tumblers arranged in a circle. The lock is often used on vending machines and coin-operated washing machines. The lock is operated by a tubular key.

UL abb. Underwriters Laboratories.

VATS abb. Vehicle Anti-theft System. An electromechanical system that uses a special key to start an ignition. The system is used in many General Motors vehicles manufactured after 1989. VATS is sometimes called *pass-key*.

Volt A unit of voltage.

Voltage (also called electromotive force) The force that pushes current. Voltage is measured in volts.

Watt A unit of power.

Index

Other Bestsellers of Related Interest